STATE OF ILLINOIS
WILLIAM G. STRATTON, *Governor*

DEPARTMENT OF REGISTRATION AND EDUCATION
VERA M. BINKS, *Director*

NATURAL HISTORY SURVEY DIVISION
HARLOW B. MILLS, *Chief*

Volume 26	BULLETIN	Article 1

The Mayflies, or Ephemeroptera, of Illinois

B. D. BURKS

Printed by Authority of the State of Illinois

URBANA, ILLINOIS

May 1953

Reprinted by Arrangement
ENTOMOLOGICAL REPRINT SPECIALISTS
P. O. Box 77224, Dockweiler Station
Los Angeles, California 90007

WILLIAM MADISON RANDALL LIBRARY UNC AT WILMINGTON

First reprinting, 1975

ISBN: 0-911836-06-3

Library of Congress Cataloging in Publication Data

Burks, Barnard De Witt, 1909-
    The mayflies, or Ephemeroptera, of Illinois.

    Reprint, with a new pref., of the 1953 ed. published
by the Illinois Natural History Survey Division, Urbana,
which was issued as v. 26, article 1 of its Bulletin.
    Bibliography: p.
    1. May-flies—Illinois. 2. Insects—Illinois.
I. Title. II. Series: Illinois. Natural History
Survey. Bulletin ; v. 26, article 1.
[QL505.2.U6B87 1975]   595.7′34′09773   75-2296
ISBN 0-911836-06-3

QL505
.2
.U6
B87
1975

Preface to the Reprint Edition

More than 20 years have passed since the original of this work was published in 1953, and more than 25 years since most of the study was completed in May, 1949. During this time, there has been a great growth in interest in aquatic insects, but new keys and revisions for the geographic area are notably lacking. The only significant study including the area treated by this book was the revision of the large genus *Ephemerella* by Allen and Edmunds.

There have been a number of major changes in the higher classification as noted in the columns below and the paragraphs that follow. The few changes in the specific names are also noted, but a catalogue of the additional species that might be found in the area was not attempted, because new lists of species and their general distributions will be available in a 1975 publication in press by Edmunds, Jensen, and Berner.

Higher Classification of the Ephemeroptera

Burks (1953)   vs.   Edmunds, Jensen, Berner (1975)

Burks (1953)	Edmunds, Jensen, Berner (1975)
Ephemeridae	Ephemeroidea
Campsurinae	Campsurinae (of Polymitarcyidae)
Potamanthinae	Potamanthidae
Ephoroninae	Polymitarcyinae (of Polymitarcyidae)
Ephemerinae	Ephemeridae
Neoephemeridae	Neoephemeridae
Caenidae	Caenidae + Tricorythidae
Ephemerellidae	Ephemerellidae
Baetiscidae	Baetiscidae
Oligoneuriidae	Oligoneuriidae
Leptophlebiidae	Leptophlebiidae
Baetidae	Baetidae + Siphlonuridae
Siphlonurinae	Siphlonurinae (of Siphlonuridae)
Isonychiinae	Isonychiinae (of Siphlonuridae)
Baetinae	Baetidae
Ametropidae	Ametropodidae + Metretopodidae + Pseudironinae (of Heptageniidae) + Acanthametropodinae (of Siphlonuridae)
Heptageniidae	Heptageniidae

The taxonomic changes are discussed below for genera and higher categories for North America north of Mexico and for species in the general region covered by this reprint (North Central and Northeastern United States and adjacent Canada).

Ephemeridae (Ephemeroidea). The family Behningiidae is now known from South Carolina to Florida (the genus *Dolania*).

Ephemerinae (Ephemeridae). The genus *Pentagenia* is now placed in the subfamily Pentageniinae of Ephemeridae. *Hexagenia recurvata* has been placed in the separate genus *Litobrancha*.

Caenidae. The family has been split into Caenidae *(Caenis* and *Brachycerus)* and Tricorythidae *(Tricorythodes* and *Leptohyphes)*. *Leptohyphes* is now known as far north as Maryland and Utah.

Ephemerellidae. In *Ephemerella* the *walkeri* group is now the subgenus *Drunella,* the *serrata* group is the subgenus *Serratella,* the *invaria* group is the nominal subgenus *Ephemerella* sensu stricto, the *bicolor* group is now the subgenus *Eurylophella, Attenella* (originally *Attenuatella)* (for *attenuata)* and *Dannella* (for *simplex* and *lita)*. Other subgenera in North America are *Timpanoga* and *Caudatella*.

Oligoneuriidae. The species *ammophila* Spieth is now assigned to *Homoeoneuria* and the wings figured as *Oligoneuria* (fig. 183) are those of a South American oligoneuriid, and not referable to *Homoeneuria*.

Leptophlebiidae. The genus *Homothraulus* is now known to occur in Texas (south to Argentina). The genus *Traverella* is now known from Indiana.

Baetidae. The Siphlonurinae and Isonychiinae are now placed in the separate family Siphlonuridae. The genus *Metreturus*, placed by Burks in the Ametropidae, has proved to be a synonym of a Siberian genus *Acanthametropus*, therefore the correct species name is *A. pecatonica*. This and a related genus, *Analetris* (Utah to Saskatchewan) are now placed in the sub-family Acanthametropodinae of Siphlonuridae. An additional genus of Siphlonurinae, *Edmundsius*, has been described from California.

In the Baetinae *(Baetidae)*, three genera have been described, *Apobaetis* from California, *Dactylobaetis* from Oregon, Saskatchewan and Oklahoma south to Argentina and Uruguay, and *Paracloeodes* from California, Wyoming, Minnesota, Mississippi and Georgia. *"Pseudocloeon" minutum* from Minnesota, mentioned by Burks in the discussion of *Pseudocloeon*, is now placed in the genus *Paracloeodes*.

Neocloeon is now regarded as a synonym of *Cloeon*.

The genus *Heterocloeon* has been placed in the synonymy of *Baetis* but its status is being re-evaluated on the basis of new studies by W. P. McCafferty (in press).

Baetis herodes Burks appears to be identical with the formerly "unidentifiable" *Baetis hageni* Eaton *(Baetis unicolor* Hagen).

Ametropidae. The correct form of the family name is Ametropodidae; the family now contains the single genus *Ametropus*. *"Metreturus"* is now in Siphlonuridae. *Metretopus* and *Siphloplecton* are now in a separate family Metretopodidae and the genera are distinguished as nymphs by simple gills in *Metretopus* and double gills on segments 1-3 in *Siphloplecton*. *Pseudiron* has been placed in the Heptageniidae.

Heptageniidae. *Stenonema* has been divided into two genera. Those species referred to by Burks as the *interpunctatum* group are now in the genus *Stenacron*.

The subgenus *Ironodes* of *Epeorus* is now regarded as a full genus. The genera *Arthroplea* and *Anepeorus* are each now placed in a separate subfamily of Heptageniidae (Arthropleinae and Anepeorinae). The nymph questionably referred to as *Anepeorus* is still believed to be the nymph of that genus. A new genus, *Spinadis*, is now known from Georgia, Wisconsin and Indiana (Spinadinae).

Papers of taxonomic importance likely to be significant in the area generally covered by Burks **The Mayflies, or Ephemeroptera, of Illinois** have been published by the following: R. K. Allen; Allen and G. F. Edmunds, Jr.; L. Berner; Edmunds, Berner and J. R. Traver; Edmunds; Edmunds, S. L. Jensen and Berner; W. Flowers; F. Harper; W. L. Hilsenhoff; F. P.

Ide; Jensen; Jensen and Edmunds; R. W. Koss; J. W. and F. A. Leonard; P. A. Lewis; W. P. McCafferty; W. L. Peters; D. C. Scott, Berner and A. Hirsh; and J. R. Traver.

A bibliography of papers published on North American Ephemeroptera from 1935 to date, and for most other areas of the world from about the date of the Burks bulletin can be found in *Eatonia, A Newsletter for Ephemeropterists*. It is currently (1974) being published twice a year. Complete sets (1954 to 1974) can be purchased through the editor, Janice G. Peters, University P. O. Box 111, Florida A & M University, Tallahassee, Fla. 32307 (as of 1974, $5.00 per set, paid to Florida A & M Univ., account #04-3501-000).

George F. Edmunds, Jr.
Salt Lake City, Utah
November, 1974

FOREWORD

ALMOST a century ago, Benjamin D. Walsh, the first State Entomologist of Illinois, became interested in the mayflies or shadflies of the area about his home in Rock Island. Since his first observations and writings concerning these insects, a great deal has happened to the Illinois environment, and great advances have been made in our knowledge of this interesting part of our native fauna. The following treatise by Dr. B. D. Burks, formerly of the Natural History Survey, brings our knowledge of this group up to date.

Mayflies are of importance to people in many ways. One of the most obvious is the swarming and massing of some species in such strategic places as river bridges, where at times the bodies of countless millions form barriers or hazards to traffic. But these dramatic occurrences are far from the most important aspects of the lives of mayflies.

As a part of the biological complex of our waters, for all mayflies are aquatic in their developmental stages, these insects find their most important place in human economy and interest. They are an important link in converting microscopic food organisms and vegetable detritus into units large enough and of proper character to be of value to our predatory fishes. This fact has been employed by fly tiers in the design of certain artificial lures intended to be attractive to certain fishes. Furthermore, mayflies may be characteristic of certain types of waters. Dr. Burks has listed a dozen different habitats, with species typical of each. This association of insect with habitat is of importance to all workers interested in our fresh waters, as the mayfly species which are present in a given body of water may indicate the condition of that water and therefore its usefulness for domestic or industrial purposes. When Walsh studied the mayfly populations in the Rock Island area, he found a considerably different species complex from that existing in the area now. In spite of much careful collecting in the Rock Island vicinity, Dr. Burks could recover only 8 of the 31 species which Walsh recorded from there. This recovery represents only about 26 per cent of the mayfly fauna present before the damming, dredging, siltation, and addition of pollutants which characterize these waters now. It was necessary to extend the search into less modified waters in order to rediscover some of the other species with which Walsh was familiar.

And then there is the peculiarness, the uniqueness, of this archaic group of insects, of interest to all who profess a delight in nature. In this group, as seldom found in an aggregation of related animals, there is a great divergence from that which we consider

to be the accepted pattern. The morphology of mayflies is reminiscent of the morphology of insects which disappeared many millions of years ago and which are now known only through fossil remains. The adult stage has become a mere vestige, lasting usually but a very few days at the most. The life span is consumed almost entirely by the developmental stages under water. All of the eating is done during this growth period; the adults have useless mouthparts and digestive systems. Of all of the winged insects, mayflies are the only ones which shed their skins after they have developed wings with which they can fly. And a few species reproduce their kind without ever attaining the true adult stage!

Thus, we find that mayflies are important not alone to harassed highway maintenance men and press photographers. They are much more so to those entrusted with the well-being of our fish populations, those interested in the public health and other values in inland waters, and those interested in the peculiarities of nature.

Dr. Burks, who prepared the following treatise, obtained the B.A. degree in 1933, the M.A. in 1934, and the Ph.D. in 1937, all from the University of Illinois. On July 1, 1937, he joined the staff of the Natural History Survey. But for short leaves during which he assisted at the United States National Museum, and leave for military service, he remained in the employ of this Survey until May 21, 1949. At that time, he resigned to take a position in the Division of Insect Identification with the United States Department of Agriculture in Washington, D. C.

A number of people have been of assistance in the preparation of this manuscript. We are especially indebted to Dr. Carl O. Mohr for his excellent illustrations, to Mrs. Elizabeth Maxwell for the preparation of many of the line drawings, and to Mrs. Leonora K. Gloyd for her careful and painstaking work in the later stages of preparation of the manuscript for publication.

For permission to use figs. 88, 188–192, 218, and 300, most of them redrawn without appreciable change from Traver in *The Biology of Mayflies*, we are indebted to the Comstock Publishing Company. For loan of critical material for study, we are grateful to Dr. Henry Dietrich of Cornell University, Dr. Joseph C. Bequaert of the Museum of Comparative Zoology, Harvard University, Dr. C. E. Mickel of the University of Minnesota, and Mr. W. J. Brown and Mr. G. P. Holland of the Canadian National Museum.

HARLOW B. MILLS, *Chief*
Illinois Natural History Survey
Urbana, Illinois

CONTENTS

Kankakee River at Momence. Habitat of *Ephemerella needhami, Baetisca bajkovi, Siphloplecton interlineatum, Stenonema lepton.*

The Mayflies, or Ephemeroptera, of Illinois

B. D. BURKS

MAYFLIES or shadflies are a group of insects constituting the order Ephemeroptera. In the young or nymphal stages, they live in the water of ponds, lakes, or streams, where they can be found under rocks or logs, in the mud at the bottom, or occasionally swimming about. When the nymphs are full grown, they come to the surface of the water and transform into free-flying aerial insects. As such, they are familiar to many fishermen and nature lovers.

In Illinois, a few large, conspicuous forms come to general attention every year when they emerge on warm midsummer evenings in enormous numbers from our larger lakes and rivers. However, these constitute only a relatively small part of the mayfly fauna of the state. Other forms are to be found emerging at various times of the year from all the relatively permanent and unpolluted bodies of water, including ponds, lakes, brooks, creeks, and rivers.

The mayfly may be distinguished readily from all other aquatic insects. The nymph has a definite head, thorax, and abdomen. It has three pairs of well-developed legs, a pair of gills on each of the middle abdominal segments, and either two or three long "tails" (called caudal filaments) extending from the posterior end of the body. It more closely resembles the stonefly nymph than any other nymph but differs from it in having gills on the middle abdominal segments.

Unlike most insects, the mayfly typically has two winged stages. It is the only existing insect that molts after getting functional wings. The first winged stage is called the *subimago,* which is actually a subadult stage; soon after it is formed this subimago (in most species) molts to form the true *adult* or reproductive stage, sometimes called the *imago.* In a very few species, noted later, which never develop to the adult, the female lays her eggs while in the subimago stage.

The subimago is very similar to the adult in appearance, but the body and all appendages are incased in a transparent skin or pellicle. The adult has its mouthparts and alimentary system represented by only minute, distorted vestiges; it usually has two pairs of extremely thin and papery, triangular wings (the posterior pair being much smaller, or lacking in a few species), which are held upright and not folded above the back when the insect is at rest. As in the nymph, the adult has two or three long, well-developed caudal filaments; if the median one appears to be lacking, it will be found on close examination to be represented by at least a small rudiment.

There are over 550 different species of mayflies known for North America north of Mexico. This report includes 48 genera and 222 species, with Illinois records of 126 species, 15 of which are described as new.

Importance of Mayflies

Although, at times of unusual abundance, the adults of a few species may swarm to lights and become an expensive nuisance in towns and cities near rivers and lakes, mayflies are, on the whole, harmless and gentle creatures. Some species apparently do not even indulge in the activity of swarming. The nymphs are likewise innocuous except in two exotic species of *Povilla.* One species of this genus, found in the East Indies, bores into wood submerged in fresh water, often seriously weakening or destroying piles and other wood structures; the other, occurring in the Belgian Congo, bores into freshwater sponges.

Occasionally, over a period of years, adult mayflies have caused damage in certain local areas. Unusual hordes of these insects may leave the water on the first suitable day after adverse weather conditions, or other

PHOTOGRAPH FROM WORLD WIDE PHOTOS.

Fig. 1.—The picture above was published in the *Chicago Daily News* on July 8, 1946. It carried the legend: "May flies stop motorists. Lawrence Rutz stops his truck on the west channel bridge at La Crosse, Wis., yesterday to clear May flies from the front of the vehicle. The insects got so thick they obstructed the view of the driver, clogged radiators and made roadway of the bridge slippery."

rare circumstances, have interfered with the normal rhythm of successive-day emergence. A few of the larger species have

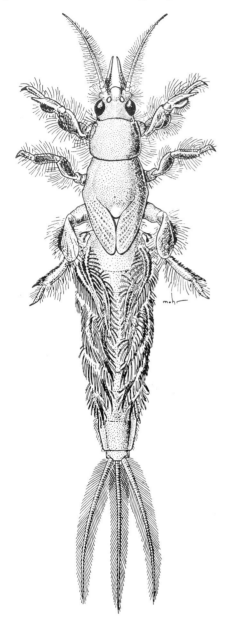

Fig. 2.—Nymph of *Hexagenia limbata*.

been known to form dense clouds and to settle in great drifts over roads, bridges, and streets. These fragile insects die within a few hours, and when occurring in such hordes their dead bodies may clog ventilator ducts and sewers and may also cause temporary traffic difficulties, fig. 1. On July 23, 1940, the Associated Press carried the following dispatch:

"Sterling, Ill.—Shadflies that in some places piled to a depth of four feet blocked traffic over the Fulton-Clinton highway bridge for nearly two hours last night.

"Fifteen men in hip boots used shovels and a snow plow to clear a path. The bridge appeared to be covered with ice and snow. Trucks without chains were unable to operate until most of the flies had been shoveled into the Mississippi river."

In both aquatic and terrestrial stages, figs. 2 and 3, mayflies achieve their chief importance as food for other animals. They are preyed upon by birds, fish, amphibians (frogs and salamanders), spiders, and many predaceous insects. It is as a natural food for fish that they are of primary economic value.

The first extensive observations on the role of mayflies as fish food were made by Forbes (1878–1888). They were based on examinations of the stomachs of Illinois fishes. Since the observations by Forbes, many contributions to this subject have been made by many authors, but no attempt is made here to collate the material in the limnological literature.

In general, it has been found that the diet of fishes consists of the most readily available suitable food. Consequently, fish of the same species in a body of water will be found to have quite different organisms in their stomachs at different seasons of the year. At certain times of year, mayflies are abundant in lakes or streams, and at these times are readily eaten by fish. Both adult and nymphal mayflies are eaten by the fish, the adults either when molting at the surface of the water, or when alighting later to lay eggs.

Mayflies have been found in the stomachs of most species of the larger Illinois fishes, including all the sport fishes, such as crappies, bass, and various other types of sunfish. There is little doubt that fish of many species feed extensively on mayflies, and that, at times of great mayfly emergence, the fish of a considerable number of species subsist chiefly on these insects.

That fish will consume mayfly nymphs readily was shown in an interesting way in New Zealand. The New Zealand mayflies had evolved over a long period in streams which did not contain game fishes. As a result, the nymphs of such forms as *Oniscigaster wakefieldi* McLachlan, instead

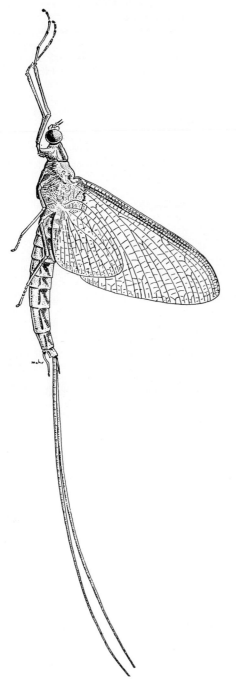

Fig. 3.—Adult male of *Hexagenia limbata*.

of remaining concealed beneath rocks and debris as our American mayflies do, swam freely in the open water. Shortly after the turn of the century, brown trout and rainbow trout were introduced into the New Zealand streams to provide game fishing. The trout ate the mayfly nymphs in such numbers that within about 20 years the once extremely abundant mayflies had become almost extinct. Tillyard (1926:64) states, "The introduced trout have greatly reduced this once abundant fauna [of mayflies] and some species are now extinct, or nearly so . . . . The mayfly fauna of Australia and New Zealand is not specialized to hold its own against the introduced brown and rainbow trout and is rapidly being reduced to a minimum."

Habitat Preference

Some species of mayflies may develop in a variety of situations, but most species are restricted to definite types of aquatic habitats. The various types of mayfly habitats found in Illinois and the species which have been observed to be characteristic of them are as follows:

1. Large, relatively slow rivers, such as the Mississippi, Ohio, and Illinois: *Hexagenia bilineata* and *limbata*, *Pentagenia vittigera*, *Tortopus primus*, *Potamanthus myops* and *verticis*, and *Ephoron leukon* and *album*.

2. Moderate-sized, fairly rapid rivers, such as the Kankakee, fig. 4, and Rock: *Hexagenia rigida* and *limbata*, *Potamanthus myops* and *verticis*, *Ephoron leukon* and *album*, *Ephemerella dorothea*, *invaria*, and *simplex*, *Baetisca bajkovi*, *Pseudiron centralis*, *Siphloplecton interlineatum*, *Stenonema* spp., and *Heptagenia* spp.

3. Small rivers or creeks with fairly rapid flow, such as the Salt Fork River and Lusk Creek: *Ephemera simulans*, *Hexagenia atrocaudata*, *Ephemerella frisoni* and *needhami*, *Baetisca laurentina*, *obesa*, and *bajkovi*, *Oligoneuria ammophila*, *Paraleptophlebia praepedita* and *ontario*, *Ameletus lineatus*, *Isonychia sicca*, *bicolor*, and *rufa*, *Centroptilum walshi* and *rufostrigatum*, *Baetis spinosus*, *propinquus*, *harti*, *pygmaeus*, *frondalis*, *pallidulus*, and *intercalaris*, *Pseudocloeon punctiventris*, *dubium*, *parvulum*, and *veteris*, *Stenonema* spp., and *Heptagenia* spp.

4. Sluggish creeks or small rivers with a great deal of silt, constituting the majority

of the streams in central Illinois: *Stenonema tripunctatum*.

5. Permanent or semipermanent brooks with rapid flow, such as the spring-fed stream in the Botanical Gardens near Elgin or the spring-fed tributaries of Lusk Creek and Gibbons Creek in southern Illinois, fig. 5: *Paraleptophlebia moerens* and *praepedita* and *Baetis vagans*.

6. Relatively permanent ponds, many of which are found in Lake, Cook, and Du Page counties, almost always in woods: *Callibaetis skokianus, ferrugineus, and fluctuans*.

7. Temporary ponds, which are commonly found in central Illinois in the springtime: *Leptophlebia nebulosa* and *cupida*.

8. Small, temporary pools, usually along the margins of streams, which have greatly reduced current or no current: *Siphlonurus marshalli, Tricorythodes* spp., *Caenis ridens* and *hilaris, Leptophlebia cupida, Habrophlebiodes americana, Callibaetis fluctuans,* and *Cloeon mendax, rubropictum,* and *simplex*.

9. Stagnant bodies of water, such as the vegetation-choked backwaters of streams or bogs: *Caenis simulans*.

Fig. 4.—Kankakee River at Aroma Park. Habitat of *Ephemerella needhami, Baetisca bajkovi* and *laurentina, Pseudocloeon dubium, Baetis spinosus, Stenonema integrum* and *bipunctatum*.

Fig. 5.—Gibbons Creek at Herod, from which *Ameletus lineatus* emerges in early spring. In late spring, *Baetis herodes* and *Leptophlebia cupida* emerge here. *Stenonema tripunctatum* emerges here throughout spring and summer months.

10. Large lakes, represented by Lake Michigan: *Ephemera simulans*.

11. Small, relatively shallow glacial lakes, such as those found in Lake County: *Ephemerella lutulenta* and *temporalis*.

12. Small, temporary brooks, which flow into larger streams in late winter and early spring. This type of stream may furnish a habitat favorable for an occasional small nymph of *Stenonema* or *Caenis* but no mayfly has been found to mature in such a place. Stoneflies and craneflies, however, often occur abundantly in these streams.

Another category might have been made for the various bodies of impounded water which now exist in considerable numbers in central and southern Illinois, but no mayfly can be said to be characteristic of such bodies of water. Mayflies that occur in such waters indicate the ecological characteristics of the individual impoundments. Some of these impoundments are stagnant and produce large flights of *Caenis simulans* only; others apparently have the characteristics of a large, slow river, and produce flights of *Hexagenia bilineata* or *limbata*.

Life History

Mayfly nymphs require a relatively long time to develop from egg to full-grown nymph. The shortest known length of nymphal life is in species of the genus *Callibaetis* which, in summer, mature from egg to adult in 5 to 6 weeks. Some of the smaller baetines require 4 to 5 months, a length of time that results in the production each year of two waves of adults for a species, one in early spring, the other in late summer or early autumn. Some, as was shown by Murphy (1922) for *Baetis vagans*, have a complex, overlapping series of broods. The summer brood matures in 6 months, the winter brood in 9. The large ephemerids, such as *Hexagenia*, figs. 2 and 3, require 2 years to mature; the annual appearance of a given species in a locality is due to overlapping broods of the species. Many other mayflies, such as *Stenonema* and *Heptagenia*, emerging as they do year after year in the same locality at about the same dates, may be inferred to require 1 year to mature from egg to adult.

During nymphal life, the developing mayfly passes through a very large number of instars. *Baetis vagans*, which has a relatively short nymphal life of 6 to 9 months, passes through 27 instars (Murphy 1922). Other mayflies, such as *Callibaetis*, have been estimated to go through about 20 nymphal instars (Needham *et al.* 1935:15).

Stenonema possibly has 30 nymphal instars. *Hexagenia,* with its 2-year life cycle, has an unknown but quite large number of nymphal instars. Mayfly nymphs have been observed to grow relatively little during each stadium.

The developing adult wings, eggs, and genitalia can clearly be seen inside the later nymphal instars. Feeding terminates with the next to the last nymphal instar. When the last nymphal instar is reached, development of the adult structures is almost complete. During the last instar, the nymph is quiescent, and the alimentary canal degenerates rapidly. The vestigial mouthparts and short antennae of the adult can clearly be seen developed beneath the nymphal cuticle.

In such forms as *Hexagenia,* the mature nymph when ready to molt comes to the surface of the water, the nymphal skin splits rapidly, and the subimago emerges quickly. The subimago rests for a short time on the shed nymphal skin, which floats like a raft. Then it is ready to take flight for a place of safety. The whole process requires only about 2 minutes. This molt from the nymph to the subimago represents a very dangerous time in the life of the mayfly. Many mature nymphs, as they are swimming to the surface of the water to molt, and many more subimagoes while resting on their nymphal skins, are devoured by fish. Some, even as they take flight, fall prey to fish that jump out of the water to catch them. Some of the mayflies that elude their fish enemies at this time are likely, as they flutter up from the surface of the water, to be eaten by birds.

In other mayflies, such as *Isonychia, Ameletus,* and *Siphlonurus,* the nymph crawls out of the water onto stones, sticks, or other convenient objects, the nymphal skin splits, and the subimago emerges fairly slowly, the process requiring 3 to 5 minutes. The empty nymphal exoskeleton is left clinging to the support where the subimago emerged. If the shed last nymphal skin is examined, it will be found to contain, almost intact, the nymphal structures for which the adult has no use. The nymphal mouthparts are complete and still contain some of the musculature in only a partly disintegrated state. The nymphal gills also are intact, even in such forms as *Isonychia,* which retain gill rudiments in the adult.

The subimaginal stage in most mayflies normally lasts 1 or 1½ days. This is subject to some prolongation at low temperatures. *Siphlonurus marshalli* requires 2½ days when the daytime temperature is from 45 to 50 degrees F. When daytime temperatures rise to 70 degrees F., the imago appears in 1½ days. During the subimaginal stage, almost 25 per cent of the body weight is lost, probably due principally to losses of water through evaporation and respiration. It can be shown that subimagoes must lose water before the adults can emerge, as subimagoes kept in a too-moist atmosphere are never able to emerge as adults. On the other hand, water loss must not be too rapid, or the subimagoes will die without producing the adults. My experience in rearing mayflies has been that relative humidity is the most critical single factor in the maturing of subimagoes to adults.

In some mayflies, such as *Ephoron, Tortopus,* and *Caenis,* the subimago stage is greatly · abbreviated. In them, the subimaginal skin or pellicle is shed almost immediately after the emergence from the nymph. In *Ephoron* and *Tortopus,* which do not have functional legs in the adults, the males shed the subimaginal skins in flight, but the females remain as subimagoes. In *Caenis simulans,* I have observed the subimaginal pellicle to be shed in flight, but Needham (Needham *et al.* 1935:99) states that in *Caenis* sp. the subimagoes alight to shed the subimaginal skin. The observations probably were made on different species of the genus. At any rate, in the females of all species of *Caenis* which have been observed, the subimaginal pellicle is only partly shed, but in the males it is shed completely.

In the great Papuan mayfly of the East Indies, *Plethogenesia papuana* (Eaton), both males and females remain subimagoes, never attaining the adult stage. However, of the many preserved specimens of this species I have seen, the subimaginal pellicle is differentiated from the enclosed adult structures. The same is true of the subimago females of our American species of *Ephoron* and *Tortopus* which, as has already been mentioned, do not attain the ultimate adult stage.

By the time the subimago stage is reached, the eggs and sperm are already mature and can be stripped from subimagoes and mixed in normal saline solution, by which fertilization is accomplished. Nymphs can be hatched from these fertilized eggs. This has been done successfully with several species of mayflies, among them *Hexagenia limbata* and *bilineata* and *Isonychia bicolor*. During the subimago state, the degeneration of the alimentary tract becomes almost complete. This tract then becomes, in the adult male, fig. 3, only an air-filled and transparent sac which serves to make the body buoyant. In the female, the degenerated digestive tract is crowded and depressed by the eggs but, as eggs are expelled, it becomes inflated again, possibly aiding in the expulsion. It is the bursting of these inflated digestive tracts that produces the familiar popping noise living mayfly adults make when they are stepped on.

Most mayflies spend the subimago state resting in the shade among plants near water. In many species, the subimaginal pellicle is shed in the early evening hours and in others during the night. It is left behind as an extremely delicate and fragile skin adhering to the support the subimago has occupied. After this molt, adults of most species continue to rest quietly among the concealing vegetation until the next sundown; then the males swarm, mating occurs, and the females deposit all their eggs before midnight. In such forms as *Caenis* and *Ephoron*, which shed the subimaginal pellicle immediately after emerging from the nymphal stage, mating and egg laying occur within a very few hours of emergence, at most during the same night. In these forms, the total winged life is thus but a few hours. In most mayflies, however, the life in the winged stages endures for 2 or 3 days at summer temperatures. There are records in the literature of its being prolonged to 5 or 6 days, but such length of life is unusual, except in the females of *Callibaetis* and *Cloeon*. In these forms, the female adults have been observed to live from 1 to 3 weeks.

Mayflies deposit their eggs in the water in a number of different ways. In many baetines, such as *Baetis intercalaris*, the female crawls beneath the surface to oviposit on stones or other objects on the bottom. This phenomenon is often referred to in the literature, and I have observed it several times in Illinois. In many genera, such as *Ameletus, Siphlonurus,* and *Leptophlebia,* and in some species of *Stenonema,* the female

flies near the surface of the water and dips the end of the abdomen into it at intervals, by this action permitting a few eggs to be washed off at a time, much as in the oviposition of some dragonflies. In other genera, such as *Ephemerella,* the female extrudes all the eggs during flight, and temporarily holds them in a mass beneath the recurved tip of the abdomen. As she flies along near the surface of the water, she darts quickly down to break the surface film momentarily, and drops the entire packet of eggs into the water. The eggs sink instantly and adhere to rocks or other objects on the bottom.

The female of other mayflies, such as *Heptagenia,* flies a short distance above the surface of running water, and then alights on the surface for a few seconds, permitting the current to carry her a short distance downstream. During this time, she extrudes a few eggs, which are washed into the stream. Then she flies up from the surface for a few minutes, returning again to it to deposit more eggs. This alternate flying and dropping of eggs continues until the female is spent, or, as I have seen so often happen, until some bird or fish eats her.

In most of the large Ephemeridae, such as *Hexagenia,* the female simply alights flat on the surface of the water, with wings outspread, and extrudes all the eggs at once in two elongate packets. These eggs sink almost instantly, and the female remains on the surface until she drowns or, more likely, is eaten by a fish.

The eggs of most mayflies hatch in from 1 to 2 weeks, depending on temperature, and nymphal development begins at once. In a few genera, however, such as *Ameletus* and *Siphlonurus,* the eggs, deposited in the spring, do not hatch until the following February or March. The long period before hatching is due to the fact that, during the summer and fall, the breeding sites for these genera become completely dry. The eggs of these genera evidently can tolerate such desiccation. Clemens (1922) pointed out this phenomenon for *Ameletus ludens* in New York, and his findings agree well with my own field observations made in Illinois.

Mayfly eggs are of a great variety of forms (Smith 1935: 67), but, characteristically, most possess long, coiled, adhesive filaments which serve to attach them to stones or other objects in the water.

Some mayflies, such as *Ameletus ludens* and *lineatus,* normally are parthenogenetic, males being either unknown or extremely rare. Some, such as species of *Callibaetis* and *Cloeon dipterum,* are said to be ovoviviparous. Edmunds (1945:170) and Berner (1941:32) have observed a process approximating ovoviviparity in species of *Callibaetis.* They found that the eggs are retained within the abdomen of the fertilized female for several days, during which time the embryos develop. Then, when the female alights on the surface of a suitable body of water, she expels the eggs, and the nymphs hatch within a few minutes.

Food Habits

With very few exceptions, the nymphs are herbivores or scavengers, living on vegetable detritus and microscopic aquatic organisms, principally diatoms. A few, such as those of *Isonychia,* are partly predaceous, eating apparently almost anything that comes within their grasp, including other mayflies. Others, such as the supposed nymph of *Anepeorus,* fig. 394, and the nymph of *Metreturus pecatonica,* fig. 312, have long, sharp mandibles which indicate that they are entirely predaceous in habit. In the subimago and adult stages, mayflies do not feed.

Emergence Peaks

The adults of many species consistently appear year after year in the same localities on about the same dates for those localities. The species of *Callibaetis,* which develop in ponds and woodland pools, emerge continuously throughout the open growing season from April to October, but they usually have a peak of emergence in late May or early June. Some species of *Baetis,* developing in small, well-aerated streams and along the margins of small rivers, have two peaks of emergence in a season, in late April or May and in August or September. Other species of *Baetis* have three peaks of emergence, in May, July, and September.

In *Hexagenia bilineata,* which develops in large, slow rivers such as the Mississippi and Ohio, the adults emerge in late June, in July, or, rarely, in August. The swarms which appear in mid-July are the largest. *H. limbata,* which may be found in the largest rivers but which prefers somewhat smaller ones such as the Kankakee or Illinois, emerges in greatest numbers in late June or July. *H. rigida,* restricted to fairly rapid, well-aerated rivers such as the Rock and Kankakee, appears in greatest numbers in June. *Pentagenia vittigera,* which seems to prefer large, slow rivers but is also known to develop in smaller numbers in a great variety of streams, emerges in greatest numbers in July or early August but never in such tremendous numbers as does *Hexagenia. Ephemera simulans,* which inhabits lakes with considerable wave action as well as fairly sluggish streams, emerges in greatest numbers in early June. It should be noted, however, that occasional specimens of all these larger ephemerines are to be taken from April to October.

Dispersal

Mating usually occurs between males and females from the same brood of a species, and oviposition most often is carried out in the same body of water in which the individuals have developed. From this fact, it would seem that the chances are rather small that new genetic factors will be introduced into a given mayfly population by cross-breeding with other populations. However, on exceptionally warm, humid summer evenings, females still carrying their egg masses will sometimes be found late at night flying to lights that are located several miles from any body of water in which they could have developed as nymphs. At times, the flight range of fertilized females may thus be fairly long. Evidently by means of long flights of such females, mayflies are able to establish themselves in new breeding grounds fairly far removed from their places of origin. As an example, a large swarm of adults of *Hexagenia bilineata* was observed to emerge from Lake Glendale, an impounded body of water in Pope County, on June 18, 1942. *H. bilineata* ordinarily is a large-river species, common in the Ohio River, 12 air-line miles from Lake Glendale. The Ohio River is the nearest known source of this species. Adults from the Ohio evidently had flown the 12 miles to Lake Glendale in the summer of 1940 and established a colony of nymphs there. The valve in the dam which formed Lake Glendale was closed in the autumn of 1939. This artificial lake is not, however, a very suitable place for the development of *bilineata,* and that species is not known to have developed there subsequently.

Many adult females of *Tortopus primus* were taken in a trap light in Urbana, on August 13, 1943. These females still carried their eggs. There is no locality where this species is known to breed within

The length of nymphal life is so long and the nymphs are so frail that usually mayflies cannot be reared from eggs to adults without great loss of life. If mature or nearly mature nymphs whose wingpads show signs of darkening can be collected for the purpose of rearing, the number of deaths

Fig. 6.—*Siphlonurus marshalli* adult being removed from a rearing pan.

a radius of 20 miles from Urbana. The most likely source of this typically large-river species was the Vermilion River, 30 air-line miles away, or the Wabash River in Indiana, almost 50 miles away.

Rearing Mayflies

Although the classification of the mayflies is based almost entirely on the adults, it is the nymphs which are most often taken in aquatic collecting or limnological work. In many instances, the generic and specific differences are much more distinct in the nymphs than in the adults. In order to be certain of their identity, it is necessary to rear the adults from the immature forms, but for most purposes of identification the association of mature nymph and adult is adequate.

can be greatly reduced. Such nymphs are in the last instar, or at most only a molt or two removed from it.

Mayflies which spend all or the latter part of their nymphal existences in still water can be easily and successfully reared from a late instar to the adult form in shallow pans. Such mayflies are *Callibaetis, Siphlonurus, Leptophlebia, Paraleptophlebia, Stenonema, Heptagenia,* all the Caenidae, most Ephemeridae, and some species of *Baetis.*

The rearing pans may be circular, flat-bottomed, enameled pans approximately 10 inches in diameter and 4 inches deep. A large, flat rock should be put in each pan and water from the place where the nymphs were collected added to a depth of not over 1 inch. The rock should be of a size to project partly above the water and thus provide a place on which an adult may emerge, fig. 6. A few rotting leaves or other detritus from the water where the nymphs were found should also be added to each pan to provide food. Care must be taken to have the pan, rock, and water all at the same temperature as the water from which the nymphs were collected.

The nymphs may be taken from the pool or stream with a dipnet, or simply picked off rocks or other objects submerged in the water. They must be handled with the greatest care, as they are easily injured, always with fatal results. They may be kept in glass jars or cap vials partly filled with water until they can be transferred to the rearing pans. These temporary storage jars or vials must be kept cool, preferably by being partly immersed in the water where collecting is being done.

An effort should be made to sort the nymphs to species, using the obvious characters that can be seen with a hand lens. Part of each collection of a species should be preserved in alcohol at the time the other specimens are placed in the pans for rearing. An accession number should be given to each lot of specimens so that the adults, when secured, can unquestionably be associated with their nymphs.

The rearing pans containing the nymphs should be covered with a screen-wire or cheesecloth top to provide a place for the adults to rest after they have emerged and, also, to prevent their escape. The pans can be transported to any convenient and suitable place for observation, but must be kept cool and protected from the direct rays of the sun. An outdoor, open-air insectary, if cool and shaded, is an ideal place for them. A cool basement also will serve. Many of the mayflies for this report were reared at the fish hatchery of the Illinois Department of Conservation at Spring Grove, Illinois.

If more than a day or two elapses between the time the living nymphs are collected and the time subimagoes begin to appear, it is advisable to change the water in the rearing pans. This can be done by dipping out part of the water in each pan and replacing it with aerated, unchlorinated water of the same temperature.

When the subimagoes emerge, each specimen should be removed to a cap vial or other similar glass jar in which one moderate-sized green leaf has been placed. The cap of this container must not make an air-tight seal. The subimago has to lose water during the subimaginal stage, but the loss must not be too rapid. The leaf will maintain the humidity at a satisfactory level, while the loose cover on the container will permit the loss of water vapor by diffusion. If water is allowed to condense on the inside of the container, the subimago will almost certainly die without shedding the subimaginal pellicle.

The subimaginal skin will be shed usually within 24 to 36 hours, although, at very low temperatures, the subimago stage may last 2 or 3 days. When the adults emerge, they should be killed and mounted on pins. Notes should be made on the colors of the eyes and body of each specimen at the time of its death. The shed nymphal and subimaginal exuviae should be preserved in alcohol and be given the same lot accession numbers as the respective preserved nymphs.

The reared adults should be studied carefully to determine if all the specimens in each lot actually are of but a single species. The nymphs associated with each lot should also be studied critically at this time to determine if a pure culture of a single species is represented. The nymphs, the nymphal and subimaginal exuviae, and the adults should now be clearly and permanently labeled in such a way that there will never be doubts, in the future, as to the correct association of nymphal and adult specimens.

Some mayflies, such as *Isonychia, Ephemerella, Ameletus,* some species of *Baetis,* and *Baetisca,* cannot be reared successfully in pans. However, they can be reared in screen-wire cages partly submerged in the waters in which the nymphs live. When the subimagoes appear in these cages, they can be removed to cap vials, where the adults will emerge. The only real disadvantage of this method is that it requires that much time be spent in the field and makes difficult the finding of the shed nymphal skins intact.

Phillips (1930) was able to rear most of the New Zealand mayflies in laboratory aquariums, but I have had very poor results in my attempts to rear Illinois mayflies by this method.

In the past, many associations of nymphs

and adults have been made by the process of relating adults taken by sweeping vegetation around bodies of water with mature nymphs found in those waters at the same time. This method has led to so many misassociations that it should be followed only as a last resort, and the results should always be viewed with suspicion.

Collecting and Preserving

Both nymphs and adults must be collected very carefully if intact specimens are to be secured. The most valuable adult specimens are those reared from nymphs. Rearing not only yields a definite association of the nymph and the adult form but makes possible a collection of well-preserved specimens.

Collecting at lights will yield the largest number of adults. Although usually specimens taken at light traps are in very poor condition, some very worth-while Illinois records have been secured by these devices. Careful sweeping of the vegetation near bodies of water will yield much valuable adult material. Subimagoes secured by the same means should be placed in cap vials so that the adults can emerge a day or so later.

Experience has shown that adult mayfly specimens are best preserved dry, on pins. Each specimen may be pinned through the thorax from dorsum to venter with the wings spread in the conventional manner for entomological specimens, or it may be pinned on the side, with the wings to the left, the pin being inserted through the pleura of the thorax. The latter method is the more rapid, and the specimens, although perhaps not so neat in appearance, are more easily handled for study and can be stored in smaller space.

The wings may be spread during the drying process, without the use of a conventional spreading board, in the following manner. A square piece of 50-pound ledger paper, as large as the maximum wing expanse, when placed on the pin above the specimen will serve to hold the wings outspread at the proper angle during the 2 or 3 days required for drying. Another piece of the same weight paper of equivalent width, but twice as long, pushed up on the pin from below, will serve to hold the caudal filaments and fore legs at the proper extension. Extremely small specimens, such

as those of most species of *Caenis,* may be mounted on card points.

If a long series of adult mayflies is available, it is desirable to preserve some dry and some in ethyl alcohol. Dry specimens retain their color characters longer than those preserved in alcohol. Specimens I have seen that were collected by Benjamin D. Walsh of Rock Island, Illinois, and preserved dry for more than 80 years show most of the color characters fairly well. In alcohol, the colors fade so rapidly that 10-year-old specimens in many genera are almost impossible to identify. Specimens in 85 per cent alcohol retain their color longer than those in 70 per cent alcohol, but, after storage in 85 per cent alcohol for several years, specimens become so hardened that it is almost impossible to make satisfactory slide mounts of the genitalia. Although alcohol is not satisfactory in some respects, it provides an easy method of preservation and permits compact storage of large series taken at one time and place, and, in some instances, is better for the study of structural characters.

No matter how preserved, the adult specimens are extremely fragile and must be handled with the greatest care. Specimens, whether dry or in alcohol, should be stored so that they are not exposed to the direct rays of sunlight, as light quickly bleaches them.

If it is ever desirable to preserve subimagoes, they may be preserved in 70 per cent ethyl alcohol, as the colors of subimagoes are usually unimportant. Then, at any later date, the adult genitalia of such specimens may be dissected out of the subimaginal pellicles and be cleared and mounted in the same fashion as those from any adult specimens.

Nymphal specimens may be picked from rocks or other objects in the water or collected from the water itself by careful dipping with an aquatic dipnet. In running water, large numbers of specimens may be secured if a seine or dipnet is held in the current and the rocks, gravel, and other objects on the bottom upstream from the net are carefully turned over. The nymphs will release their holds on these objects and the current will carry them into the net.

Nymphs and exuviae are best preserved in 70 per cent ethyl alcohol. Preserving mixtures containing acetic acid or glycerin should be avoided, as they eventually make the specimens so soft that the parts will not hold together. Glycerin also eventually turns mayflies almost pitch black. Formalin hardens specimens so much that it should not be used.

Collections of specimens in alcohol, if in vials with cork stoppers, will almost certainly have many losses due to evaporation of the alcohol. The use of red rubber stoppers in the vials will greatly reduce evaporation, but, even with the best care, specimens will occasionally be found to have dried out. Many dried specimens can, however, be partially restored by careful treatment with trisodiumphosphate (Van Cleave & Ross 1947). This method of restoration often makes it possible to identify nymphs which formerly had to be discarded.

Study Preparations

For a study of the male genitalia, slide mounts of these structures must be prepared with special care. I have used the following procedures in material examined in this report.

The entire mayfly is first relaxed and the caudal filaments removed but saved in case it should be necessary to study them later. Then the apex of the abdomen is cut off and cleared in cold 10 per cent potassium hydroxide for a period of 6 to 12 hours. After this treatment the dissection is placed in distilled water and any undisintegrated muscle or abdominal contents are carefully teased out with fine dissecting needles. The preparation is transferred first to 50 per cent ethyl alcohol and then through two changes of 70 per cent alcohol, being left in each not less than 15 minutes. A few drops of acidulated acid fuchsin are added to the last change of 70 per cent alcohol, and after a minimum of 15 minutes the preparation is removed to 95 per cent alcohol. In this alcohol the tenth tergite and the bases of the caudal filaments are dissected off so that they will not obscure the structure of the penis lobes in the finished slide. The genitalia are then left in the 95 per cent alcohol until all excess stain has been washed away.

The preparation is mounted, directly from the 95 per cent alcohol, in balsam. The latter is a special medium made by diluting standard, filtered Canada balsam with 10 per cent turpentine. A small drop of this medium is put on the slide and the stained genitalia preparation placed in it; the transferring is done with a hooked dissecting needle, not a pipette. This turpentine mounting medium dries slowly enough to allow ample time for orienting the dissection correctly on the slide. Each specimen is mounted with the genital forceps down, care being taken to mount all dissections as nearly as possible in the same position, for the structures of the penis lobes of a single species may look quite different if seen from different angles. The preparation is then placed in a dust-tight box to dry. It should be examined at intervals over a period of several days and the dissection straightened, if necessary, with a dissecting needle. The coverslip is not put on until the preparation is almost completely dry.

In this paper, the drawings of the male genitalia were, except in Leptophlebiidae, made from the dorsal aspect, that is, with the penis lobes above the genital forceps. In the Leptophlebiidae, the genitalia were drawn from the ventral aspect, so as better to show the ventral appendages of the penis lobes.

In many instances, it is necessary to make dry mounts of the adult wings so that the venation can be studied critically. If the adult specimen is preserved in alcohol, the mount can be made directly, but, if the specimen is dry, it must first be relaxed. The wings from the alcoholic specimen or relaxed dry specimen are then carefully dissected off with dissecting needles or knives. As these wings are extremely fragile, the operation must be done with great care. The wings can easily be removed by severing the muscles at the base of each wing.

The detached wings are carefully washed in a watchglass containing 70 per cent alcohol to remove any dust or debris. It may be necessary to use a fine camel's-hair brush to remove all the dust; the brushing must be done very carefully, else the fragile wings will be torn.

Next a drop or two of 70 per cent alcohol is placed on a clean microscope slide and the wings are floated onto it. They should be spread and arranged symmetrically, dissecting needles being used for the manipulation. Then, before the alcohol evaporates, a square, No. 1 thickness coverslip is put on. Coverslips three-fourths inch or seven-eighths inch square will serve for most mayfly wings. The weight of this coverslip will serve to hold the wings flat and in place as the alcohol evaporates.

When most of the alcohol has evaporated, but the wings are still slightly damp, a narrow strip of gummed paper is moistened and affixed along each of the lateral margins of the coverslip. The strips of paper serve to hold the coverslip in place during the drying process. After the mount is completely dry (it is best to let the preparation stand 24 hours to be sure it is dry), strips of gummed paper are affixed to the top and bottom margins of the coverslip to make a permanent mount. Gummed paper having animal glue should be used, as it can be relied upon to adhere tightly to the polished glass surfaces for years. These slide preparations may then be studied and stored, as are the balsam mounts of the genitalia. They must, however, always be handled carefully, as the wings are most fragile after they are thoroughly dry.

Literature

The mayflies, in contrast with many other insect groups, are fairly well known for the world as a whole. The great majority of specimens of adults from anywhere in the world can be placed generically with the keys of Ulmer (1933, 1939). Generic keys to the nymphs of the world are not so readily available, but Ulmer (1940) keyed out many nymphs, and most of the faunal papers cited below include keys to nymphs as well as adults. The excellent monograph of Eaton (1883–1888) still serves for the generic and specific identification of the adults and nymphs of many forms from all parts of the world.

A great many papers have been published which include keys and descriptions of the species of a single country or region. Klapálek (1909), Ulmer (1924a), and Schoenemund (1930) published keys and descriptions for the German and central European species of mayflies. Kimmins (1942) treated the British species, Perrier (1934) the French, Grandi (1941–1951) the Italian, and Lestage (1928 et seq.) the Belgian. Chernova (1940) described and keyed the nymphs of the Russian species.

Barnard (1932, 1940) published on the South African species, Lestage (1925b) cataloged and described part of the North African species, and Ulmer (1930) published a paper on some Abyssinian species. Chopra (1927) published a comprehensive paper on the Indian species, while Traver (1939) considered species endemic to the Himalayan region. Phillips (1930) published a revision of the New Zealand species. Ulmer (1924c, 1939–1940) described the East Indian and Philippine species, Lestage (1921, 1924b) treated the species of Indo-China, while Ulmer (1926a) published a large paper on the species of China, and Uéno (1931) treated those of the Japanese fauna.

Traver (1938) published a work on the mayflies of Puerto Rico; she also (1944) treated many of the species of Brazil. Ulmer (1938) treated the Chilean species. Needham & Murphy (1924), Ulmer (1942–1943), and Traver (1946–1947) published major contributions on the South American and Central American species. Spieth (1943) made some taxonomic notes on several species from Surinam and other Neotropical localities.

The majority of the species of mayflies occurring in North America north of Mexico can be identified with the keys and descriptions of Traver (1935a). Despite the fact that some of the keys and many of the descriptions in her paper were compiled from the literature, and that various other workers have not agreed with her conclusions on a number of points, her work nevertheless remains the greatest single contribution yet made to the study of North American mayflies. This paper is especially valuable in that it brings together the great number of descriptions of North American species that appeared in the nearly 50-year period between 1888 (the time that Eaton's monograph was published) and 1935. McDunnough (1921–1943) published a long series of extremely valuable papers containing descriptions and illustrations of a great many of the North American species of mayflies. Berner (1940–1950) studied and keyed out the Florida species.

This list of works just cited does not, of course, constitute a complete bibliography of the world literature on mayfly classification, but it will serve to indicate the present extent of the comprehensive literature.

The Walsh Species

Benjamin D. Walsh became interested in mayflies and related insects about the year

Table 1.—Mayfly species described by Benjamin D. Walsh in 1862 and 1863, and status of lectotypes designated by Nathan Banks from specimens collected by Walsh and deposited by Herman A. Hagen in the Museum of Comparative Zoology, Harvard University, Cambridge, Massachusetts.

Walsh Species	MCZ Type No.	Condition	Label
Species Described in 1862			
Baetis sicca	11248, ♂	Good	B. sicca ♂ Rock Island Walsh 672
Potamanthus odonatus*	—	—	—
Palingenia vittigera*	—	—	—
Palingenia flavescens	11252, ♂	Good	Rock Island Walsh 675
Palingenia pulchella	11251, ♂	Bleached; genitalia gone	Rock Island Walsh 677
Palingenia terminata	11253, ♂	Good	Rock Island Walsh 679
Ephemera flaveola	11210, ♂	Good	E. flaveola Rock Island Walsh 688
Ephemerella excrucians	11213, ♂	Good; genitalia on slide	Ephemerella excrucians ♂ Rock Island Walsh 691
Ephemerella consimilis*	—	—	—
Cloe ferruginea*	—	—	—
Cloe fluctuans*	—	—	—
Cloe dubia	11214, ♂	Genitalia broken	Rock Island Walsh 700
Cloe mendax	11215, ♀	Faded	Rock Island Walsh 697
Species Described in 1863			
Baetis interlineata	Undesignated	Good	Baetis femorata Say Rock Island 1863
Pentagenia quadripunctata*	—	—	—
Heptagenia simplex	11250†	—	—
Heptagenia cruentata*	—	—	—
Heptagenia maculipennis	11249, ♂	Badly broken	maculipennis ♂ Rock Island Walsh 683
Ephemera myops	11209**	—	—
Cloe propinqua	11218, ♂	Good	Cl. propinqua Walsh Rock Island, not Hagen 667

* Type material missing.
† Should be disregarded; see McDunnough 1929:179 and page 184 below.
** Should be disregarded, as specimen was collected in 1864.

1860. He amassed a large collection of mayflies, principally from and near his home at Rock Island, Illinois. He also collected a few specimens from Coal Valley Creek, a tributary of the Rock River, in Rock Island County; from the Des Plaines River near Chicago; and from southern Illinois, along the Ohio River. He preserved all the specimens dry, on pins, and made careful notes on the colors of the specimens while they were still alive or very shortly after their death.

By 1862, Walsh had segregated these mayfly specimens to genera and species, basing his determinations principally on Hagen's *Synopsis of the Neuroptera of North America* (1861). He published the results of these studies in 1862. In his paper, he described 26 species of mayflies, of which he considered 13 to be new; 7 he identified as species previously described by Say or Pictet, and the remaining 6 he questionably referred to species described by Say, Walker, or Hagen. He also described two new genera of mayflies in his paper.

Shortly after the publication of this work, Walsh sent duplicate specimens of most of these species to Hagen, then living and working in Koenigsberg, Prussia. Hagen examined these specimens and returned critical notes to Walsh concerning them. These notes were published (Hagen 1863) along with Walsh's further observations on the species, in the light of Hagen's comments (Walsh 1863). In this latter paper, Walsh described three additional new genera and five new species; he also provided new names for two species he had misidentified in his first paper.

Hagen did not return Walsh's duplicate specimens to him, but added them to his own collection. Walsh sent specimens of additional species to Hagen in 1864; these Hagen added to his collection. Later, in 1870, Hagen brought his collection to the United States and deposited it in the Museum of Comparative Zoology, Harvard University. In the meantime, Walsh had died, in 1869, and his collection, containing the types of all his mayflies, was deposited in the Chicago Academy of Sciences. In 1871, this collection was completely destroyed in the great Chicago fire. As a result, the specimens of Walsh's species in the Hagen collection, still preserved at the Museum of Comparative Zoology, became the sole remaining authentic representatives of those species.

It should be noted that, although the true types of Walsh's species are destroyed, most of the specimens of Walsh material now in the Museum of Comparative Zoology may be considered to be cotypes. Some specimens, however, were collected by Walsh after the descriptions were published, as is shown by the specimen labels, and these specimens are only autotypes. Lectotypes for most of Walsh's species of mayflies have been designated by Nathan Banks. A number of workers, myself included, have studied these specimens. Table 1 gives a list of the species described by Walsh and the present status of the lectotypes.

In addition to the species listed in table 1, the Museum of Comparative Zoology collection includes specimens determined and labeled by Walsh as *Baetis arida* Say, *B. alternata* Say, *Potamanthus cupidus* Say, *Palingenia limbata* Pictet, *Palingenia bilineata* Say, *Palingenia interpunctata* Say, and *Baetisca obesa* Say. All were collected by Walsh at Rock Island.

In collecting material for the present Illinois report on mayflies, we made a great effort to secure good series of all the Walsh species from the type locality. The results were somewhat disappointing, as, of the 31 species described or identified by Walsh, we were able to secure only 8 at Rock Island, even with intensive collecting. This is not surprising when it is realized that the rivers around Rock Island are now quite different than they were in Walsh's time. In the 1860's the Mississippi and Rock rivers at Rock Island were large, rapid rivers (Walsh

1863: 202). Since Walsh's time, extensive dredging, channel straightening, and damming operations have greatly reduced the rapidity of flow of these rivers. As a result, the Mississippi River at Rock Island is now ecologically more like a lake than a rapid river, and the Rock River is extremely sluggish. Changes in the rivers have produced a corresponding change in the local mayfly fauna.

In Walsh's time, also, it was the annual practice, in spring, to float log rafts down the Rock and Mississippi rivers to the saw mills located in Rock Island (Walsh 1862: 372; 1863: 202). These rafts originated in the pine forests of Wisconsin and Minnesota. Walsh noticed that large numbers of mature mayfly nymphs were brought down the river to Rock Island with these rafts. The nymphs, in the accumulated debris of the rafts, crawled out on the logs to molt. Walsh collected much of his material from these log rafts. Needless to say, this source of specimens long since has disappeared from Rock Island.

Although we collected only 8 of Walsh's species at Rock Island, we ultimately secured 27 of his 31 species by searching in other localities. Most of the specimens we took at various points on the Rock River upstream from Rock Island, notably at Prophetstown, Dixon, Sterling, Oregon, and Rockford. Some also we took in southern Illinois at Mount Carmel, on the Wabash River.

Of the Walsh species which have not been collected again in Illinois, *Ephemerella excrucians* has recently been reared in northern Michigan (Leonard 1949:158). *E. consimilis,* based on a very brief, comparative description, and unrepresented in the Museum of Comparative Zoology collection, has of necessity remained unrecognized. Walsh identified Rock Island specimens as questionably belonging to two species described by other authors: *Baetis debilis* Walker and *Cloe unicolor* Hagen. The former has been shown by Eaton (1885: 253) to be a misidentification, but exactly what species Walsh had cannot now be determined. The latter was an identification of a Hagen species which is itself unrecognizable today. As we lack the original material, what species Walsh ascribed to the name *unicolor* cannot be ascertained with any degree of certainty.

External Morphology

The external features of both nymph and adult mayfly are described in detail in *The Biology of Mayflies* (Needham *et al.* 1935). Only those characters used in identification of Illinois species are mentioned here.

Many of the morphological structures which have proved useful in classification are shown in fig. 7 of a generalized mayfly adult and in fig. 8 of a generalized nymph. The terminology for the adult thorax is essentially the same as that used by Velma Knox (1935). The location of the various margins and areas of the wings is illustrated in figs. 7 and 9. The disc of the wing is the middle part, bounded roughly anteriorly by R_1 and posteriorly by Cu, but not including the marginal or extreme basal areas. The spines on various parts of the penis valves which have proved of diagnostic value, and their locations, are shown in fig. 10, of the generalized male genitalia.

All mayfly nymphs are strictly aquatic and respire by means of gills, which vary greatly in size and form in different species. Typically, there is one pair of gills on each of the first seven abdominal segments. They may be lamelliform (platelike), filiform (threadlike), or a combination of these two forms, figs. 51–55, 91*a–d*, 96, 113, 172, 199, and others.

When mayflies first emerge as winged insects, in the subimago stage, the surfaces of the semiopaque wings are covered with microtrichia; the wing margins and the caudal filaments are clothed with numerous long, slender setae. Beneath the subimaginal pellicle the adult eyes, legs, genitalia, and caudal filaments can be seen, contracted and wrinkled. Most mayflies remain in this stage about a day before shedding the subimaginal pellicle. Immediately following, the legs in most species and the caudal filaments become greatly lengthened and the eyes of the males greatly expanded; the wings are clear and hyaline, and they lack microtrichia and marginal setae. In a few mayfly adults, the legs are aborted, while, in some others, they are so greatly reduced as to be useless. Although mayflies typically have two pairs of wings, the anterior pair being much the larger, there are some species in which the hind wings are lacking.

In most adults, the compound eyes occupy the greater part of the head and, in the males of most species, the eyes are larger and closer together on top of the head than in the females. In the males of some genera, the eyes are greatly expanded and each is divided into an upper and lower portion, which may be further differentiated by size and color, figs. 255–257. In the nymphs, the compound eyes show a considerable range in size and position but none shows the high degree of development exhibited in adults.

The colors of the compound eyes of the male as given in the species descriptions and keys are those of the eyes while the specimens were still alive or had been dead only a short time. In mayflies, such as *Baetis,* which have divided eyes, the color of the upper portion of each eye is the significant one; the lower portion, as is well known, is a day-eye, with pigmentation so extensive that this part of the eye appears black. In such forms as *Stenonema* and *Heptagenia,* occasional specimens are to be found in which the entire eye is completely black, before death as well as after. These specimens should be disregarded, as this blackening is apparently due to the corneal layer of the eye separating from the hypodermis. After specimens have been killed and are thoroughly dry, the colors the eyes had in life disappear completely.

The other color characters as given in the descriptions in this report are, wherever possible, those of fresh, dry specimens. All colors fade or change somewhat within a few days after death of the specimens, regardless of the method of preservation; however, this loss of color is relatively slight in dry specimens stored away from the light. Specimens preserved in 70 per cent alcohol soon change color almost completely, becoming, after a few months, only a dull tan or yellow, even though the color in life may have been of various shades from dark brown to white. On dry specimens, the salient features of the color pattern are relatively permanent, but the delicate tints of red, green, gray, or yellow of fresh specimens eventually disappear or become scarcely discernible.

The system of nomenclature for wing veins used by Traver (1935*a*:119) has been followed in this report. It may be noted that this differs from the system of Tillyard (1923) in several respects. For instance, the branched, convex vein in the center of the fore wing called MA by Tillyard is called R_{4+5} in this report. The branches have been named M_1 and M_2 by Ulmer (1933, pls. I, II) and 6 and 6^1 by Eaton (1883, pls. I–XXIII). The vein R_{4+5} is of greatest importance, as it is the best landmark that can be used in identifying the veins of the mayfly fore wing. Once it has been located (except in the greatly specialized Oligoneuriidae), all other veins anterior and posterior to it can quickly be located.

Traver considered R_{4+5} to be a branch of R_s, whereas Tillyard believed it to be a

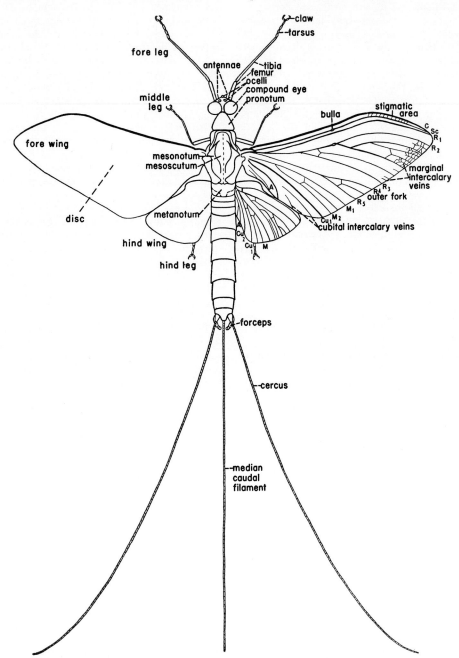

Fig. 7.—Generalized mayfly adult male, showing structures used in classification.

branch of M. Actually, as good a case can be made for considering this vein a branch of R_s as of M. It is not clearly joined to either, but sometimes appears to arise nearer the stem of R than of M. The poorly developed axillary sclerites of the mayfly wing give no conclusive evidence either way. Such being the case, I have decided to follow Traver's system, in which this vein is considered to be a branch of R_1, as her system was used consistently in *The Biology of Mayflies* (Traver 1935a). This work has

been, and is most likely to continue for many years to be, the standard reference work on North American mayflies.

It should further be noted that the veins called M_1 and M_2 by Traver have been called Cu_1 and Cu_2 by Ulmer, MP by Tillyard, and branches of vein 7 by Eaton. The veins Cu_1 and Cu_2 of Traver's system correspond to 1st and 2nd anal veins of Ulmer's, Cu_1 and Cu_2 of Tillyard's, and veins 8 and 9 of Eaton's.

In mayfly literature, it is frequently stated

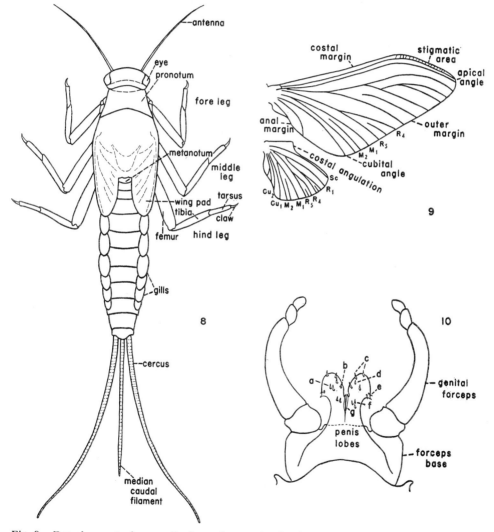

Fig. 8.—Dorsal aspect of generalized mayfly nymph, showing structures used in classification.
Fig. 9.—Generalized mayfly wings, showing principal veins and areas used in classification.
    Fig. 10.—Generalized mayfly male genitalia, showing structures used in classification: *a*, discal spines; *b*, subapical spines; *c*, apical spines; *d*, apicomesal spines; *e*, lateral spines; *f*, basal spines; *g*, mesal spines.

that the median caudal filament is wanting, but actually this filament is simply reduced to a minute, unsegmented or partly segmented vestige. In all adult mayflies there is at least some indication of it. The outer caudal filaments are the cerci.

An arbitrary system for the designation of sizes of specimens has been followed in this report. The size of an adult specimen is taken to be the length of the body without the head or fore legs and without the caudal filaments. Specimens having such a body length of 5 mm. or less are called small, those from 6 to 10 mm. long are called medium, and those over 1 cm. long are designated as large. The length of the body in nymphs is without the head or caudal filaments, and the sizes given refer to mature nymphs, that is, ones with dark wingpads.

Identification of Sexes

The adult male mayfly is easily recognized by the presence of a pair of forceps, or claspers, near the apex of the ninth sternite, as well as by two shorter structures, the penes, between the arms of the forceps. The forceps, except in the subfamily Campsurinae, are segmented, and both forceps and penes vary greatly in size and form, figs. 60–67. These same structures can be seen through the pellicle of the subimago and of the mature nymph, although in these two stages they are soft and less sclerotized than in the adult. They are not fully expanded in the subimago, and much less developed in the nymph.

In the female, the posterior margin of the ninth abdominal sternite is rounded, simple, and without prominent processes or additional structures. In a few genera, the female possesses a rudimentary ovipositor.

Classification

The order Ephemeroptera is one of the most archaic of winged insect groups. It is not closely related to any other order but, as Tillyard (1917) points out, certain points of resemblance give some slight evidence of a very ancient connection with the Odonata. These points of resemblance, not common to other orders, are the presence of only one wing axillary, the inability to fold the wings backward or downward over the abdomen, and the retention of abdominal gills in Ephemeroptera and in a few primitive

nymphs of Odonata (*Cora* and *Pseudophaea*). The fossil record indicates that the two orders were already differentiated in the Upper Carboniferous Period but were then more closely related than they are today (Tillyard 1917:6).

At the time Eaton wrote his *Revisional Monograph* (1883–88), the mayflies were considered to constitute the single family Ephemeridae. However, several years earlier Eaton (1869:132) had indicated that the family Ephemeridae could be subdivided into three major divisions, based on the habits and structures of the nymphs: (1) the burrowing forms with tusked mandibles, (2) the flat, crawling forms, and (3) the rather long, slender, free-swimming forms. In the *Revisional Monograph,* he divided the adults into three groups; these groups somewhat paralleled the divisions he had previously suggested, based on the nymphs. He further subdivided these three groups into 13 generic types; the three groups were not named as taxonomic categories.

Banks (1900:246) published a classification of the mayflies, considering them to represent but the one family Ephemeridae, but dividing this family into seven tribes: Baetiscini, Polymitarcini, Leptophlebini, Siphlurini, Ephemerini, Baetini, and Caenini.

Needham (1901:419) published a key to the nymphs of the family Ephemeridae and indicated that this family could be divided into three subfamilies, the Ephemerinae, Heptageninae, and Baetinae. Needham mentions (1905:29, footnote) that this key indicated subdivisions of the Ephemeridae into subfamilies which were very similar to those given in a manuscript key prepared earlier by C. A. Hart for use by students at the University of Illinois.

A few years later, Needham (1905:22) published a revised key to the family Ephemeridae in North America. In this key, which included both nymphs and adults, he again divided the Ephemeridae into the three subfamilies Ephemerinae, Heptageninae, and Baetinae. These three subfamilies corresponded only very roughly to the three groups into which Eaton had divided the family Ephemeridae in his classification.

Klapálek (1909) then published a greatly expanded classification of the mayflies, based on German species, dividing the order into

10 families; Palingeniidae, Polymitarcidae, Ephemeridae, Potamanthidae, Leptophlebiidae, Ephemerellidae, Caenidae, Baetidae, Siphlonuridae, and Ecdyonuridae. These families he placed in the order Ephemerida. It may be noted that Klapálek's classification bears considerable resemblance to the earlier classification of Banks (1900:246), and, as Ulmer (1920a: 98) remarks, the 10 families in Klapálek's classification correspond almost exactly to 10 of the unnamed subdivisions of the family Ephemeridae that Eaton had indicated in his monograph.

Ulmer (1914), in a new classification of the German mayflies, added the family Oligoneuriidae.

Bengtsson (1917) added the family Ametropidae to the classification and changed the name of the Ecdyonuridae of Klapálek's classification to the Heptageniidae. The same year, Lestage (1917) published a classification of the Palearctic mayfly nymphs in which he used only the families Ephemeridae, Heptageniidae, Baetidae, Oligoneuriidae, and the new family Prosopistomatidae.

Three years later, Ulmer (1920a: 99) published a revised classification for the mayflies of the world in which he combined the essential features of all the preceding classifications. He elevated the three subfamilies of Needham's classification to suborders, with 14 families beneath them, but no subfamilies. Ulmer's classification was arranged as follows:

Suborder Ephemeroidea
　Family Palingeniidae
　Family Polymitarcidae
　Family Ephemeridae
　Family Potamanthidae
Suborder Baetoidea
　Family Leptophlebiidae
　Family Ephemerellidae
　Family Caenidae
　Family Baetidae
　Family Oligoneuriidae
　Family Prosopistomatidae
Suborder Heptagenioidea
　Family Baetiscidae
　Family Siphlonuridae
　Family Ametropodidae
　Family Ecdyonuridae
　　　(Heptageniidae)

A few years later, Handlirsch (1925: 415) published a much more conservative classification for the mayflies of the world. In this he placed all the forms in one family,

but employed a number of subfamilies and tribes, as follows:

Family Ephemeridae
　Subfamily Siphlurinae
　　Tribe Siphlurini
　　Tribe Ametropodini
　　Tribe Ecdyurini
　Subfamily Baetiscinae
　Subfamily Prosopistomatinae
　Subfamily Baetidinae
　Subfamily Caenidinae
　Subfamily Leptophlebiinae
　　Tribe Ephemerellini
　　Tribe Leptophlebiini
　Subfamily Ephemerinae
　　Tribe Ephemerini
　　Tribe Potamanthini
　　Tribe Polymitarcini
　　Tribe Palingeniini
　Subfamily Oligoneuriinae

Ulmer (1933) later revised his classification of 1920 but made no changes in the arrangement of suborders and families. The same year, Spieth (1933) published a paper on the phylogeny of mayflies. His conclusions, based on a study of adult wings and male genitalia and nymphal gills and mouthparts, were that the North American mayflies represented four superfamilies and eight families, as follows:

Superfamily Siphlonuroidea
　Family Siphlonuridae
　Family Heptageniidae
　Family Baetidae
Superfamily Ephemeroidea
　Family Leptophlebiidae
　Family Ephemeridae
　Family Ephemerellidae
Superfamily Caenoidea
　Family Caenidae
Superfamily Baetiscoidea
　Family Baetiscidae

Two years later, *The Biology of Mayflies* (Needham *et al.* 1935) appeared. In this work the mayflies were divided into three families (the three subfamilies of Needham's earlier classification) and 17 subfamilies:

Family Ephemeridae
　Subfamily Palingeniinae
　Subfamily Ephoroninae
　Subfamily Ephemerinae
　Subfamily Potamanthinae
　Subfamily Campsurinae
　Subfamily Neoephemerinae
Family Heptageniidae
　Subfamily Heptageninae

Family Baetidae
   Subfamily Oligoneurinae
   Subfamily Ametropinae
   Subfamily Metretopinae
   Subfamily Siphlonurinae
   Subfamily Baetiscinae
   Subfamily Ephemerellinae
   Subfamily Leptophlebiinae
   Subfamily Caeninae
   Subfamily Baetinae
   Subfamily Prosopistomatinae

The supergeneric classification of the mayflies which I have adopted here does not coincide exactly with any of the previous classifications, although it more closely agrees with the classification of Banks (1900) than with that of any other author. I do not believe that all mayflies can be divided into only three main categories without introducing an unjustifiably large number of exceptions into the characterization of each of those three main divisions. Although the mayflies are an extremely archaic group, we still have living representatives of many of the diverse branches that have arisen within the order during its long history. As has been shown by Tillyard (1925, 1932), the mayflies reached their maximum abundance in the Permian Period and have declined since. In the long period from the Permian to the present, many quite distinct types of mayflies have arisen, all of which evidently represent considerable divergence from the Permian mayfly prototype. Characteristics of this ancestral mayfly type are discussed by Tillyard (1932) and Carpenter (1933).

In our present-day mayfly fauna, we retain representatives of many of these diverse lines of mayfly evolution. The family classification should, as much as possible, reflect these degrees of divergence from the ancestral mayfly prototype. Certainly there are more than 3, and, in my opinion, there are at least 11 distinct lines of descent. I have accordingly distributed our Illinois mayflies among 10 families; the eleventh, Prosopistomatidae, is not represented in North America. This family classification has been arrived at through an evaluation of all available characteristics in nymphs and adults. The conspectus of the supergeneric classification followed here is given below:

Order Ephemeroptera
   Family Ephemeridae
      Subfamily Campsurinae
      Subfamily Potamanthinae
      Subfamily Ephoroninae
      Subfamily Ephemerinae
      Subfamily Palingeniinae
   Family Neoephemeridae
   Family Caenidae
   Family Ephemerellidae
   Family Baetiscidae
   Family Prosopistomatidae
   Family Oligoneuriidae
   Family Leptophlebiidae
   Family Baetidae
      Subfamily Siphlonurinae
      Subfamily Isonychiinae
      Subfamily Baetinae
   Family Ametropidae
   Family Heptageniidae

It must be admitted that varying degrees of relationship are indicated among some of these families. The Caenidae almost certainly arose from an ephemerid ancestor, as is shown by the still existing but extremely rare interstitial forms which are placed here in the Neoephemeridae. The point of divergence, however, of the Caenidae from the ephemerid stem must have been quite remote. The Ephemerellidae, Oligoneuriidae, Baetiscidae, and Prosopistomatidae apparently have no near relatives in the recent fauna. The Leptophlebiidae possibly arose from the same stem which produced the Baetidae, although the similarity between the two is slight. On the other hand, the Heptageniidae and the Baetidae must have arisen from an ancestor common to both of these families; the rare but still existing interstitial forms between these two families indicate this probable relationship.

The forms here included in the Baetidae admittedly represent rather widely divergent types. The Siphlonurinae contain quite archaic forms that (Spieth 1933: 329) probably arose very early from the Permian mayfly prototype, while the Baetinae contain greatly reduced adult forms that probably arose much later. The baetine and siphlonurine nymphs are, however, quite similar, and most adult baetine structures can be derived by simple reduction from corresponding adult structures in the Siphlonurinae. The divided compound eyes of the adult male baetines, on the other hand, seem to be strikingly different from the eyes of the siphlonurines, but actually the beginnings of the development of this divided eye

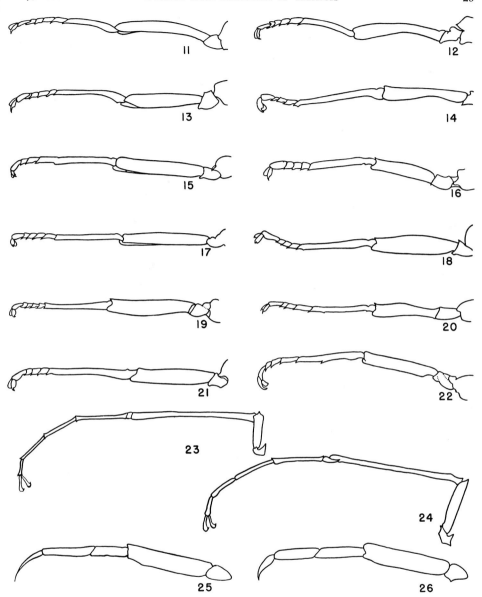

Fig. 11.—*Caenis simulans*, hind leg of adult male.

Fig. 12.—*Brachycercus lacustris*, hind leg of adult male.

Fig. 13.—*Tricorythodes atratus*, hind leg of adult male.

Fig. 14.—*Neoephemera purpurea*, hind leg of adult male.

Fig. 15.—*Callibaetis fluctuans*, hind leg of adult male.

Fig. 16.—*Ephemerella needhami*, hind leg of adult male.

Fig. 17.—*Stenonema pulchellum*, hind leg of adult male.

Fig. 18.—*Isonychia sicca*, hind leg of adult male.

Fig. 19.—*Heptagenia* sp., hind leg of adult female.

Fig. 20.—*Siphlonurus alternatus*, hind leg of adult female.

Fig. 21.—*Leptophlebia nebulosa*, hind leg of adult male.

Fig. 22.—*Siphloplecton interlineatum*, hind leg of adult male.

Fig. 23.—*Ephoron leukon*, fore leg of adult male.

Fig. 24.—*Ephemera simulans*, fore leg of adult male.

Fig 25.—*Siphloplecton interlineatum*, hind leg of mature nymph.

Fig. 26.—*Ameletus lineatus*, hind leg of mature nymph.

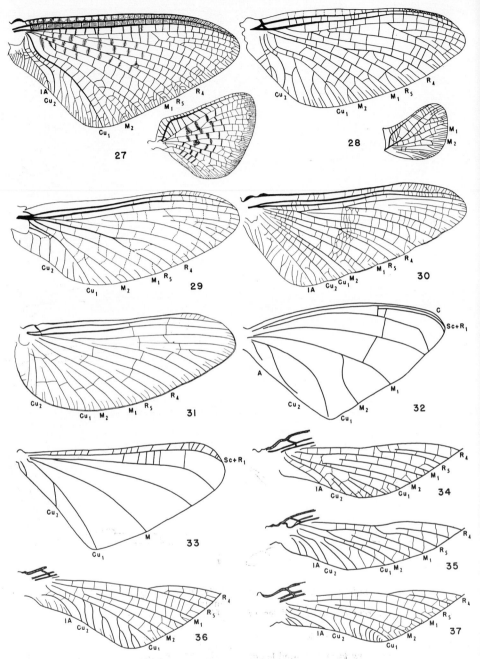

Fig. 27.—*Hexagenia bilineata,* wings.
Fig. 28.—*Neoephemera purpurea,* wings.
Fig. 31.—*Baetis propinquus,* fore wing.
Fig. 32.—*Lachlania saskatchewanensis,* fore wing. (After Ide.)
Fig. 33.—*Oligoneuria anomala,* fore wing. (After Eaton.)
Fig. 34.—*Siphloplecton basale,* posterior half of fore wing.

Fig. 29.—*Ephemerella lutulenta,* fore wing.
Fig. 30.—*Baetisca obesa,* fore wing.
Fig. 35.—*Paraleptophlebia praepedita,* posterior half of fore wing.
Fig. 36.—*Isonychia rufa,* posterior half of fore wing.
Fig. 37.—*Siphlonurus quebecensis,* posterior half of fore wing.

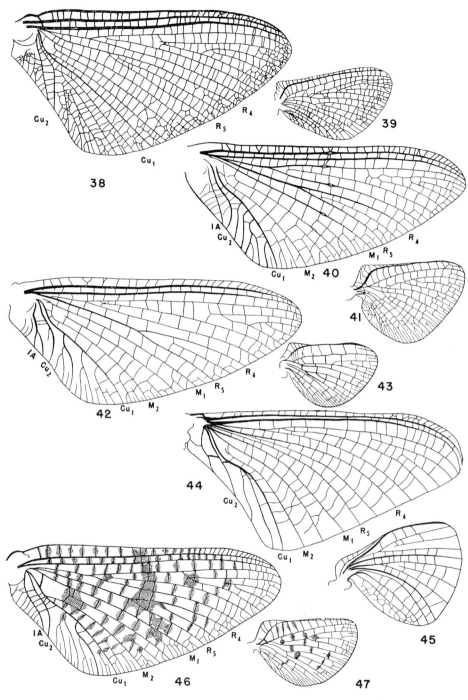

Fig. 38.—*Ephoron leukon*, fore wing.
Fig. 39.—*Ephoron leukon*, hind wing.
Fig. 40.—*Pentagenia vittigera*, fore wing.
Fig. 41.—*Pentagenia vittigera*, hind wing.
Fig. 42.—*Potamanthus verticis*, fore wing.

Fig. 43.—*Potamanthus verticis*, hind wing.
Fig. 44.—*Tortopus primus*, fore wing.
Fig. 45.—*Tortopus primus*, hind wing.
Fig. 46.—*Ephemera simulans*, fore wing.
Fig. 47.—*Ephemera simulans*, hind wing.

can be seen clearly in the living adult male siphlonurines.

KEY TO FAMILIES

Adults

1. Lateral ocellus in both sexes extremely large in comparison with a compound eye; each ocellus approximately one-half as large as a compound eye . . . . **Caenidae**, p. 43
   Lateral ocellus in males never more than one-tenth as large as a compound eye; in females lateral ocellus not more than one-fourth as large as a compound eye 2
2. Vein Sc of fore wing wanting or united with R_1, figs. 32, 33 . . . . **Oligoneuriidae**, p. 79
   Vein Sc of fore wing present as a separate vein extending from base to apex of wing, figs. 27–31 3
3. Cubital intercalary veins entirely absent, crossveins in disc of fore wing weak and netlike, fig. 30 . . . . . . **Baetiscidae**, p. 75
   Cubital intercalary veins present, crossveins in disc of wing not netlike, figs. 27–29, 31 4
4. Hind tarsus with five clearly differentiated segments, true first tarsal segment not fused with tibia, figs. 17, 19 . . . . **Heptageniidae**, p. 151
   Hind tarsus with only three or four clearly differentiated segments, figs. 14–16, 18, 20–22 5
5. Vein M_2 of fore wing sharply bent near base, figs. 27, 28, running parallel with vein Cu_1 in this area 6
   Vein M_2 of fore wing straight throughout its length, figs. 30, 31, 34–37, or curved no more than in fig. 29 7
6. Costal crossveins of fore wing in area basad of bulla partly or almost completely atrophied; veins M_1 and M_2 of hind wing separating distad of the center of wing; fig. 28 . . . . **Neoephemeridae**, p. 42
   Costal crossveins in both fore and hind wings well developed; veins M_1 and M_2 of hind wing separating proximad of center of wing; figs. 27, 38–47 . . . . **Ephemeridae**, p. 27
7. Fore wing with one or two long intercalary veins between M_2 and Cu_1, fig. 29 . . . . **Ephemerellidae**, p. 55
   Fore wing without long intercalaries between veins M_2 and Cu_1, figs. 34–37 8
8. Vein Cu_2 of fore wing angularly bent toward inner wing margin, figs. 35, 185–192 . . . . . . **Leptophlebiidae**, p. 81
   Vein Cu_2 of fore wing straight or evenly curved, figs. 31, 34, 36, 37 9
9. Cubital intercalary veins of fore wing consisting of one or two pairs of long, parallel veins, free marginal veinlets always absent, fig. 34; hind wings invariably present. **Ametropidae**, p. 144

Cubital intercalary veins of fore wing either a series of short, slightly sinuate veins extending from vein Cu_1 to inner wing margin, figs. 36, 37; or cubital intercalaries one or two long, basally detached veins accompanied by free marginal veinlets, figs. 31, 220–222; hind wing sometimes absent . . . . **Baetidae**, p. 97

Mature Nymphs

1. Gills on abdominal segments 1–6 concealed under carapace-like projection of thoracic notum, fig. 181 . . . . **Baetiscidae**, p. 75
   Abdominal gills not concealed under carapace-like projection of thoracic notum 2
2. Second abdominal segment bearing a pair of operculate or lidlike gills which cover gills on segments 3–6; gills lacking on segment 7; figs. 88, 96, 113, 114 3
   Second abdominal segment not bearing a pair of operculate gills which cover more posterior gills; operculate gills, if present, borne by segment 4; gills present on segment 7 4
3. Operculate gills fused on meson, connate; a hooklike, median spine at posterior margin of abdominal tergites 6–8, fig. 88 . . . . **Neoephemeridae**, p. 42
   Operculate gills not fused on meson; abdominal tergites 6–8 without hooklike spines, figs. 96, 113, 114 . . . . **Caenidae**, p. 43
4. Gills of first abdominal segment large and situated on the venter, fig. 184 . . . . **Oligoneuriidae**, p. 79
   Gills of first abdominal segment dorsal, or first abdominal segment without gills . . 5
5. Gills always absent from second abdominal segment and sometimes absent from third segment, also . . . . **Ephemerellidae**, p. 55
   Gills present on abdominal segments 1–7 6
6. Tibia of hind leg shorter than hind tarsal claw, figs. 301, 312; or claw at least six times as long as wide at its widest point, fig. 25 . . . . **Ametropidae**, p. 144
   Tibia of hind leg as long as, or longer than, hind tarsal claw, figs. 26, 247, 254, 266, 385; claws relatively thick 7
7. Each mandible with a projecting tusk, visible from dorsal aspect, figs. 2, 55, 59 8
   Mandibles without projecting tusks . . . 9
8. Gills relatively broad, biramous, with margins ciliate, figs. 51–54 . . . . **Ephemeridae**, p. 27
   Gills slender, filamentous, and bare . . . . **Leptophlebiidae**, p. 81
9. Head flattened dorsoventrally, prognathous, eyes dorsal, and labrum mostly or completely concealed under projecting anterior margin of head, figs. 360, 383–386, 390, 394, 395 . . . . **Heptageniidae**, p. 151
   Head hypognathous, eyes lateral, or, if prognathous, labrum completely ex-

EPHEMERIDAE

The family Ephemeridae, as defined here, corresponds to the families Ephemeridae, Polymitarcidae, and Potamanthidae in the classification of Ulmer (1933: 195), except as to be approximately parallel with Cu_1 in the basal area. The basal costal crossveins are well developed, and the stigmatic crossveins usually are not anastomosed. In many ephemerids, each compound eye in the males is divided into a dark lower portion and a light-colored upper portion, but the facets usually are the same size in each portion of the eye. The eye of the male in some species is not much larger than the eye of the female. The fore tarsus in the males has five segments, the basal one very short, figs. 23, 24. The hind tarsus in both

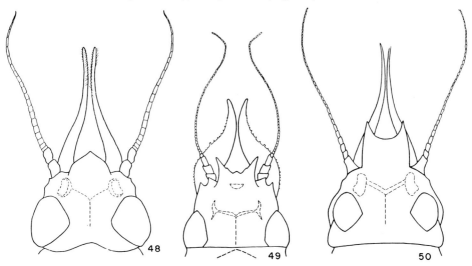

Fig. 48.—*Hexagenia bilineata*, doral aspect of head of nymph.
Fig. 49.—*Pentagenia vittigera*, dorsal aspect of head of nymph.
Fig. 50.—*Ephemera simulans*, dorsal aspect of head of nymph.

that *Neoephemera,* included by Ulmer in the Ephemeridae, is excluded from this family and placed in a new family, Neoephemeridae. The classification of the Ephemeridae used here is identical with that of Traver (1935a: 240), except for her inclusion of *Neoephemera* in this family.

The Ephemeridae include the largest mayflies occurring in Illinois. These are the mayflies that on warm summer evenings commonly emerge in enormous numbers from our larger rivers and lakes, and cover bridges and water-front buildings.

The wings of adult ephemerids have extremely numerous crossveins, and, except in *Tortopus* and *Campsurus,* a band of fine, short veinlets along the outer margin of each wing, figs. 38–43, 46, 47. In the fore wing, vein Cu_1 is sinuate near the base, and the posterior branch of vein M_2 is curved so sexes has only four clearly differentiated segments. In *Tortopus,* the middle and hind legs are nonfunctional and almost completely degenerated; in *Ephoron* the legs are reduced somewhat and are nonfunctional.

The nymphs of all ephemerids are provided with mandibular tusks and a row of completely exposed, biramous gills on either side of the abdomen, figs. 2, 55, 59. These nymphs live in the sand, gravel, or silt on the bottoms of our larger streams and lakes. Although they remain almost completely buried most of the time, they can sometimes be seen swimming freely in fairly deep water near shore. They have been observed to swim with a characteristic darting and undulating motion at about a foot beneath the surface. The periods in which the nymphs swim freely in the water may be near the time of molting. Nymphs in an aquarium were observed to leave the sand in the bottom, swim about in the water, and, after molting, re-bury themselves in the sand.

KEY TO SUBFAMILIES

ADULTS

1. Veins Sc and R_1 of fore wing curved posteriorly and continued around apical angle of wing; marginal veinlets wanting, fig. 44. Middle and hind legs of both male and female completely atrophied beyond trochanter, or femur, tibia, and tarsus degenerated to mere membranous flaps; male forceps unsegmented, fig. 60...................
   . . . . **Campsurinae,** p. 28
   Veins Sc and R_1 of fore wing straight at apexes; marginal veinlets present, figs. 38, 40, 42, 46. Middle and hind legs beyond trochanter not atrophied or degenerated to membranous vestiges; male forceps segmented, figs. 61–63, 65–67.........................2

2. Marginal veinlets of fore wing extremely numerous; cubital intercalary veins straight, not attached at bases to Cu_1, fig. 38...........**Ephoroninae,** p. 32
   Marginal veinlets of fore wing relatively few in number; cubital intercalary veins sinuate, attached at bases to Cu_1, figs. 42, 46.......................3

3. First anal vein of fore wing forked near wing margin, fig. 42.................
   ....**Potamanthinae,** p. 30
   First anal vein of fore wing not forked, fig. 46...........**Ephemerinae,** p. 35

MATURE NYMPHS

1. Gills lateral, fig. 55.................
   ....**Potamanthinae,** p. 30
   Gills dorsal, figs. 2, 59................2
2. Head without a frontal process.........
   ....**Campsurinae,** p. 28
   Head with a frontal process, figs. 2, 59. .3
3. Mandibular tusks upcurved............
   ....**Ephemerinae,** p. 35
   Mandibular tusks downcurved........
   ....**Ephoroninae,** p. 32

CAMPSURINAE

The subfamily Campsurinae, as defined here, corresponds to the first section of the family Polymitarcidae in the classification of Ulmer (1933: 197). It includes only *Tortopus* and the very closely related genus *Campsurus* in the Nearctic region. Together, these two genera include about 50 species in South America and Central America, but only 4 species of campsurine mayflies have been described or identified from America north of Mexico. Of these, 2 occur in Texas only, another cannot be

identified at present, and the fourth is quite generally distributed in eastern North America. The nymphs are unknown for the Nearctic species.

KEY TO GENERA

ADULTS

Middle and hind legs reduced to functionless, membranous vestiges, but with all leg-parts still discernible..............**1. Tortopus**
Middle and hind legs completely aborted beyond the trochanters........**2. Campsurus**

1. *TORTOPUS* Needham & Murphy

Tortopus Needham & Murphy (1924:23).

In the adults, the fore wing has veins C and Sc recurved around the apical angle of the wing, fig. 44; the veinlets along the apical wing margin are absent, vein M is forked at the wing base, and there are two long, cubital intercalary veins. The fore leg in the males is developed normally, but the middle and hind legs, as well as all of the legs in the females, are reduced to small, nonfunctional semimembranous vestiges, which, however, have all leg-parts still discernible. The females never molt to the imago stage, but mate and lay their eggs as subimagoes. In adults of both sexes, gill stumps are retained along the lateral margins of the abdomen. The median caudal filament is vestigial in the males, but is well developed in the females.

It has, unfortunately, not been possible as yet to find the nymphs of *Tortopus* in Illinois. The nymphs are presumed to be burrowing forms, relatively close in structure to the nymphs of *Campsurus*. Characteristics for the latter have been given by Needam & Murphy (1924:13) and Ulmer (1920b:17).

Tortopus primus (McDunnough)

Campsurus primus McDunnough (1924a:7).
Campsurus incertus Traver (1935a:286).
Campsurus manitobensis Ide (1941:155).

As has been pointed out before (McDunnough 1926:185; Traver 1935a:288), this species may be a synonym of *Palingenia puella* Pictet (1843: 145). The type of *puella* has not, so far as is known, been located and compared with recently collected material. Pictet's description was apparently drawn from a single mutilated female specimen, almost certainly a sub-imago. The species *puella* can probably never be identified with certainty. For the present, the name *primus* may be used for the species.

MALE.—Length of body 10–14 mm., of fore wing 9–12 mm. Body of living specimen snow-white, with wings white and costal margin of each fore wing faintly shaded with bluish gray. Antennae and apical tarsal segments of each fore leg shaded with bluish gray.

Fore leg with femur and tibia approximately equal in length, first tarsal segment one-third as long as tibia, second to fifth tarsal segments equal in length, each twice as long as first segment; longer of two tarsal claws as long as fifth tarsal segment, shorter claw two-thirds as long as longer one. Middle and hind legs greatly reduced, but all parts still represented by small, membranous vestiges. Genitalia, fig. 60, with two characteristically broad, flat penis lobes.

FEMALE.—Length of body 13–17 mm., of fore wing 16–22 mm. Body of living specimens snow-white, but often with a faint, longitudinal band of grayish brown on meson of pronotum and on dorsum of ab-

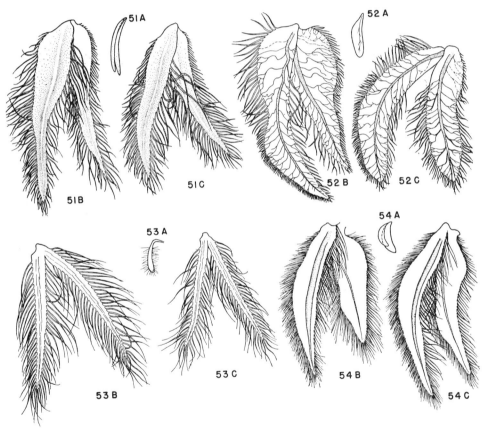

Fig. 51*A*.—*Hexagenia limbata*, gill of first abdominal segment.
Fig. 51*B*.—*Hexagenia limbata*, gill of fourth abdominal segment.
Fig. 51*C*.—*Hexagenia limbata*, gill of seventh abdominal segment.
Fig. 52*A*.—*Ephoron leukon*, gill of first abdominal segment.
Fig. 52*B*.—*Ephoron leukon*, gill of fourth abdominal segment.
Fig. 52*C*.—*Ephoron leukon*, gill of seventh abdominal segment.
Fig. 53*A*.—*Potamanthus myops*, gill of first abdominal segment.
Fig. 53*B*.—*Potamanthus myops*, gill of fourth abdominal segment.
Fig. 53*C*.—*Potamanthus myops*, gill of seventh abdominal segment.
Fig. 54*A*.—*Pentagenia vittigera*, gill of first abdominal segment.
Fig. 54*B*.—*Pentagenia vittigera*, gill of fourth abdominal segment.
Fig. 54*C*.—*Pentagenia vittigera*, gill of seventh abdominal segment.

domen. Mesoscutum of thorax with three longitudinal, obscure, brown lines which converge toward scutellum. Dorsal meson of metathorax vaguely stained with grayish

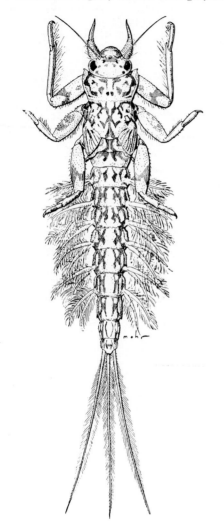

Fig. 55.—*Potamanthus* sp., mature nymph, dorsal aspect.

brown. All legs greatly reduced, semi-membranous, nonfunctional.

Known from Alabama, Arkansas, Georgia, Illinois, Kansas, Manitoba, Missouri, Nebraska, Ontario, and Tennessee. Develops in large, slow rivers.

Illinois Records.—ALTON: Aug. 29, 1913, 2♀. BLOOMINGTON: C. C. Adams, 10♀. CHAMPAIGN: Sept. 21, 1892, C. A. Hart, 1♀. ELIZABETHTOWN: at light, July 14,

1948, Mills & Ross, 1♀. GRAND TOWER: Aug. 14, 1898, C. A. Hart, 73♂. HAVANA: Aug. 10, 1889, C. A. Hart, 1♀; White Oak Creek, Aug. 14, 1896, C. A. Hart, 3♀. MOMENCE: Aug. 16, 1938, Ross & Burks, 1♀. OQUAWKA: Sept. 26, 1947, H. H. Ross, 30♀. QUINCY: Aug. 10, 1889, C. A. Hart, 1♀. SHAWNEETOWN: Oct. 3, 1942, Frison & Ross, 1♀. URBANA: at light, Aug. 23, 1943, H. B. Petty, 5♀; Sept. 20, 1909, 2♀.

2. *CAMPSURUS* Eaton

Campsurus Eaton (1868:83).

The adults of *Campsurus* differ from those of *Tortopus* mainly in the structure of the legs; the middle and hind legs in both sexes in *Campsurus* are completely aborted beyond the trochanters. The difference in the wing venation between the two, described by Needham & Murphy (1924:23) when they defined the genus *Tortopus,* is not reliable, according to Ulmer (1942:108).

Campsurus decoloratus (Hagen) (1861: 43), known from Texas and Mexico, and *circumfluus* Ulmer (1942:110), described from Texas, are the only known Nearctic species. *C. puella* (Pictet) (1843:145), described from Louisiana, has been tentatively placed in *Campsurus,* but is at present unidentifiable.

POTAMANTHINAE

The subfamily Potamanthinae corresponds to Ulmer's family Potamanthidae (1933: 199). It has only one Nearctic genus.

3. *POTAMANTHUS* Pictet

Potamanthus Pictet.(1845:208, pl. 25).

The adults are fairly large, whitish mayflies with the vertex and the dorsum of the thorax light reddish brown. The marginal intercalary veins of the wings are not net-like, fig. 42. In the fore wing, the basal part of vein M_2 is more strongly curved to the rear than is vein Cu_1, fig. 42, and the first anal vein is forked near the wing margin. In the hind wing, the costal projection is acute and veins M_1 and M_2 diverge near the wing base. The middle and hind legs of the adults are functional. The male genitalia, fig. 61, have the penis lobes broad, flattened, and almost completely fused on the meson; each arm of the forceps has three segments. The nymphs, fig. 55, are sprawlers, with the gills extended laterally. The median caudal filament is well developed in the nymphs and in the adults of both sexes.

Adult specimens of *Potamanthus* should be studied when freshly killed, as the faint

myops. In the Museum of Comparative Zoology there is a specimen labeled as the type of *myops,* but it was collected a year after the description was published and cannot, therefore, be the type. It is, however, a specimen determined as *myops* by Walsh himself, and is in agreement with the present-day concept of the species.

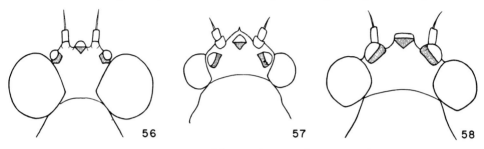

Fig. 56.—*Potamanthus verticis,* dorsal aspect of head of adult male.
Fig. 57.—*Potamanthus myops,* dorsal aspect of head of adult male.
Fig. 58.—*Potamanthus distinctus,* dorsal aspect of head of adult male.

color markings fade rapidly after death. This fading occurs much more rapidly in alcoholic than in dried specimens.

Reliable specific characters for the nymphs of this genus have not been found.

KEY TO SPECIES

ADULTS

1. Abdomen usually entirely unmarked, occasionally each abdominal segment with a faint, minute, pink spot on either side; crossveins in fore wing hyaline in each sex; compound eyes of male small, fig. 57....................**1. myops**
   Abdomen with large, well-marked, lateral, salmon-pink spots or stripes.........2
2. Abdomen with lateral stripes; fore wing in each sex with black crossveins; compound eyes of male moderate in size, fig. 58................**3. distinctus**
   Abdomen with lateral spots...........3
3. Female with crossveins of each fore wing black, male with those crossveins hyaline, or with only a few of the anterior crossveins black; compound eyes of male large, fig. 56.........**2. verticis**
   Both male and female with all crossveins of each fore wing black; compound eyes of male small, as in fig. 57.
   ....**4. neglectus**

1. *Potamanthus myops* (Walsh)

Ephemera myops Walsh (1863:207).
Potamanthus medius Banks (1908:259).

A study of the type of *medius* leaves no doubt in my mind that it represents the same species as the one we have been calling

MALE.—Length of body 10–13 mm., of fore wing 11–14 mm. Compound eyes small, fig. 57; wings completely hyaline, with no crossveins black; abdomen without lateral, salmon-pink spots or stripes, or, rarely, with small, faint, lateral spots discernible in living specimens; caudal filaments with articulations usually slightly darkened with red-brown or pink.

FEMALE.—Length of body 11–13 mm., of fore wing 12–14 mm. Eyes same size as in male; wings with no crossveins black; abdomen without lateral, salmon-pink spots or stripes; caudal filaments with articulations usually slightly darkened with red-brown or pink.

Known from Illinois, Indiana, Iowa, Kansas, Michigan, and Wisconsin. Develops in large and moderate-sized rivers.

Illinois Records.—Adult specimens, collected June 6 to August 17, are from Aurora (Fox River), Champaign, Dixon (Rock River), East Dubuque (Mississippi River), Effingham, Freeport, Galesburg, Homer, Kankakee, Monmouth, Monticello, Muncie, Oakwood, Oregon, Peoria, Rockford, Rock Island, Rockton, Sterling, and Urbana.

2. *Potamanthus verticis* (Say)

Baetis verticis Say (1839:42).
Ephemera flaveola Walsh (1862:377).

The type of Say's species is lost, but both male and female types of Walsh's species

are in the Museum of Comparative Zoology. By common consent, the identity of
this species has long been based on the
characters of Walsh's species, although Say's
name has priority. There is at present no
reason for changing this practice, as is
mentioned below under *neglectus*.

MALE.—Length of body 7–9 mm., of fore
wing 8–10 mm. Compound eyes large, fig.
56; wings with all crossveins hyaline or, in
occasional specimens, with a few anterior
crossveins of each fore wing darkened; abdomen with a row of salmon-pink spots
on either side; cerci and median caudal filament with articulations darkened.

FEMALE.—Length of body 8–10 mm., of
fore wing 9–11 mm. Compound eyes small,
each less than half the size of eye of male;
wings with crossveins darkened; abdomen
with a row of salmon-pink spots on either
side; cerci and median caudal filament with
articulations darkened.

Known from the midwestern and northeastern states. Develops in large and moderate-sized rivers.

Illinois Records.—Adult specimens, col-
lected May 2 to August 16, are from An-
tioch, Aurora, Bloomington, Champaign,
Dixon (Rock River), Foster (Mississippi
River), Hardin (Illinois River), Kankakee,
Keithsburg, Mount Carmel, Oregon, Proph-
etstown (Rock River), Quincy, Rockford,
Rock Island, Rockton, Savanna (Mississippi
River), South Beloit, Sterling, Warsaw
(Mississippi River), Wilmington, Yorkville.

3. *Potamanthus distinctus* Traver

Potamanthus distinctus Traver (1935a: 280).

The crossveins of the fore and hind wings
are black in both sexes; the fore wing is 11
mm. long; there is a reddish-tan stripe on
the vertex, pronotum, and the anterior part
of the mesonotum; the abdomen has a
salmon-pink stripe on either side; and the
articulations of the cerci and the median
caudal filament are darkened. The compound eyes of the male are moderately large,
fig. 58.

Known from New York and Ohio.

4. *Potamanthus neglectus* Traver

Potamanthus neglectus Traver (1935a: 282).

In describing this species, Traver mentioned that it might eventually prove to be a
synonym of *verticis*. There is, however,
nothing in the original description of Say's
species that would conclusively decide the
matter; if *neglectus* were to be placed as a
synonym of *verticis,* it would then be necessary to resurrect from synonymy the name
flaveola for that species at present being
called *verticis*. I prefer to follow McDunnough (1926:186) in considering *flaveola* a
synonym of *verticis,* as there is no strong
reason for not doing so, and the present
known distribution of the species involved is
in agreement with that practice. *P. verticis*
was described from Indiana, and *flaveola*
was described from Illinois; the species now
going under the name *verticis* occurs in the
midwestern and northeastern states. *P.
neglectus* is known only from the Atlantic
Seaboard.

The crossveins in the wings in both sexes
are darkened; the length of the fore wing
of the male is 8–9 mm.; there is a reddish-
brown median stripe on the vertex of the
head, on the pronotum, and on the mesonotum; the abdomen has a row of salmon-pink
spots on either side; the caudal filaments
have the articulations darkened; the compound eyes of the male are small, as in
myops, fig. 57.

Known from Maryland, New York, and
Pennsylvania.

EPHORONINAE

The subfamily Ephoroninae includes only
one genus in the Nearctic region, *Ephoron*.
As used here, this subfamily corresponds
to the second section of the family Polymitarcidae in Ulmer's classification (1933:
197).

4. *EPHORON* Williamson

Ephoron Williamson (1802:71).
Polymitarcys Eaton (1868:84).

The adults of *Ephoron* are fairly large,
snow-white mayflies with all legs of the
females and the middle and hind legs of
the males greatly reduced and functionless.
The females do not molt to the adult stage,
but mate and lay their eggs as subimagoes.
The costal and subcostal areas of the fore
wing are grayish purple; otherwise the wings
are snow-white. These wings, fig. 38, have
extremely abundant crossveins and netlike
marginal intercalaries suggesting the archaic

orthopteroid archedictyon. The cubital intercalaries of the fore wing consist of three or four long, straight veins, the posterior one of which is attached to the anal wing margin by a series of confused, short, and irregular veinlets. The hind wing has a blunt costal angulation. In the fore leg of the males, fig. 23, the tarsus is normally developed but the femur is quite short. In the females, the median caudal filament is well developed, while, in the males, it is reduced to a minute rudiment.

The nymphs, fig. 59, have prominently ‑oothed, downcurved mandibular tusks; the gills have short, relatively inconspicuous marginal ciliae, figs. 52B, 52C.

Reliable characters for separating the females to species have not yet been found.

KEY TO SPECIES

ADULT MALES

Mesonotum dark brown; apicolateral angle of each penis lobe rounded, fig. 66..........
....**1. leukon**
Mesonotum light yellow, shaded with tan; apicolateral angle of each penis lobe acute, fig. 67........................**2. album**

MATURE NYMPHS

Gills on abdominal segments 2–6 with lateral tracheal branches pigmented, figs. 52B, C, 59.............................**1. leukon**
Gills on abdominal segments 2–6 with lateral tracheal branches hyaline........**2. album**

1. *Ephoron leukon* Williamson

Ephoron leukon Williamson (1802:71).
Polymitarcys albus of authors,
   misidentification.

Rearing work and field observations carried on here in Illinois show that the mature nymphs of this species, when ready to transform, migrate to the shores of the large rivers in which they develop. At dusk, they congregate in the shallow water or even in the wet mud at the edge of the water. The subimagoes emerge there, leaving their cast nymphal skins floating on the shallow water or partly submerged in the mud. These subimagoes take flight at once, and the males molt to the adult stage almost immediately. Molting occurs in the air, during flight, as the legs are nonfunctional. The adults then disperse to mate and lay their eggs. All the adults that emerge during one evening are, apparently, dead by the following morning. The length of adult

life I observed for this species in Illinois was about one hour. My observations do not agree with those of Howard (Needham 1905:60; 1920:285), also made in Illinois. However, Howard's observations were made on a mixture of individuals belonging to the genera *Ephoron* and *Potamanthus*. My observations on *leukon* agree closely with those made by Ide (1937a:25) on this species in Ontario.

MALE.—Length of body 12–14 mm., of fore wing 11–13 mm. Vertex of head light yellow, shaded with very dark gray between ocelli; anterodorsal area of mesonotum dark brown, metanotum a lighter brown; legs light yellowish to snow-white (in freshly killed specimens), with each fore femur and tibia stained with purplish gray; abdomen white, with variable areas of gray shading

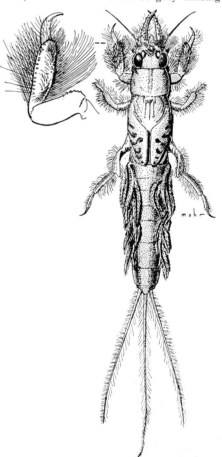

Fig. 59.—*Ephoron leukon*, mature nymph, dorsal aspect. Small figure at left represents enlargement of fore leg to show detail.

on apical tergites; penis lobes, fig. 66, with lateral angles relatively blunt, slightly down-curved; each cercus stained with gray-lavender near base, color paling to snow-white toward apex.

The nymph, fig. 59, was described by Ide (1935a:113).

Known from central, eastern, and north-eastern states and southeastern Canada. Most of the older published records of *album*

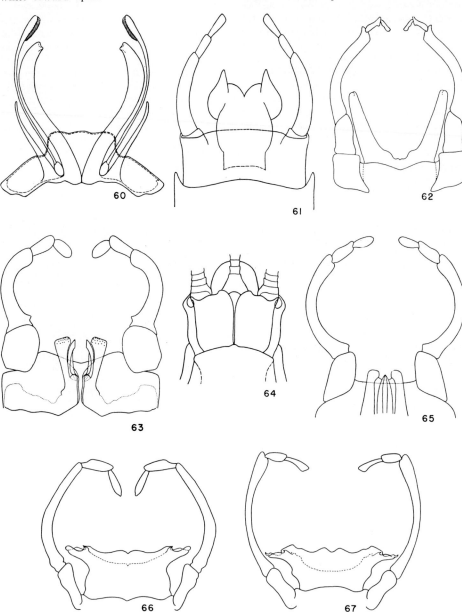

Fig. 60.—*Tortopus primus*, male genitalia.
Fig. 61.—*Potamanthus myops*, male genitalia.
Fig. 62.—*Pentagenia vittigera*, male genitalia.
Fig. 63.—*Ephemera simulans*, male genitalia.

Fig. 64.—*Pentagenia vittigera*, apex of fe-male abdomen, ventral aspect.
Fig. 65.—*Ephemera varia*, male genitalia.
Fig. 66.—*Ephoron leukon*, male genitalia.
Fig. 67.—*Ephoron album*, male genitalia.

for the eastern states should be referred to *leukon.* Develops in large and moderate-sized rivers.

Illinois Records.—Specimens, collected June 9 to September 24, are from Bloomington, Champaign, Cleveland (Rock River), Como (Rock River), Dixon (Rock River), Golconda, Havana, Homer, Kankakee, Lyndon (Rock River), Mahomet (Sangamon River), Milan, Muncie, Oakwood, Prophetstown, Rock Island, Rockton (Rock River), Roscoe (Rock River), Savanna (Mississippi River), Sterling (Rock River), Urbana, West Salem, and Wilmington.

2. *Ephoron album* (Say)

Baetis alba Say (1824:305).
Polymitarcys albus (Say) in part.
    Eaton (1883:47).

MALE.—Length of body 9–12 mm., of fore wing 8–11 mm. Vertex of head light yellowish, shaded with purplish gray between ocelli; entire dorsum of thorax light cream colored, almost white, sutures faintly shaded with tan; legs white to pale yellow, with each fore femur and tibia stained with purplish gray; abdomen snow-white, apical tergites faintly yellowish; genitalia, fig. 67, with penis lobes having lateral angles acute, slightly upcurved. Each cercus stained with gray near base, blending into snow-white toward apex.

The nymph was described by Edmunds (1948a: 12).

Known from the north-central and northwestern states. Develops in large and moderate-sized rivers.

Illinois Records.—FOSTER: Mississippi River, July 11, 1939, B. G. Berger, 2 ♂. KANKAKEE: at light, Aug. 4, 1936, Frison & Burks, 2 ♂; Aug. 15, 1938, H. H. Ross, 1 ♂; Aug. 16, 1938, Ross & Burks, 3 ♂. MOMENCE: at light, Aug. 5, 1938, Burks & Boesel, 1 ♂; Aug. 16, 1938, Ross & Burks, 25 ♂. PROPHETSTOWN: July 7, 1925, T. H. Frison, 5 ♂. ST. CHARLES: Fox River, July 8, 1948, Ross & Burks, 1 ♂. WILMINGTON: at light, Aug. 3, 1937, Ross & Burks, 18 ♂.

EPHEMERINAE

The subfamily Ephemerinae, as defined here, corresponds to the family Ephemeridae in Ulmer's classification (1933:198). It in-cludes the genus *Hexagenia,* whose members are the largest and commonest of Illinois mayflies. They are also the most important mayflies in the state when considered as food for the fishes of our larger lakes and streams.

In the adults, the middle and hind legs are well developed and functional. The marginal intercalary veins of the wings are not netlike. The cubital intercalary veins of each fore wing consist of two to four long, slightly sinuate, forked veins attached to Cu_1 and extending to the wing margin, figs. 27, 40, 46. The median caudal filament may be well developed, greatly reduced, or vestigial.

In the nymphs, figs. 2, 48–50, the frontal process is well developed and the mandibular tusks are strong. When the nymph is alive, the gills are held curved over the abdominal tergites.

KEY TO GENERA

ADULTS

1. Fore wing with crossveins at and posterior to bulla crowded and darkened so as to form a path extending half way across wing, fig. 46 . . . .**5. Ephemera**
    Fore wing with crossveins in region of bulla not so arranged as to form a path extending across wing, figs. 27, 40 . . . 2
2. Median caudal filament reduced but relatively well developed; in female, median caudal filament four-fifths to five-sixths as long as each cercus; in male, one-fifth to one-sixth as long as cercus . . . .**6. Pentagenia**
    Median caudal filament vestigial: in each sex, reduced to only four to nine small, poorly defined segments .**7. Hexagenia**

NYMPHS

1. Head with a more or less dome-shaped anterior projection between bases of antennae, figs. 48, 68–72 . . . .**7. Hexagenia**
    Head with a two-pronged anterior projection between bases of antennae, figs. 49, 50 2
2. Mandibular tusk with dorsolateral angle smooth, rounded, fig. 50 . .**5. Ephemera**
    Mandibular tusk with dorsolateral angle carinate and toothed, fig. 49 . . . .**6. Pentagenia**

5. *EPHEMERA* Linnaeus

Ephemera Linnaeus (1758:546).

The adults are large, relatively slender-bodied mayflies, usually with spotted wings and an abdominal color pattern made up of dark longitudinal stripes and blotches on a very pale yellowish background. In the males, the fore leg is almost as long as the body. The fore wing, fig. 46, has the cross-veins at and posterior to the bulla crowded together so as to form a path extending halfway across the wing. The median caudal filament in both sexes is as long as each cercus. The cerci and the median caudal filament are extremely long—each more than twice as long as the body.

In the nymphs, the frontal process of the head has a conspicuous, sharply projecting angle at each lateral margin, fig. 50. The mandibular tusks are long, slender, and smooth, with a few small, toothlike rasps on the outer side near the base. The apex of the labial palp is broad and truncate.

KEY TO SPECIES

ADULTS

1. Abdomen creamy white, without dark markings...............**1. guttulata**
   Abdomen yellowish or tan, with longitudinal, dark brown markings.......2
2. Hind wing with small, dark clouds surrounding discal crossveins, making wing appear spotted; genitalia with penis lobes relatively broad, fig. 63........
   ....**2. simulans**
   Hind wing not spotted, discal crossveins not surrounded by dark clouds; genitalia with penis lobes relatively narrow, fig. 65.....................**3. varia**

MATURE NYMPHS

1. Abdomen without color markings on venter.................**1. guttulata**
   Abdomen with longitudinal color markings on venter..........................2
2. Hind wingpad with dark color pattern indicated..............**2. simulans**
   Hind wingpad without indication of color pattern....................**3. varia**

1. Ephemera guttulata Pictet

Ephemera guttulata Pictet (1843:135).

The wings appear to be almost solid brown because of the dense, dark red-brown pigmentation around and between the crossveins; abdomen uniformly pale cream, without a darker color pattern, contrasting markedly with the very dark wings; cerci and median caudal filament tan to brown, with articulations very dark brown. Distinctive genitalia of male figured by Needham (1921, pl. 81, fig. 58) and others.

Known from New York, Pennsylvania, and Quebec.

2. Ephemera simulans Walker

Ephemera simulans Walker (1853:536).
Ephemera natata Walker (1853:551).
Ephemera decora Walsh (1862:376).

MALE.—Length of body 10–12 mm., of fore wing 11–13 mm. Thorax mostly dark red-brown, with relatively small, light tan areas on pro- and mesopleuron; each fore leg dark red-brown, middle and hind legs tan to yellow; fore wing and hind wing, figs. 46, 47, each with dark brown spots in discal area. Abdomen yellow to tan, with dark red-brown markings: each tergite with a pair of lateral dark blotches; each of apical three tergites with a pair of submedian, longitudinal, lateral stripes; sternites with longitudinal, lateral stripes, and a pair of submedian, longitudinal spots on each sternite; penis lobes, fig. 63, relatively wide; caudal filaments dark yellow-brown, with articulations darkened.

FEMALE.—Length of body 11–13 mm., of fore wing 12–14 mm. Color pattern same as in male, but dark areas and spots relatively less extensive; legs light tan or yellow, fore legs shaded with light red-brown; caudal filaments light yellow, with brown articulations.

Known from the northeastern and central states and eastern Canada. This is one of the commonest mayflies in the Chicago region. It emerges along the lake front in enormous numbers every summer. The nymph is found near the shores of lakes having considerable wave action and in moderate-sized rivers and creeks.

Illinois Records.—CEDAR LAKE: Oct. 21, 1882, 3 nymphs. CHICAGO: July 31, 1887, C. A. Hart, 3 ♂, 3 ♀; July, 1916, 1 ♂; July 8, 1937, Frison & Ross, 43 ♂, 18 ♀. EDDYVILLE: Lusk Creek, May 16, 1947, B. D. Burks, 1 ♀. EVANSTON: July 17, 1938, G. T. Riegel, 2 ♂; July 9, 1939, G. T. Riegel, 1 ♂; July 22, 1942, J. S. Ayars, 2 ♂. HOMER: June 30, 1925, T. H. Frison, 1 ♂; June 10, 1926, T. H. Frison, 9 ♀; June 30, 1927, Frison & Glasgow, 9 ♂, 11 ♀; Oct. 3, 1946, L. J. Stannard, 2 nymphs. KANKAKEE: June 12, 1931, Frison & Mohr, 2 ♂, 3 ♀; June 5, 1932, Frison & Mohr, 2 ♂; May 31, 1938, Burks & Mohr, 3 ♂, 1 ♀;

June 15, 1938, Ross & Burks, 1 ♀. Mc-
HENRY: June 3, 1943, Ross & Sanderson,
1 ♀. MUNCIE: June 29, 1919, 1 ♂ ; May
25, 1941, Ivabel Johnson, 1 ♀. OAKWOOD:
June 6, 1925, T. H. Frison, 18 ♂, 1 ♀ ; June
9, 1926, Frison & Auden, 81 ♀ ; June 14,
1935, C. O. Mohr, 1 ♀ ; June 5, 1948, Burks
& Sanderson, 5 ♂, 1 ♀ ; June 23, 1948, B. D.
Burks, 3 ♂. SOUTH BELOIT: July 2, 1931,
Betten, Frison, & Ross, 1 ♂, 1 ♀.

3. *Ephemera varia* Eaton

Ephemera decora Hagen (1861:38),
    not Walker. Misidentification.
Ephemera varia Eaton (1883:69).

The types of this species are in the
Museum of Comparative Zoology.

The fore wing in the adults has the same
fundamental color pattern as that in
simulans, although the dark-colored areas
are relatively smaller; the hind wing in
varia, however, lacks the dark spots entirely. The thorax is relatively lighter
colored than the thorax of *simulans;* the
fore leg of the male is dark yellow, the
base and apex of the tibia brown; the
middle and hind legs are light yellow, almost
white. The penis lobes, fig. 65, are relatively
narrower than those in *simulans,* fig. 63.

Known from Connecticut, Maine, Michigan, New Hampshire, New York, Ontario,
and West Virginia.

6. *PENTAGENIA* Walsh

Pentagenia Walsh (1863:196).

The adults of *Pentagenia* are large,
cream-colored mayflies, each with a conspicuous, dark brown, longitudinal band on
the dorsum of the thorax. In the males,
the eyes, which are almost contiguous on
the meson, are larger than those of any
other member of the family Ephemeridae.
Each fore leg in the males is relatively short,
about one-half as long as the body, and only
slightly longer than the middle and hind
legs; the legs in the females are quite similar
to those in the males. The wing venation,
figs. 40, 41, is typical for the subfamily. The
median caudal filament in both sexes is
reduced in length.

In the nymphs, fig. 49, the frontal projection of the head has a stout prong at
each lateral margin, the mandibular tusks
are short and stout, bearing irregular teeth

along each dorsolateral margin; and the
apical segment of the labial palp is a broad,
somewhat scoop-shaped triangle.

KEY TO SPECIES

ADULTS

Caudal filaments uniformly light yellowish or
  white; abdominal sternites unmarked.......
  ....**1. vittigera**
Caudal filaments brown, with narrow, yellowish band at each articulation; abdominal
  sternites marked with brown lines.........
  ....**2. robusta**

1. *Pentagenia vittigera* (Walsh)

Palingenia vittigera Walsh (1862:373).
Pentagenia quadripunctata Walsh (1863:198).

MALE.—Length of body and of fore wing
15–18 mm. Head and body generally cream
colored, with dorsum of thorax marked
with brown and entire dorsum of abdomen
occupied by a conspicuous, dark brown,
longitudinal band; legs light yellowish, almost white, with variable, vague gray shading at articulations; wings hyaline, C, Sc,
and R of fore wing light yellow, and crossveins in this area yellowish, other veins
and crossveins hyaline; fore wing often with
four black dots in a row extending across
wing from bulla toward posterolateral angle
of wing; abdominal sternites unmarked;
genitalia, fig. 62, and caudal filaments very
light yellow, almost white.

FEMALE.—Length of body 18–25 mm., of
fore wing 18–23 mm. Colored as in male;
apical abdominal sternite with a median,
V-shaped notch on apical margin, fig. 64.

Known from Arkansas, Illinois, Iowa,
Kansas, Minnesota, Missouri, Tennessee,
and Texas. Although this species in Illinois
is never so abundant as the species of
Hexagenia, the adults of *vittigera* occur in
considerable numbers throughout the summer along our larger rivers. Apparently,
vittigera develops only in large rivers.

· **Illinois Records.**—Specimens, collected
from June 6 to September 20, are from
Anna, Bloomington, Cairo, Carbon Cliff,
Carbondale, Carlyle, Centralia, Champaign,
Chicago, Dixon, Elizabethtown, Freeport,
Gibson City, Golconda, Grafton (Illinois
River), Harrisburg, Havana, Keithsburg,
McConnell (Rock River), Meredosia (Illi-
nois River), Mount Carmel, Murphysboro,
Oquawka, Peoria (Illinois River), Pere

Marquette State Park, Poplar Bluff, Quincy (Mississippi River), Rock Island, Rockton (Rock River), Rosiclare, Shepherd, Urbana, and Waukegan.

2. *Pentagenia robusta* McDunnough

Pentagenia robusta McDunnough (1926: 185).

All crossveins in wings tan; no pigmented dots in fore wing. Abdominal sternites marked with longitudinal, brown lines, and terminal two sternites mostly brown; caudal filaments brown, with yellowish articulations.

Known only from Ohio.

7. *HEXAGENIA* Walsh

Hexagenia Walsh (1863:197).

The various members of this genus are the commonest Illinois mayflies, as well as the largest. The eyes in the adult males are large, but never quite so large as in *Pentagenia vittigera,* and are always separated on the meson by at least a small space. Each fore leg in the males is approximately as long as the body. In the fore wing, the crossveins near the bulla are not crowded, fig. 27, as they are in *Ephemera,* fig. 46. The median caudal filament is reduced to a mere vestige in both sexes.

In the nymphs, fig. 2, the frontal process is dome-shaped, conical, or truncate; the lateral angles are never produced anteriorly as spines or prongs. The upcurved mandibular tusks are long, slender, and smooth, entirely lacking rasplike teeth. The apical segment of the labial palp is broad, with a median apical point.

My views of specific limits in this genus correspond in the main with those of Spieth (1941b: 233); I have not, however, followed him in recognizing subspecific segregates within the species *limbata* and *munda.*

It is not advisable to attempt to name single female specimens, unassociated with males; the characters given below for females apply to typical specimens only.

KEY TO SPECIES

ADULTS

1. Males.............................2
   Females............................7
2. Apex of each penis lobe reflexed, fig. 73; first segment of forceps shorter than penis lobe..............**6. recurvata**

Apex of each penis lobe not reflexed, figs. 74–78; first segment of forceps longer than penis lobe....................3
3. An anteapical protuberance on mesal margin of each penis lobe (dorsal view), fig. 78..................**2. bilineata**
   No anteapical protuberance on mesal margin of each penis lobe, figs. 74–77....4
4. Apexes of penis lobes sharply incurved, fig. 75...................**3. limbata**
   Apexes of penis lobes arcuate, figs. 74, 76, 77........................5
5. Penis lobes short and blunt, fig. 77......
   ....**1. atrocaudata**
   Penis lobes elongate, figs. 74, 76.......6
6. Penis lobes tapered gradually to apexes, fig. 74....................**5. rigida**
   Penis lobes abruptly constricted at about midlength, fig. 76.........**4. munda**
7. Membrane of wings uniformly stained with brown pigment, and lacking darker spots or areas at costal margin of fore wing, posterior margin of hind wing, or in disc of either of these wings.......
   ....**6. recurvata**
   Wing membrane hyaline, yellowish, or very light tan, and always with some darker areas or spots in wings.......8
8. Hind wing with a broad, reddish or purplish-brown band at posterior margin, fig. 83...............**1. atrocaudata**
   Hind wing either without darkened posterior margin or with a relatively narrow, sometimes discontinuous, dark brown band at posterior margin, figs. 82, 84–87........................9
9. Hind wing with membrane hyaline, veins hyaline, and discal crossveins black, figs. 86, 87; abdomen white or cream colored, with dorsal color pattern of dull red markings..........**5. rigida**
   Hind wing with membrane hyaline, yellowish, or light tan, with veins partly or entirely yellowish or tan, and crossveins black or brown both in discal and marginal areas of wing; abdomen yellow, tan, or light brown, with brown color pattern......................10
10. Hind wing with a continuous, dark brown band at posterior margin and two relatively large, brown spots in disc of wing; each discal crossvein surrounded by a small, brown cloud, fig. 82.......
    ....**2. bilineata**
    Posterior margin of hind wing without brown band, fig. 85, or often with an irregular, discontinuous one, fig. 84; no large, brown spots in disc of wing...
    ....**3. limbata, 4. munda**

MATURE NYMPHS

1. Each gill of first pair composed of a single filament...............**6. recurvata**
   Each gill of first pair composed of two or more filaments 2
2. Frontal process of head truncate at apex and with a small mesal indentation, fig. 70..............**1. atrocaudata**
   Frontal process of head rounded or angled on meson, figs. 68, 69, 71, 72........3

3. Frontal process of head dome shaped, fig.
   69...........................**3. limbata**
   Frontal process of head angled on meson
   at apex, figs. 68, 71, 72 4
4. Mid-tarsal claw slender, long, fig. 79;
   frontal process of head bluntly angled
   on meson, broad at base, fig. 72......
   .... **5. rigida**
   Mid-tarsal claw broadened near base, figs.
   80, 81; frontal process of head as in
   figs. 68, 71......................5
5. Frontal process of head relatively narrow,
   and with straight lateral margins, fig.
   71; mid-tarsal claw slender near tip,
   fig. 81.................**4. munda**
   Frontal process of head wider, and with
   curved lateral margins, fig. 68; mid-
   tarsal claw thick near tip, fig. 80......
   .... **2. bilineata**

1. *Hexagenia atrocaudata* McDunnough

Hexagenia atrocaudata McDunnough
   (1924*b*: 92).

MALE.—Length of body 22–24 mm., of
fore wing 23–25 mm. Eyes large, almost
contiguous on meson of head. Dorsum and
sternum of thorax mostly very dark red-
brown, pleura with yellow and light red
areas. Fore wing hyaline, with all veins
and crossveins very dark red-brown, costal
interspace red-brown, no conspicuous, discal,
dark spots present; hind wing, fig. 83, hyaline, veins and crossveins dark, no discal,
dark spots present, outer margin with a
broad, purplish or reddish-brown band.
Abdominal tergites mostly dark red-brown,
with small, yellowish streaks or spots on
dorsal meson and at lateral margins; genitalia, fig. 77, with short and blunt penis
lobes.
   FEMALE.—Length of body 23–25 mm., of
fore wing 24–25 mm. Colored much as in
male, but brown areas slightly smaller and
lighter colored; wings as in male.
   NYMPH.—Length of mature specimen 25
mm. Frontal process of head, fig. 70, truncate at apex and with a small mesal indentation.
   Known from Georgia, Illinois, Indiana,
Maryland, Michigan, Missouri, New York,
North Carolina, Ohio, Ontario, Pennsylvania, Virginia, and West Virginia. Develops in relatively cool, rapid creeks and
smaller rivers.
   Illinois Records.—MOMENCE: Aug. 21,
1936, Ross & Burks, 1 ♀; Aug. 22, 1936, 1 ♂.
RICHMOND: Aug. 15, 1938, Ross & Burks,
1 ♂. WILMINGTON: at light, Aug. 6, 1947,
Burks & Sanderson, 2 ♀.

2. *Hexagenia bilineata* (Say)

Baetis bilineata Say (1824: 303).
Hexagenia bilineata falcata Needham
   (1921: 292; pl. 81, fig. 62).

MALE.—Length of body 14–20 mm., of
fore wing 14–18 mm. Eyes separated on
meson of head by a space one-half as wide
as width of one eye. Thorax mostly red-
brown; fore wing hyaline, veins red-brown,
crossveins darker, several prominent, brown
spots in disc of wing, costal interspace tinted
with red-brown; hind wing, fig. 82, with
several prominent, discal, brown spots. Abdomen, when viewed from side without
magnification, appears to have two parallel,
longitudinal, dark brown bands; penis lobes,
fig. 78, beaked.
   FEMALE.—Length of body 18–22 mm., of
fore wing 20–22 mm. Color of body and
wings as in male, but generally slightly
lighter; parallel, longitudinal, color bands
of abdomen not so obvious.
   NYMPH.—Length 25–35 mm. Frontal
process of head, fig. 68, relatively broad,
with curved margins; mid-tarsal claw thick
near tip, fig. 80.
   Known from the District of Columbia,
Mississippi River valley, Maryland, and
Virginia. Develops usually only in large,
relatively slow rivers, but sometimes develops also in impounded bodies of water.
   Illinois Records.—Specimens, collected
from June 6 to September 16, are from
Alton, Cairo, Decatur, Elizabethtown, Fulton, Glendale, Grafton, Hardin, Harrisburg, Havana, Kankakee, Mahomet, Meredosia, Milan, Monticello, Mound City,
Murphysboro, Oregon, Peoria, Pontiac,
Prophetstown, Quincy, Ripley, Rockford,
Rock Island, Rockton, Rosiclare, Savanna,
Springfield, Venedy Station, and Wilmington.

3. *Hexagenia limbata* (Serville)

Ephemera limbata Serville *in* Guérin
   (1829: 384; pl. 60, figs. 7–9).
Palingenia occulta Walker (1853: 564).
Palingenia viridescens Walker (1853: 550).
Baetis angulata Walker (1853: 564).
Hexagenia variabilis Eaton (1883: 55).
Hexagenia venusta Eaton (1883: 54).
Hexagenia mingo Traver (1931*b*: 597).
Hexagenia pallens Traver (1935*a*: 271).

MALE.—Length of body 16–21 mm., of
fore wing 13–19 mm. The compound eyes
separated on meson by a space slightly nar-

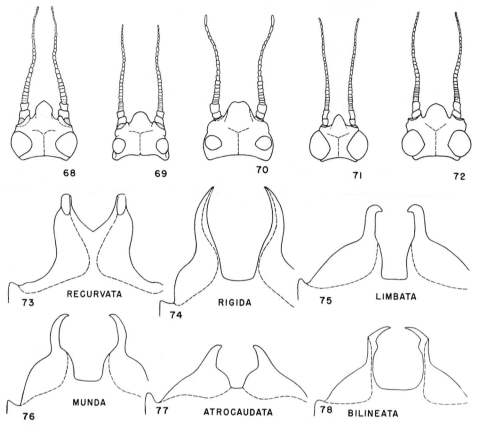

Fig. 68.—*Hexagenia bilineata,* dorsal aspect of head of nymph.

Fig. 69.—*Hexagenia limbata,* dorsal aspect of head of nymph.

Fig. 70.—*Hexagenia atrocaudata,* dorsal aspect of head of nymph.

Fig. 71.—*Hexagenia munda,* dorsal aspect of head of nymph.

Fig. 72.—*Hexagenia rigida,* dorsal aspect of head of nymph.

Fig. 73.—*Hexagenia recurvata,* penis lobes.

Fig. 74.—*Hexagenia rigida,* penis lobes.

Fig. 75.—*Hexagenia limbata,* penis lobes.

Fig. 76.—*Hexagenia munda,* penis lobes.

Fig. 77.—*Hexagenia atrocaudata,* penis lobes.

Fig. 78.—*Hexagenia bilineata,* penis lobes.

rower than diameter of one eye. Thorax typically mostly dark red-brown, varying to light brown; fore wing hyaline or faintly stained with yellow or tan, costal interspace shaded with red-brown, usually several prominent, brown spots in disc of wing; outer margin of hind wing usually with a conspicuous brown band. Abdominal tergites typically marked with dark brown lateral triangles and lighter brown mesal ones, sternites varying from almost completely unmarked to having a conspicuous row of dark mesal triangles; genitalia, fig. 75, with penis lobes hook shaped at apexes.

FEMALE.—Length of body 22–24 mm., of fore wing 20–22 mm. Colored much as in male, but generally lighter, some individuals almost completely yellow, darkest individuals never so dark as darkest males; fore wing usually lacking discal spots; hind wing, figs. 84, 85, without discal spots, posterior margin with or without brown band, this band, when present, discontinuous.

NYMPH.—Length 23–30 mm. Frontal process of head dome shaped, fig. 69.

Known from the Mississippi and Missouri river valleys, eastern states, and southern Canadian provinces. Develops in a great variety of waters: small to large creeks, small to large rivers, small lakes, and bodies of impounded water.

Illinois Records.—Specimens, collected from June 4 to September 15, are from Anna, Antioch, Beardstown, Bloomington, Carbondale, Champaign, Channel Lake, Decatur, Dixon, Elgin, Fisher, Fox Lake,

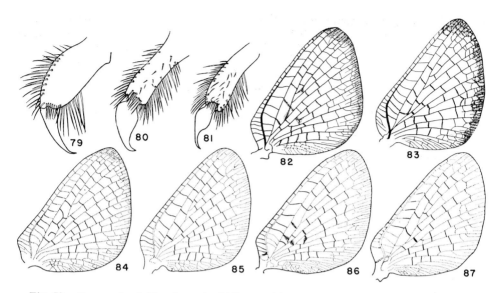

Fig. 79.—*Hexagenia rigida*, claw of middle leg of mature nymph.

Fig. 80.—*Hexagenia bilineata*, claw of middle leg of mature nymph.

Fig. 81.—*Hexagenia munda*, claw of middle leg of mature nymph.

Fig. 82.—*Hexagenia bilineata*, hind wing of female.

Fig. 83.—*Hexagenia atrocaudata*, hind wing of female.

Fig. 84.—*Hexagenia limbata*, dark phase, hind wing of female.

Fig. 85.—*Hexagenia limbata*, light phase, hind wing of female.

Fig. 86.—*Hexagenia rigida*, dark phase, hind wing of female.

Fig. 87.—*Hexagenia rigida,* light phase, hind wing of female.

Freeport, Giant City State Park, Gilman, Glen Ellyn, Grayslake, Havana, Homer, Kankakee, Kickapoo State Park, Laclede, La Salle, McHenry, Mahomet, Momence, Monticello, Morris, Mount Carmel, Murphysboro, New Milford, Oakwood, Oregon, Peoria, Poplar Bluff, Quincy, Rantoul, Richmond, Rockford, Rock Island, Roodhouse, Rosecrans, Springfield, Spring Grove, Urbana, and Waukegan.

4. *Hexagenia munda* Eaton

Hexagenia munda Eaton (1883:53).
Hexagenia affiliata McDunnough (1927c:119).
Hexagenia carolina Traver (1931b:601, 616).
Hexagenia elegans Traver (1931b:594).
Hexagenia marilandica Traver (1931b:599).
Hexagenia orlando Traver (1931b:608).
Hexagenia rosacea Traver (1931b:607).
Hexagenia weewa Traver (1931b:605).
Hexagenia kanuga Traver (1937:29).

This species is distinguishable from *limbata* only in the male and the nymph. Male genitalia, fig. 76, have penis lobes in form of elongate, shallow hooks. Nymph has a relatively narrow, straight-margined frontal process, fig. 71; mid-tarsal claw slender near tip, fig. 81.

Known from the northeastern, central, and southeastern states; it is extremely rare in Illinois, possibly an adventive. Apparently normally develops in small lakes.

Illinois Record.—Monticello: April 11, 1934, 1 ♂.

5. *Hexagenia rigida* McDunnough

Hexagenia bilineata falcata Needham
    (1921:292; pl. 81, fig. 65).
Hexagenia rigida McDunnough (1924b:90).

MALE.—Length of body 18–24 mm., of fore wing 16–20 mm. Eyes separated on meson by a space equal to diameter of one eye. Thorax mostly dark red-brown; fore wing hyaline, costal interspace stained with brown, veins and crossveins dark brown; hind wing with prominent discal spots, outer margin usually with brown band. Abdomen with extensive, dark red-brown color pattern on yellow- or red-brown background; penis lobes, fig. 74, elongate, almost straight.

FEMALE.—Length of body 18–28 mm., of fore wing 18–24 mm. Background color of body white or light cream; dorsal abdominal pattern dull red, venter entirely white; hind wing with membrane and veins hyaline, and crossveins black, figs. 86, 87. This is the only species of *Hexagenia* so colored.

NYMPH.—Length of body of mature specimen 22–28 mm. Apex of frontal process of head bluntly angled on meson, broad at base, fig. 72; mid-tarsal claw elongate, slender, fig. 79.

Known from Illinois, Iowa, Kansas, Manitoba, Michigan, Missouri, New Brunswick, New York, Ohio, Oklahoma, Ontario, Pennsylvania, Quebec, and Vermont. Develops in moderate-sized rivers that have a fairly rapid flow.

Illinois Records.—Specimens, collected from April 26 to September 6, are from Aroma Park, Aurora (Fox River), Beardstown, Chicago, Dixon (Rock River), Homer, Kankakee (Kankakee River), Libertyville, Mahomet, Momence, Oakwood, Oregon, Prophetstown, Rockford, Rockton (Rock River), Roscoe (Rock River), Rosiclare, South Beloit, White Heath, and Wilmington.

6. *Hexagenia recurvata* Morgan

Hexagenia recurvata Morgan (1913:395).

This species has the wing membranes so heavily tinted with dark brown that freshly killed specimens appear to have almost black wings. The dorsum of the thorax is mostly very dark brown, but the abdomen is mostly light yellow, with relatively small, darker markings; the male genitalia are distinctive, fig. 73, as the penis lobes have their apexes recurved. The nymph differs from all others in the genus in having the first pair of abdominal gills single.

Known from Maine, Massachusetts, Michigan, New York, North Carolina, Ontario, Quebec, and West Virginia.

NEOEPHEMERIDAE new family

This family includes a single North American genus, *Neoephemera*, which I am segregating from the Ephemeridae. Although the adults of *Neoephemera* show considerable similarity to the typical ephemerids, the nymphs are very similar to the caenid type. This indicates that *Neoephemera* is an interstitial form. Rather than include it in either the Ephemeridae or the Caenidae, necessitating the inclusion of a number of

exceptions in the characterization of each family, I believe *Neoephemera* can better be considered as representing a distinct family. Ulmer (1933:199) placed *Neoephemera* in the family Ephemeridae, as did Traver (1935a:289). In describing the new Javanese genus *Neoephemeropsis,* Ulmer (1939:481, 483; 1940:606) pointed out that it is a near relative of the American genus *Neoephemera* and placed it in his family Potamanthidae. *Neoephemeropsis* almost certainly belongs in the family Neoephemeridae. The European form that is known as *Caenis maxima* Joly (1870:144) probably also belongs here.

8. *NEOEPHEMERA* McDunnough

Neoephemera McDunnough (1925b:168).
Oreianthus Traver (1931a:103).
    New synonymy.

In the adults of this genus, the fore wing, fig. 28, has the basal costal crossveins weak or entirely absent, the stigmatic crossveins partly anastomosed, vein M_2 slightly curved toward vein Cu_1 near the wing base, two long, forked, cubital intercalary veins, and vein 1A with one to three crossveins extending from it to the anal wing margin. The hind wing has an acute marginal projection, and veins M_1 and M_2 diverge near the center of the wing. The median caudal filament is well developed in both sexes. It may be noted that, in *Neoephemeropsis,* the median caudal filament is rudimentary in both sexes.

The nymphs, fig. 88, are typically caenid in type, although each possesses two pairs of wingpads. Each tarsal claw is long, slender, and edentate, fig. 92. There is a minute, median, backward-projecting spine on the posterior margin of dorsum of the metathorax and on each of abdominal tergites 1–2 and 6–8. Each gill of the pair on the first abdominal segment is minute, single, and filiform, fig. 89B; on second segment elytroid, connate, covering the gills of the four following abdominal segments, and bearing a ventral tuft of filaments, fig. 89A; on segments 3–6, fig. 89C, similar in form to the corresponding gills in *Caenis,* fig. 91D, but differing in that each dorsal, platelike gill has a ventral tuft of filaments near the base. The median caudal filament is well developed and all three caudal filaments are uniformly clothed with short spines. In

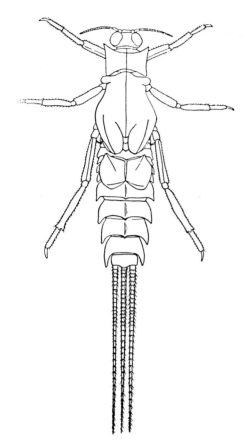

Fig. 88.—Nymph of *Neoephemera purpurea*. (From Traver. Figure used by permission of the Comstock Publishing Company, Ithaca, New York.)

the Javanese *Neoephemeropsis,* each cercus is also clothed, on the mesal side, with a dense comb of long setae, and the median caudal filament bears such a comb of long setae on either side.

There are two described species of *Neoephemera* in North America. *N. bicolor* McDunnough (1925b:168) is known from Michigan, Quebec, and Georgia; *purpurea* (Traver) (1931a:103; 1937:34) is known from Florida, Georgia, North Carolina, and South Carolina.

CAENIDAE

This family corresponds to the subfamily Caeninae of the family Baetidae in Traver's classification (1935a:629), and to the family Caenidae in Ulmer's classification (1933:206).

The members of the Caenidae are among the most distinct of the mayflies. In most species, the individuals are small, but some attain a body length of 6 or 7 mm. The adult females are slightly larger than the males, but otherwise they are almost identical in appearance; even the compound eyes of one are not larger than those of the other. In both sexes, each lateral ocellus is at least one-half as large as one of the compound eyes. The thorax is greatly developed, while the abdomen is relatively small and contracted, giving these mayflies a rather thickset appearance. The fore wing is white, being quite cloudy or milky, the costal margins tinged with grayish lavender; the marginal ciliae are numerous, even in the imago stage. In all genera, the wings characteristically have very few crossveins, and, except in *Leptohyphes,* they are quite broad in the anal region, figs. 97, 98. The hind wing is wanting except in the subimago stage of the genus *Leptohyphes.* Each tarsus has five segments, figs. 11–13. The abdomen is somewhat broad and flattened dorsoventrally, with the posterolateral angles of each segment obliquely produced. There are three well-developed caudal filaments, and the individual segments making up these filaments are relatively longer than in most mayflies. In the subimagoes, the filaments bear prominent setae, but in the adult males they are bare. The adult females retain, partly or completely, the subimaginal filaments. Especially in the females, the subimaginal exuviae are often only partly shed.

In the hairy nymphs, figs. 96, 113, 114, the head is hypognathous and the body somewhat flattened dorsoventrally. The tarsal claws are relatively large and long, figs. 93–95. The lateral margins of the abdominal segments are produced as spines or plates. The first abdominal segment has either a pair of single, filamentous gills or none. The gills borne by the second abdominal segment are operculate, completely covering the gills on segments 3–6. There are three well-developed caudal filaments; the cerci bear setae on all sides.

KEY TO GENERA

ADULTS

1. Fore wing with very few crossveins and with median intercalary vein extending to wing base, fig. 97; male genital forceps with only one segment, fig. 108..2

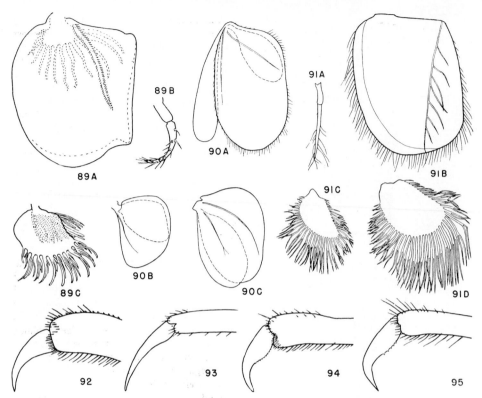

Fig. 89A.—*Neoephemera purpurea*, elytroid gill.
Fig. 89B.—*Neoephemera purpurea*, gill of first abdominal segment.
Fig. 89C.—*Neoephemera purpurea*, gill of fourth abdominal segment.
Fig. 90A.—*Tricorythodes atratus*, elytroid gill.
Fig. 90B.—*Tricorythodes atratus*, gill of sixth abdominal segment.
Fig. 90C.—*Tricorythodes atratus*, gill of fourth abdominal segment.
Fig. 91A.—*Caenis* sp., gill of first abdominal segment.
Fig. 91B.—*Caenis* sp., elytroid gill.
Fig. 91C.—*Caenis* sp., gill of sixth abdominal segment.
Fig. 91D.—*Caenis* sp., gill of fourth abdominal segment.
Fig. 92.—*Neoephemera purpurea*, claw of middle leg of mature nymph.
Fig. 93.—*Brachycercus lacustris*, claw of middle leg of mature nymph.
Fig. 94.—*Caenis* sp., claw of middle leg of mature nymph.
Fig. 95.—*Tricorythodes atratus*, claw of middle leg of mature nymph.

Fore wing with relatively numerous cross-veins and with median intercalary vein extending only halfway to wing base, fig. 98; male genital forceps with three segments, fig. 101 3
2. Prosternum twice as broad as long, fore coxae being widely separated on venter . . . .**10. Brachycercus**
Prosternum at least twice as long as broad, fore coxae being close together on venter . . . .**11. Caenis**
3. Fore wing broadest in the anal region, fig. 98; subimago without hind wing . . . .**9. Tricorythodes**
Fore wing more elongate and narrow, broadest in the center; subimago with hind wing . . . .**12. Leptohyphes**

MATURE NYMPHS

1. First abdominal segment without gills; each operculate gill borne by second abdominal segment triangular, fig. 96, or elongate-oval; operculate gills well separated on mid-dorsal line of abdomen 2
First abdominal segment bearing a pair of single, filamentous gills; each operculate gill borne by second abdominal segment semiquadrate, these operculate gills meeting or overlapping on mid-dorsal line of abdomen, figs. 113, 114 3
2. Operculate gills triangular, fig. 96 . . . .**9. Tricorythodes**

Operculate gills elongate-oval..........
.................**12. Leptohyphes**
3. Head bearing occipital and frontal tubercles, fig. 113........**10. Brachycercus**
Head without tubercles, fig. 114........
........................**11. Caenis**

9. *TRICORYTHODES* Ulmer

Tricorythodes Ulmer (1920a: 51).

The species of *Tricorythodes* consist of small, fragile mayflies which resemble *Caenis* in habitus, but differ considerably in diagnostic characteristics. *Tricorythodes* is a New World genus related to *Tricory-thus* of the Palearctic and African regions. *Tricorythodes* and *Tricorythus* are distin-

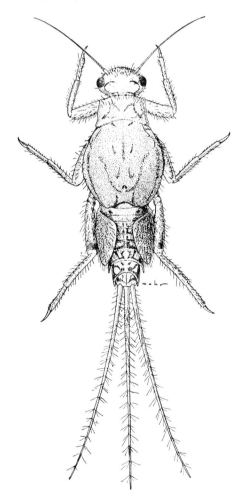

Fig. 96. — *Tricorythodes atratus,* mature nymph, dorsal aspect.

guished in the adults by the characteristics of the legs. In the former, the legs are long and slender; in the latter, they are short and less slender.

In the Illinois species of *Tricorythodes,* each adult has a pair of tubercles near the posterior margin of the vertex of the head. Each antennal pedicel in both sexes is two or three times as long as the scape; the flagellum is enlarged near the base and is four or five times as long as the pedicel. In the adult males, the fore leg is as long as the body, the fore femur is one-half as long as the fore tibia, and the fore tibia is one and one-half times as long as the fore tarsus; the second tarsal segment is one-half as long as the tibia and as long as tarsal segments 3–5 combined. In the fore wing, fig. 98, there are relatively numerous crossveins; veins R_4 and R_5 diverge in the center of the wing, and vein M_2 and the median intercalary arise some distance distad of the wing base. The posterior margin of the male genital forceps base has a wide, median excavation. Each arm of the forceps, figs. 99–101, has three segments; segment 1 is columnar, segment 2 has a bulbous, median enlargement at the base, and segment 3 is minute. The penis lobes are fused on the meson almost to the apexes, much as in some species of *Ephem-erella.* In the adult females, the caudal filaments are as long as those of the males, and these female caudal filaments usually retain the subimaginal setae only at the apexes.

The nymphs, fig. 96, have smooth heads lacking tubercles. Each antenna is almost twice as long as the head and pronotum, when measured in dorsal aspect. The legs are relatively longer than in *Caenis,* fig. 114, but shorter than in *Brachycercus,* fig. 113; each claw is relatively long, hooked at its apex, and has a ventral row of denticles, fig. 95. Abdominal segment 1 lacks gills; segment 2 bears a pair of subtriangular, operculate gills, each of which has an additional, ventral, membranous plate; each of segments 3–6 bears a pair of double, plate-like gills. All gills, fig. 90, have the margins entire. The three caudal filaments are relatively long and stout, and they have a whorl of setae at each articulation.

I have observed the subimagoes of *Tri-corythodes atratus* to shed the subimaginal

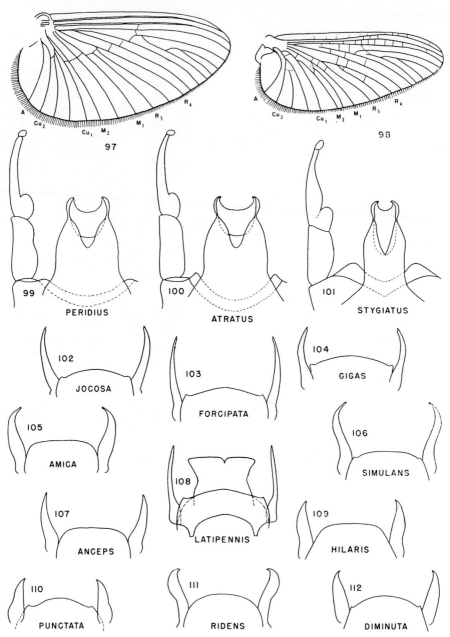

Fig. 97.—*Brachycercus lacustris*, fore wing.
Fig. 98.—*Tricorythodes atratus*, fore wing.
Fig. 99.—*Tricorythodes peridius*, male genitalia.
Fig. 100.—*Tricorythodes atratus*, male genitalia.
Fig. 101.—*Tricorythodes stygiatus*, male genitalia.
Fig. 102.—*Caenis jocosa*, penis lobes.
Fig. 103.—*Caenis forcipata*, penis lobes.

Fig. 104.—*Caenis gigas*, penis lobes.
Fig. 105.—*Caenis amica*, penis lobes.
Fig. 106.—*Caenis simulans*, penis lobes.
Fig. 107.—*Caenis anceps*, penis lobes.
Fig. 108.—*Caenis latipennis*, male genitalia. (After McDunnough.)
Fig. 109.—*Caenis hilaris*, penis lobes.
Fig. 110.—*Caenis punctata*, penis lobes.
Fig. 111.—*Caenis ridens*, penis lobes.
Fig. 112.—*Caenis diminuta*, penis lobes.

pellicle during flight; it is quite possible that all species of the genus do likewise.

The first-described Nearctic species in this genus, *allectus* (Needham) (1905: 47), is now virtually unidentifiable. It cannot be placed from the characters given in the original description alone. The types are, furthermore, either lost, or represented by only a few fragments. These fragments of specimens are not certainly the types, but might be; in any event, they are not in good enough condition to serve as a basis for an identification of the species.

Reliable characteristics for the separation to species of females and nymphs of this genus have not yet been found.

KEY TO SPECIES

ADULT MALES

1. Wings with all crossveins hyaline; antennal scape black......**1. stygiatus**
   Wings with anterior crossveins brown; antennal scape white or yellow, tinged with red-brown....................2
2. Vertex of head black........**2. atratus**
   Vertex of head mostly light yellow......
   ....**3. peridius**

1. *Tricorythodes stygiatus* McDunnough

Tricorythodes stygiatus McDunnough (1931e:267).

MALE.—Length of body 2.5–3.0 mm., of fore wing 3.5–4.0 mm. Head black, compound eyes and ocelli black; each antennal scape black, pedicel tan, flagellum yellow. Pronotum black, mesonotum very dark brown, metanotum yellow-brown; meso- and metapleura dark red-brown, with vague, black markings near wing bases; thoracic sternum dark red-brown; all coxae black; femora black, with red shading and with vague, longitudinal, yellow streaks; tibiae white, mottled with black and red-brown over basal three-fourths; tarsi white or gray; wings hyaline, anterior longitudinal veins brown. Abdominal tergum black; tergites 3–7 lighter near lateral margins: sternites dull yellow, suffused with gray, and with a median, black mark at posterior margin of each sternite; caudal filaments white, shaded with gray near bases. Genitalia, fig. 101, yellow to white, with penis lobes relatively narrow and median indentation at posterior margin of forceps base relatively narrow and shallow.

Known from Illinois, Michigan, New Brunswick, and Quebec. Develops in almost stagnant eddies along large streams.

Illinois Record.—WILMINGTON: at light, Aug. 6, 1947, Burks & Sanderson, 4 ♂.

2. *Tricorythodes atratus* (McDunnough)

Tricorythus atrata McDunnough (1923:39).

It is quite possible, as McDunnough (1931e:265) has said, that *atratus* is the same as *allectus* (Needham 1905:47), although the characters given in the original descriptions are not quite identical. As was remarked above, the specimens at present taken for the types of *allectus* are not in good enough condition to serve as a basis for an identification of the species. It is preferable at present to use *atratus* for the species, as that name is based on a detailed description, and the types are well preserved and available for study.

• MALE.—Length of body 3.0–3.5 mm., of fore wing 3.5–4.5 mm. Head black, compound eyes black, each antenna yellow or white, with faint, gray shading at base of pedicel. Pronotum black, becoming brown at lateral margins; mesonotum and metanotum dark brown, pleura lighter brown, sternum dark brown; coxae and trochanters dark gray; femora gray, with black shading, subapical area of each with red-brown shading; fore tibia brown, with median, black shading, middle and hind tibiae white, with black shading in middle, fore tarsus gray; middle and hind tarsi white; wings hyaline, veins Sc and R_1 shaded with gray, anterior veins and crossveins brown. Abdomen yellow-gray or white; tergites 1 and 2 completely shaded with black, tergites 3–7 black only on meson and at posterolateral angles, apical tergites covered by black shading; sternites yellowish gray, with a median, black mark at posterior margin of each sternite; caudal filaments white, shaded with gray near bases. Genitalia, fig. 100, white or light yellow, with penis lobes relatively wide and apical, median excavation of forceps base relatively wide and deep.

Known from Illinois, Michigan, and Quebec. Develops in almost stagnant eddies of larger streams.

Illinois Records.—MILAN: Rock River, June 4, 1940, Mohr & Burks, 1 ♂. OREGON: at light, July 18, 1927, Frison & Glasgow,

1 ♂. ROCKTON: Rock River, June 11, 1948, Burks, Stannard, & Smith, 1 ♂. WILMINGTON: at light, Aug. 3–4, 1937, Ross & Burks, 70 ♂ ; Aug. 6, 1947, Burks & Sanderson, 3 ♂.

3. *Tricorythodes peridius* new species

This species agrees with *atratus* in having the antennae almost completely white, in having a subapical, red-brown band on each of the femora, which are extensively shaded with black, and in having abdominal segments 3–7 shaded with black only on the dorsal meson and at the posterolateral angles of the tergites. The two differ in that *peridius* is larger and generally much lighter in color, with the vertex of the head mostly yellow instead of entirely black, as in *atratus*; in *peridius*, also, the apical margin of the genital forceps base has a relatively broad, shallow median excavation.

MALE.—Length of body 4.0–4.5 mm., of fore wing 4.5–5.5 mm. Head yellow-brown, shaded with black near posterior margin of vertex; eyes and ocelli black; antennae white, each usually shaded with tan and black on pedicel. Pronotum completely shaded with dark gray, mesonotum and metanotum brown, pleura slightly lighter brown, venter same color as mesonotum at lateral margins, color paling to white on meson; all coxae and trochanters dark brown; femora light brown, shaded with black, and with a subapical, red-brown band on each; fore tibia gray, middle and hind tibiae light yellow or white, with black shading in middle; fore tarsus light gray, middle and hind tarsi white; wings hyaline, veins Sc and R_1 shaded with gray, veins and crossveins in anterior half of each wing brown. Abdomen mostly light yellow or white, tergites 1 and 2 washed with gray. Tergites 3–7 shaded with gray on meson and at lateral margins, apical tergites uniformly washed with gray; sternites faint gray-tan, often almost white; caudal filaments white, basal four to six segments of each filament shaded with gray. Genitalia, fig. 99, white, with vague, gray shading along lateral margins of structures, apical margin of forceps base with a broad, shallow median indentation.

Holotype, male.—Wilmington, Illinois, at light, Aug. 6, 1947, Burks & Sanderson. Specimen in alcohol.

Paratypes.—ILLINOIS: Same data as for holotype, 11 ♂. ST. CHARLES: at light, July 8, 1948, Ross & Burks, 1 ♂. All specimens in alcohol.

10. *BRACHYCERCUS* Curtis

Brachycercus Curtis (1834: 122).
Oxycypha Burmeister (1839: 796). In part.
Eurycaenis Bengtsson (1917: 186).

The species of *Brachycercus* consist of small mayflies with broad thoraxes. All species have the mesonotum and metanotum various shades of brown, and the head, pronotum, and abdomen white, with black or gray markings. Each antennal pedicel is markedly long, three times as long as the scape. In both sexes, each lateral ocellus is two-thirds as large as one of the compound eyes. The wing venation, fig. 97, does not

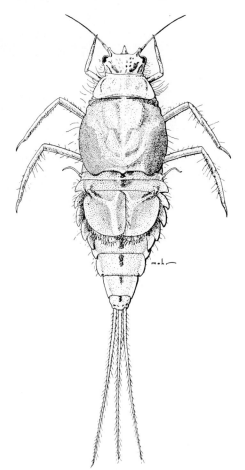

Fig. 113.—*Brachycercus lacustris*, mature nymph, dorsal aspect.

differ significantly from that of *Caenis*: veins R_4 and R_5 diverge relatively close to the wing base, veins M_1 and M_2 and the median intercalary vein all arise at the wing base, and there are relatively few crossveins. The legs are long and slender, with the femora very little stouter than the tibiae; legs conspicuously more slender than in *Caenis*. The abdomen in the males is about as long as the thorax, while, in the females, the thorax is slightly shorter than the abdomen. The segments of the abdomen bear long and filamentous, lateral projections. The male genital forceps base bears a pair of accessory lateral projections; otherwise the male genitalia are similar to those of *Caenis*. The posterior margin of the terminal abdominal sternite in the females is evenly rounded from margin to margin. The caudal filaments of the males are much longer than those of the females.

The nymphs, fig. 113, are flattened, heavy-bodied forms with conspicuously long, slender legs and tuberculate heads. Each antennal pedicel is three times as long as the scape, and the flagellum is as long as the head and pronotum combined. The tarsal claws are long, slender, and entirely without ventral denticles, fig. 93. The lateral margins of the abdominal tergites are produced as broad, flat, and thin projections. The first abdominal segment bears a pair of single, filamentous gills; the gills of abdominal segment 2 are operculate and semiquadrate; and the gills borne by segments 3–6 are single, the margins of each gill being provided with a fringe of long filaments.

Reliable characteristics for the separation to species of females and nymphs of this genus have not yet been found.

KEY TO SPECIES

ADULT MALES

Abdominal tergites white, entirely without
　　black markings . . . .**1. prudens**
Abdominal tergites white; a black, transverse
　　line at posterior margin on each of the apical
　　tergites, and traces of a median, longi-
　　tudinal, black line on each of the middle and
　　apical tergites . . . .**2. lacustris**

1. *Brachycercus prudens* (McDunnough)

Eurycaenis prudens McDunnough
　　(1931*e*: 264).

MALE.—Length of body and of fore wing 3 mm. Head below antennae yellow; vertex shaded with brown, and with a median, longitudinal, black line, lateral areas near ocelli dark gray-brown. Pronotum light yellow, almost white, with dark brown shading on median area of posterior margin; fore coxa vaguely shaded at base and apex with brown, fore femur gray-brown, fore tibia gray at base; mesonotum yellow-brown, with faint, dark brown shading on dorsal sutures; each pleuron and sternum of meso- and of metathorax light yellow-brown; metanotum yellow-brown; middle and hind legs white, with coxae sometimes slightly shaded with gray-brown. Abdomen entirely light yellow, almost white, either entirely without darker markings or with faint, pinkish-brown shading at posterior margins of caudal tergites; genital forceps long and flat, faintly gray in color; caudal filaments white.

FEMALE.—Length of body and of fore wing 3.5 mm. Coloration identical with that of male except that mesonotum is a slightly darker brown, and middle and hind coxae are extensively shaded with brown.

Known from Illinois, Kansas, and Saskatchewan. Apparently develops in large rivers.

Illinois Records.—SAVANNA: at light, July 20, 1892, Hart, Forbes, & McElfresh, 7 ♂. SHETLERVILLE: at light, Aug. 10, 1898, C. A. Hart, 4 ♂, 1 ♀.

2. *Brachycercus lacustris* (Needham)

Caenis lacustris Needham (1918: 249).
Eurycaenis pallidus Ide (1930*b*: 218),
　　　　not Tschernova. Name preoccupied.
Brachicercus idei Lestage (1931*a*: 119).
　　　　New name for *pallidus* Ide.

Caenis lacustris was described from nymphs only, and the types have been lost. I follow Lyman (1944: 3) in placing *idei* as a synonym of *lacustris*.

MALE.—Length of body 4–5 mm., of fore wing 4.5–6.0 mm. Head, pronotum, and abdom. n white, meso- and metathorax light brown; each of the apical abdominal tergites marked with a transverse, black line at posterior margin, and traces of a median, longitudinal, dorsal, black line present on middle and posterior tergites; fore leg shaded with brown, other legs white; claspers of genitalia tan, and penis lobes white; caudal filaments white.

Known from Michigan, Minnesota, New York, Ohio, and Ontario; should eventually be found to occur in Illinois.

11. *CAENIS* Stephens

Caenis Stephens (1835:61).
Oxycypha Burmeister (1839:796). In part.
Ordella Campion (1923:513). New name,
unnecessarily proposed.

The species of *Caenis* consist of small, predominantly white mayflies with shadings of purplish gray. These mayflies often emerge in enormous numbers, filling the air like snowflakes.

Each antennal pedicel in the adults of both sexes is approximately twice as long as the scape. The vertex of the head lacks tubercles. The fore coxae are close together on the venter; each fore leg in the males is as long as the body, with the tibia twice as long as the femur and slightly longer than the tarsus. In most species, there is a pair of submedian, dark brown spots on the pronotum. In each wing, there are very few crossveins, as in fig. 97 of *Brachycercus lacustris;* vein M_2 and the median intercalary vein extend to the wing base, and the cubital intercalary veins are long and relatively straight. The marginal ciliae are well developed in the adult wings; the subimaginal pellicle covering the wings is often only partly shed. The wings are whitish hyaline, with gray-purple shading in the first three interspaces and on veins Sc and R_1. Each abdominal segment bears a pair of long, lateral ciliae in the subimago; these ciliae are reduced to small, lateral projections in the adult. The posterior margin of the male genital forceps base is slightly convex; each forceps has only one segment. The penis lobes are fused on the meson; each lobe is broad and flat, and slightly widened at the apex, fig. 108. In both sexes, the three caudal filaments are well developed, those of the males being the longer; these filaments are entirely white in all Illinois species.

In the nymphs, the body is quite flat, with the pronotum narrower than the mesonotum. The nymph shown in fig. 114 to illustrate this genus has been drawn somewhat distended in order to show the structure of the abdomen; when the nymph is alive, the abdomen is more compact than it is represented in this figure. The head is smooth, lacking tubercles. As measured in dorsal aspect, each antenna is twice as long as the head and pronotum combined. The legs are relatively short and stout; the claws are small and slender, and have extremely mi-

nute ventral denticles, fig. 94. The first abdominal segment bears a pair of prominent, single, filamentous gills, fig. 91*A*; the gills borne by the second segment are single, quadrate, and operculate, fig. 91*B*. The gills borne by segments 3–6 are single and platelike, each gill having the margins deeply fissured to produce a marginal fringe of long filaments; each filament is secondarily divided near the tip to produce two or three smaller filaments, figs. 91*C*, 91*D*. The three caudal filaments are relatively stout, with a whorl of three to five setae at each articulation.

I have observed the subimagoes of *simulans* to shed the subimaginal exuviae during

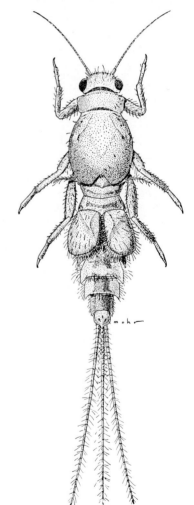

Fig. 114.—*Caenis simulans*, mature nymph, dorsal aspect.

flight, almost immediately after emerging from the nymphs.

Most adult specimens of *Caenis* intended for study should be killed with cyanide and mounted dry on pins or points; a few specimens of a series may also profitably be preserved in alcohol for ease in studying the genitalia. The colors of the specimens in alcohol fade seriously, but the characters of the male genitalia can be seen plainly in uncleared specimens preserved in alcohol. The male genitalia of dry specimens can be cleared in KOH for study, but the clearing technique is difficult. Specimens of *Caenis* are both small and fragile, which makes them difficult to handle without breakage during staining and clearing operations.

A revision, based on adult specimens of this difficult genus, was made by McDunnough (1931e); all subsequent North American workers have leaned heavily on his work. Specific characteristics for females and nymphs of *Caenis* have not been found.

KEY TO SPECIES

Adult Males

1. Vertex of head entirely and uniformly shaded with dark gray 2
   Vertex of head partly or completely white or pale yellow 5
2. Hind femur freckled with numerous, minute, black dots . . . . **1. punctata**
   Hind femur without numerous, minute, black dots; white, with an apicodorsal black spot or subapical black band . . 3
3. Apex of hind femur completely encircled by a black band . . . . **2. diminuta**
   Apex of hind femur with only a dorsal black spot 4
4. Genital forceps short and stout, fig. 105; body and fore wing each only 2 mm. long . . . . **3. amica**
   Genital forceps moderately long and curved, fig. 106; body and fore wing each 3.5–4.0 mm. long . . . . **4. simulans**
5. Abdomen lacking spiracular dots or streaks 6
   Abdomen with spiracular dots or streaks 7
6. Mesonotum light red-brown; genital forceps short, fig. 107, apexes straight; vertex of head almost entirely white, shaded with purplish gray only at margins and near ocelli . . . . **5. anceps**
   Mesonotum light yellow; genital forceps moderately long, fig. 111, apexes hooked; vertex of head heavily shaded with purplish gray near ocelli, light only in center . . . . **6. ridens**
7. Abdominal segments 7–9 only with gray-brown spiracular streaks; segments 1–6 without spiracular dots or spots . . . . **7. hilaris**

Abdominal segments 1–6 with spiracular dots or spots 8
8. Vertex of head almost entirely white, with purplish-gray shading present only at anterior and posterior margins and occasionally near lateral ocelli; length of body 2 mm. . . . . **8. gigas**
   Vertex of head dark gray-purple in entire anterior two-thirds, posterior third light yellow or white; length of body 3 mm. or more 9
9. Length of body 3 mm.; genital forceps of relatively moderate length and stout, fig. 102 . . . . **9. jocosa**
   Length of body 4 mm.; genital forceps relatively long and slender, fig. 103 . . . . **10. forcipata**

1. *Caenis punctata* McDunnough

Caenis punctata McDunnough (1931e:259).

Male.—Length of body and of fore wing 3 mm. Vertex of head completely covered with purplish gray; mesonotum yellow-brown; fore femur with dark purplish-gray shading in middle and near apex; middle and hind femora peppered with minute, black dots, each also with a dark gray band near apex; abdomen pale yellow, almost white, with extensive gray shading on tergites; spiracular marks present on all abdominal segments; each sternite with a pair of black, lateral dashes, sternites 1–6 each with a pair of sublateral, black dots, and sternites 7–9 each with two pairs of sublateral, black dots; genital forceps, fig. 110, short and stout.

Known from New York, Ontario, Quebec, and Wisconsin.

2. *Caenis diminuta* Walker

Caenis diminuta Walker (1853:584).

Male.—Length of body and of fore wing 3 mm. Vertex of head slightly but uniformly shaded with gray; mesonotum dark chestnut brown; fore femur shaded with gray and with a darker, transverse band near apex; fore tibia uniformly shaded with faint gray; middle and hind femora white, with a black band encircling each near apex; abdomen white, with extensive gray shading on tergites 1–6; spiracular dots present on segments 1–8 or –9; sternites 2–6 each with a pair of slender, lateral, black lines; two pairs of minute dots sometimes present on each abdominal sternite, those on sternites 7–9 always darker than anterior ones; genital forceps, fig. 112, short, stout, and markedly divergent.

Known from Florida, Georgia, Ontario.

3. *Caenis amica* Hagen

Caenis amica Hagen (1861:55).

MALE.—Length of body and of fore wing 2.0 mm. Vertex of head completely covered by dark gray shading; mesonotum yellow-brown; fore femur extensively shaded with dark gray, middle and hind femora each with a black, subapical, dorsal spot; abdominal tergites lightly shaded with gray, and spiracular dots present on abdominal segments 1–7 or –8; genital forceps, fig. 105, short and relatively slender.

Known from Maine, Maryland, Missouri, North Carolina, New York, and West Virginia.

4. *Caenis simulans* McDunnough

Caenis simulans McDunnough (1931e:263).

This is by far the commonest Illinois species of *Caenis*.

MALE.—Length of body and of fore wing 3.5–4.0 mm. Head below antennae yellow, vertex completely covered with dark purplish-gray shading; antennae yellow. Pronotum light yellow, with dark purple-gray shading at margins and on dorsal meson; each fore coxa shaded with grayish brown, fore femur gray, with a black apicodorsal spot, fore tibia and tarsus light gray; meso- and metanotum chestnut brown, darker brown shading present on median dorsal suture or mesoscutum, just dorsad of wing bases, and at posterior ends of outer parapsides; apex of mesoscutum shaded with gray; gray shading present near either lateral margin of metanotum; each pleuron chestnut brown, with dark gray shading around coxal cavities and light gray shading over prealar bridge; middle and hind legs white, with a black streak on outer side of each trochanter, and on dorsal side of each femur at base and near apex. Abdomen white or faintly stained with yellow, tergites 1–6 heavily shaded with dark gray, tergites 7–9 with gray shading in basolateral areas; black stigmatic dots or spots usually present on abdominal segments 1–9, these markings sometimes obsolete on segments 8 and 9; each sternite typically with a black spot near either lateral margin and another black spot on meson, in addition to a pair of minute, sublateral, black dots on each of sternites 1–7 and a pair of sublateral, black dots on each of sternites 8–10; genital forceps, fig. 106, long and slightly bowed.

FEMALE.—Length of body and of fore wing 4.0–5.5 mm. Coloration almost identical with that of male except that fore femur is light yellow, without gray shading, but with apicodorsal spots preserved; fore tibia and tarsus light yellow rather than gray, and median, black spots of abdominal sternites wanting or only faintly indicated.

Known from the northern states and Canada. Develops in nearly or quite stagnant water; it evidently tolerates considerable pollution.

Illinois Records.—Specimens, collected May 3 to August 19, are from Antioch, Banner, Beach, Chester, Fox Lake, Golconda, Havana, Herod, Kankakee, Momence, Oakwood, Palos Park, Prophetstown, Richmond, Rosecrans, Serena, Spring Grove, Sterling, Wadsworth, and Zion.

5. *Caenis anceps* Traver

Caenis anceps Traver (1935a:645).

MALE.—Length of body and of fore wing 2 mm. Vertex of head mostly white, with anterior and posterior margins edged with gray-purple, and with gray-purple shading near lateral ocelli; mesonotum light red-brown; fore femur shaded with gray at apex, and fore tibia shaded with gray at base; middle and hind femora white, with a minute, black, dorsal dot near apex of each; abdomen entirely white, without markings of any kind; genital forceps, fig. 107, short and straight.

Known from Missouri and New York.

6. *Caenis ridens* McDunnough

Caenis ridens McDunnough (1931e:256).

MALE.—Length of body and of fore wing 2 mm. Head white, vertex lightly shaded with gray-purple, this shading darker near lateral and anterior ocelli, median area of vertex relatively pale; antennae white. Pronotum white, with minute, purplish gray-shaded area at each anterolateral angle; gray-purple shading present around each fore coxal cavity and on fore coxa; fore leg white, fore femur faintly shaded with gray at apex, fore tibia gray; meso- and metanotum pale yellow, with light purplish brown shading on median, longitudinal line of mesoscutum and on apex of mesoscutellum; pleuron light yellow, with purplish brown shading around coxal cavities; all thoracic sternites white; middle and hind legs entirely white. Abdomen white, without spiracular spots or streaks; genital forceps, fig. 111, stout and moderately long.

FEMALE.—Length of body and of fore wing 2.5–3.0 mm. Head almost entirely white, with faint gray shading extending across middle of vertex only; fore femur with gray shading at apicodorsal angle; abdominal tergites 1 and 2 each with a pair of sublateral, transverse dark spots.

Known from Illinois, Kansas, Michigan, Ontario, and Wisconsin. Develops apparently in eddies in small rivers.

Illinois Records.—AURORA: at light, July 17, 1927, Frison & Glasgow, 22 ♂, 4 ♀. LAKE FOREST: J. G. Needham, 1 ♀ (Traver 1935a: 653). OAKWOOD: at light, July 10, 1927, Frison & Glasgow, 53 ♂, 21 ♀.

7. Caenis hilaris (Say)

Ephemera hilaris Say (1839:43).

MALE.—Length of body and of fore wing 2.0–2.5 mm. Head white, faint lavender shading present on vertex near lateral ocelli; antennae white. Pronotum white, with faint gray-brown shading at each anterolateral angle; a fairly large, gray-brown spot present on outer side of each fore coxa; fore leg white, with gray shading at apex of femur; meso- and metanotum light yellow, with a prominent, gray-brown spot near apex of mesoscutellum, this spot often extending forward onto mesoscutum; meso- and metapleura white, with vague gray-brown shading around coxal cavities; all thoracic sternites white; middle and hind legs white, with a black spot on dorsal side near apex of each femur. Abdomen white, with gray shading on first tergite and prominent, gray-brown spiracular lines on tergites 7–9; genital forceps, fig. 109, short and stout.

FEMALE.—Length of body and of fore wing each 2.5–3.0 mm. Coloration identical with that of male except that vertex of head is slightly darker, each fore femur has a dorsoapical, black spot, and spiracular lines may be present on tergites 5 and 6 as well as on apical tergites.

Known from the eastern and central states. Develops in eddies of moderate-sized and large rivers.

Illinois Records.—DIXON: June 27, 1935, DeLong & Ross, 17 ♂. ELIZABETHTOWN: June 22, 1932, Dozier & Park, 9 ♀.

FOSTER: Mississippi River, July 22, 1939, B. G. Berger, 8 ♀. FULTON: July 20, 1927, Frison & Glasgow, 2 ♂. HOMER: Aug. 10, 1925, T. H. Frison, 2 ♀. JACKSON ISLAND, in Mississippi River opposite Hannibal, Mo.: Sept. 6, 1940, G. T. Riegel, 1 ♂, 12 ♀. KANKAKEE: Aug. 16, 1938, Ross & Burks, 3 ♂, 5 ♀. MACKINAW: July 4, 1939, Ivabel Johnson, 3 ♀. MOMENCE: Aug. 16, 1938, Ross & Burks, 3 ♂. MONTEZUMA: at light, Oct. 9, 1931, C. O. Mohr, 36 ♀. OQUAWKA: Sept. 26, 1947, H. H. Ross, 18 ♂, 36 ♀. OREGON: at light, July 18, 1927, Frison & Glasgow, 290 ♂, 21 ♀; July 2, 1946, Burks & Sanderson, 54 ♂, 70 ♀. ROCK ISLAND: 1 ♂ (Walsh 1862: 381). SAVANNA: Mississippi River, July 23, 1892, S. A. Forbes *et al.,* 1 ♀.

8. Caenis gigas new species

This species resembles *hilaris* in being very small and having the vertex of the head almost entirely white. The two differ in that *gigas* has spiracular dots or streaks on abdominal segments 2–9 rather than on segments 7–9 only, as in *hilaris*; *gigas* also has longer and more slender genital forceps than does *hilaris*.

MALE.—Length of body and of fore wing each 2 mm. Head white, vertex white, with a narrow, gray-purple line at anterior and posterior margins, occasionally faint, gray-purple shading present on vertex near lateral ocelli; antennae white, each pedicel twice as long as scape, flagellum five times as long as pedicel. Pronotum white, with purplish shading at margins; propleura and sternum white; each fore coxa shaded with purplish gray at base, fore femur almost completely shaded with gray, fore tibia gray, tarsus white, with purplish gray shading at articulations; mesonotum light yellow-brown and with a narrow line of gray shading at principal sutures; mesopleura yellow-brown, with fairly extensive, gray-purple shading, mesosternum white; wings hyaline, costal shading typical for genus; metathorax white, with gray shading around coxal cavities; middle and hind legs white, shaded on coxae and on dorsal sides, near apexes, of femora. Abdomen white, tergites with extensive, dark gray shading on median area and at posterior margins of tergites 1–6; tergite 7 with gray shading covering median and anterolateral areas; tergite 8 with gray shading in anterolateral areas; tergite 9 with a black median spot; spiracular marks present on segments 2–9, those on segment 7 largest; a black mark present near middle of each lateral margin of sternites 2–9; genital forceps, fig. 104, slender, almost straight; caudal filaments white.

FEMALE.—Length of body 2.5–3.0 mm., of fore wing 3.0–3.5 mm. Color as in male, except that fore femur is gray, shaded only on ventral side near base and dorsally near apex; fore tibia white.

Holotype, male.—Giant City State Park, Illinois, at light, August 6, 1946, Mohr & Sanderson. Specimen in alcohol.

Allotype, female.—Same data as for holotype. Specimen in alcohol.

Paratypes.—ILLINOIS: Same data as for holotype, 7 ♂, 33 ♀; same locality: July 5, 1944, Sanderson & Leighton, 46 ♀; Aug. 22, 1944, 2 ♂, 15 ♀. HAMILTON: Aug. 30, 1931, Ross & Mohr, 62 ♂. HEROD: July 8–11, DeLong & Ross, 25 ♂, 36 ♀. All specimens in alcohol.

9. *Caenis jocosa* McDunnough

Caenis jocosa McDunnough (1931e: 260).

MALE.—Length of body and of fore wing 3 mm. Head below antennae, and on posterior third of vertex, light yellow; anterior two-thirds of vertex shaded with purplish gray; antennae pale yellow, almost white. Pronotum light yellow, with black shading at lateral margins and on dorsal meson; area around fore coxal cavities, and on outer side of each coxa, shaded with black, fore femur shaded with gray, this shading darker near base and on dorsal side near apex of femur, fore tibia light gray; mesonotum yellow-brown, shaded with dark brown on dorsal, longitudinal, median line and at apex of mesoscutum; mesoscutellum shaded with gray at apex; metanotum yellow-brown, with dark brown shading on meson; pleuron yellow-brown, shaded with black around coxal cavities; thoracic sternum light yellow, almost white; middle and hind legs light yellow, with a black streak on the outer side of each trochanter and a black apicodorsal spot on each femur. Abdomen very light yellow, tergites 1–7 shaded with gray; spiracular marks present on segments 1–7 or –8, sometimes these marks wanting on segments 3 or 4; abdominal sternites pale yellow, with a pair of lateral, black streaks on each sternite 1–5 and a pair of black

dots on each of the more posterior sternites; genital forceps, fig. 102, relatively long and slender.

FEMALE.—Length of body and of fore wing each 4 mm. Coloration identical with that of male except that fore femur is white, with gray shading in middle and near apex, and that thorax has more extensive, black or gray shading and little or no dark brown shading.

Known from the northeastern and midwestern states and southeastern Canadian provinces. Possibly a pond species.

Illinois Records.—ANNA: May 6, 1925, T. H. Frison, 2 ♂, 22 ♀. HOMER: at light, Aug. 10, 1925, T. H. Frison, 1 ♂, 3 ♀.

10. *Caenis forcipata* McDunnough

Caenis forcipata McDunnough (1931e: 257).

This species, described from Ontario and Quebec, may eventually be shown to be a synonym of *latipennis* Banks, described from the state of Washington. The male genitalia of the two species are virtually identical, figs. 103, 108.

MALE.—Length of body and of fore wing 4 mm. Head below vertex light yellow, almost white; anterior two-thirds of vertex dark gray, posterior third light yellow; antennae white, apex of each pedicel slightly darkened. Pronotum light yellow, shaded with purplish gray at margins; prosternum white, with black lateral margins; each fore femur shaded with dark gray, with a black apicodorsal spot, fore tibia gray; meso- and metanotum dark yellow-brown; pleuron dark yellow-brown; sternum white; middle and hind femora white, with a subapical, dorsal, black spot on each. Abdomen pale yellow; tergites 1–7 heavily shaded with dark gray, stigmatic marks present on segments 1–7 or –8; sternites pale yellow, with paired, gray dots faint; genital forceps, fig. 103, long and slender.

FEMALE.—Length of body and of fore wing 4.0–4.5 mm. Coloration identical with that of male except that the yellow-brown of the mesonotum is lighter and the gray shading of the fore femur is less extensive.

Known from Illinois, Manitoba, Michigan, New York, Ontario, Quebec, and Wisconsin. Possibly a pond species.

Illinois Records.—ANTIOCH: Channel Lake, June 15, 1928, T. H. Frison, 1 ♂. SHAWNEETOWN: May 11, 1935, C. O. Mohr,

2 ♂. STERLING: at light, May 21–22, 1925, D. H. Thompson, 5 ♂.

12. *LEPTOHYPHES* Eaton

Leptohyphes Eaton (1882:208).

This genus formerly was thought to be restricted to South America and Central America; I have, however, seen specimens, referable to *Leptohyphes,* which were collected at lights in San Antonio, Texas.

Species of the genus *Leptohyphes* consist of small or medium-sized, fragile mayflies which closely resemble the other members of the family Caenidae. The fore leg in the males is shorter than the body, and the hind leg is slightly longer than the fore leg. The fore wing is constricted in the cubitoanal region, as in mayflies with two pairs of wings; veins R_4 and R_5 diverge at or near the center of the wing, and vein M_2 and the median intercalary vein arise at a point some distance distad of the wing base. The hind wing, persisting in the subimagoes, usually has a very long, thin costal projection; there always is a pair of thin, membranous projections arising near the wing bases and extending along the lateral margins of the mesoscutellum. The male genitalia consist of a pair of 3-segmented forceps and a pair of slender, apically diverging penis lobes. There are two long caudal filaments; the middle one is vestigial.

In the nymphs, which are flattened and sprawling forms having relatively short, thickset legs and edentate claws, the pronotum typically is rectangular, with the anterolateral angles acute. Two pairs of wingpads are present. The abdominal gills are borne at the lateral margins of segments 2–6; the first pair is elongate-oval and elatroid, covering the following pairs of gills; the lateral margins of the abdominal segments are produced as broad, shelflike projections having the posterolateral angles acute. There are three well-developed caudal filaments.

EPHEMERELLIDAE

The family Ephemerellidae, as it is here treated, is identical with the subfamily Ephemerellinae of the family Baetidae of Traver (1935a:562) or the family Ephemerellidae of Ulmer (1933:204). In my opinion, this group of mayflies is sufficiently different from all others to require segregation as a distinct family. The ephemerellids are quite a homogeneous group and as far diverged from the fossil ephemerid prototype (Tillyard 1932) as are the Ephemeridae or Heptageniidae. It may be noted that, although the ephemerellids are world-wide in distribution, they were first recognized as a distinct group by Walsh in 1862, on the basis of a study of specimens he had collected at Rock Island, Illinois.

Although ephemerellids are one of the commonest mayfly groups in the northern and western states and are represented there by a large number of interesting species, they are relatively uncommon in Illinois. They seem to require for their development either somewhat rapid, clear streams which are cool throughout the year or small, clear lakes. Such bodies of water are rather rare in this state, but a few suitable habitats are provided by Nippersink Creek in McHenry County, the Rock River in Winnebago and Ogle counties, the Kankakee River in Kankakee County, the Salt Fork River in Vermilion County, Lusk Creek in Pope County, and some of the lakes in Lake County. A few ephemerellids have been taken in the Wabash River at Mount Carmel, but, in general, the Wabash River does not provide a habitat favorable for their development. They seem to have disappeared completely from the Mississippi and Rock rivers at Rock Island, although they obviously were fairly common there in the early 1860's when Walsh collected and described the genus *Ephemerella* and two species from that locality.

The ephmerellids offer an excellent example of a common mayfly phenomenon: the nymphs are more easily collected and more readily separated to species than the adults. The nymphs can be found at almost any time during the spring or summer months by turning over rocks and debris in the streams or lakes in which they live. The adults have a length of life of only 3 or 4 days, are strong fliers and rather difficult to net, and have a high mortality rate when being reared from the nymphs. Opportunities for obtaining adults of a given species are thus relatively few. As a consequence, most American mayfly collections include a great many more nymphal than adult specimens of ephemerillids. The Illinois Natural History Survey collection, for example, has only a few dozen adults but hundreds of nymphs.

In the Nearctic region, the family Ephemerellidae includes one genus, *Ephemerella*.

13. *EPHEMERELLA* Walsh

Ephemerella Walsh (1862:377).
Drunella Needham (1905:42).
Chitonophora Bengtsson (1908:243; 1909:6).
Torleya Lestage (1917:366).
Timpanoga Needham (1927:108).
Eatonella Needham (1927:108).

This genus now includes more than 80 Nearctic species, only 11 of which have, so far, been taken in Illinois. Quite a few more species of *Ephemerella* might, however,

basal three of which are short and approximately equal in length; the apical segment is usually as long as the basal three combined. Each arm of the male genital forceps has three segments, the basal one short, broad, and usually indistinctly set off from the distal ones; the second segment is long and somewhat bowed; and the third segment is usually minute, although it is fairly long in a few species. The penis lobes invariably are fused at the base on the meson; in most species, this mesal fusing extends almost to the tips of the penis lobes. There are three long caudal filaments.

The nymphs, figs. 172 and 173, are more or less flattened dorsoventrally. In each, the

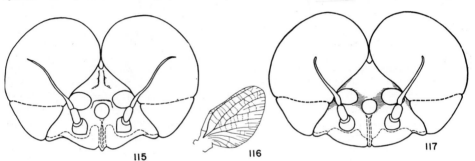

Fig. 115.—*Ephemerella argo*, head of adult male, anterior aspect.
Fig. 116.—*Ephemerella lutulenta*, hind wing of adult.
Fig. 117.—*Ephemerella ora*, head of adult male, anterior aspect.

eventually be found to occur in the smaller streams in the northern tier of Illinois counties. For that reason, a number of extralimital, northern species of *Ephemerella* have been included in the keys and discussions given below.

In each fore wing of the *Ephemerella* adults, the costal crossveins are entirely wanting or a few apical ones are very weakly indicated, the stigmatic crossveins are weak and usually slightly anastomosed, and all other crossveins are weak, fig. 29. All the longitudinal veins are well indicated. Vein Cu_2, at mid-length, is bent at right angles toward the wing margin. The hind wing, fig. 116, has a subangulate costal projection and weak crossveins.

In adult males, the large compound eyes are almost or quite contiguous on the dorsal meson and each is rather indistinctly divided, figs. 115 and 117, with the ventral portion smaller and darker than the dorsal portion. In both sexes, the hind tarsus has four clearly differentiated segments, the

head is relatively small, and the body is often as broad across the middle abdominal segments as across the thorax. The integument in the *Ephemerella* nymphs is more heavily sclerotized than it is in most mayfly nymphs. In many species of *Ephemerella*, the nymphs are provided with prominent dorsal spines and tubercles. In some western species, the nymphs have each an ingenious, ventral, abdominal sucker-disc that permits them to cling to rocks in the rapid streams in which they live. In various species, the abdomen bears either four or five platelike gills; the second segment is invariably without gills. The first pair of platelike gills sometimes forms a sort of operculum which covers the more caudal pairs of gills. Structures of the individual gills of two species of *Ephemerella* are shown in figs. 170 and 171. There are three well-developed caudal filaments.

In the figures of male genitalia included here, ventral spines borne by the penis lobes are shown by broken lines; dorsal spines are shown by unbroken lines. These stout spines borne by the penis lobes occur in three positions on the lobes: apical, near or at the apex near the meson; lateral, at or near the outer lateral margin of each penis lobe a short distance from the apex; and basal, approximately midway between base and apex of the penis lobe; see fig. 10.

The following keys to the genus *Ephemerella* do not include the species *consimilis* Walsh, described in 1862, from Rock Island, Illinois. No authentic Walsh material of this species is known to exist, and the species cannot be identified from the description alone. Extensive collecting in and around Rock Island has failed to turn up any speci-

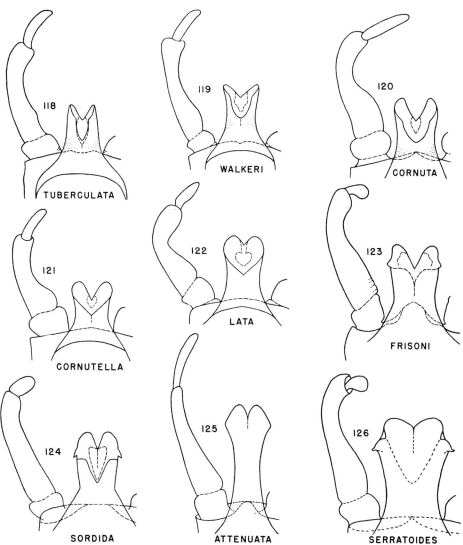

Fig. 118. — *Ephemerella tuberculata,* male genitalia.

Fig. 119.—*Ephemerella walkeri,* male genitalia.

Fig. 120.—*Ephemerella cornuta,* male genitalia.

Fig. 121.—*Ephemerella cornutella,* male genitalia.

Fig. 122.—*Ephemerella lata,* male genitalia.

Fig. 123.—*Ephemerella frisoni,* male genitalia.

Fig. 124.—*Ephemerella sordida,* male genitalia.

Fig. 125.—*Ephemerella attenuata,* male genitalia.

Fig. 126.—*Ephemerella serratoides,* male genitalia.

mens that could be assigned to this species. It may be noted, however, that the streams around Rock Island are today quite different than they were in Walsh's time. The genotype, *excrucians,* also described from Rock Island and not found to occur there now, has recently been located and reared in northern Michigan.

Examination of the lectotype of *Cloe quebecensis* Provancher (1876:267), now deposited in the Provincial Museum, Quebec, shows that the species was described from a female specimen of the genus *Ephemerella.* This species was transferred by Provancher to *Heptagenia* the year following its description (1878:127). At the present state

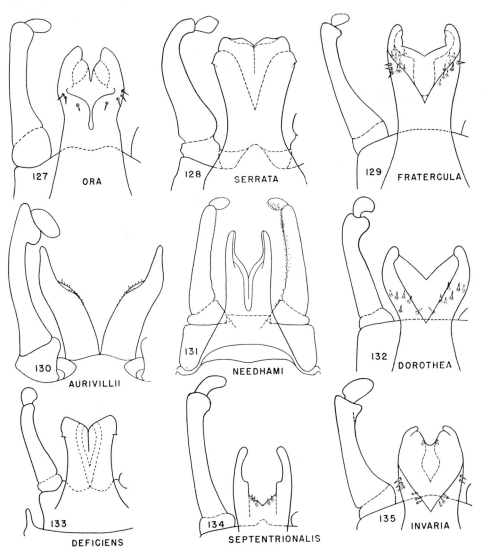

Fig. 127.—*Ephemerella ora,* male genitalia.
Fig. 128.—*Ephemerella serrata,* male genitalia.
Fig. 129.—*Ephemerella fratercula,* male genitalia.
Fig. 130.—*Ephemerella aurivillii,* male genitalia.
Fig. 131.—*Ephemerella needhami,* male genitalia.

Fig. 132.—*Ephemerella dorothea,* male genitalia.
Fig. 133.—*Ephemerella deficiens,* male genitalia.
Fig. 134.—*Ephemerella septentrionalis,* male genitalia. (After McDunnough.)
Fig. 135.—*Ephemerella invaria,* male genitalia.

of our knowledge of the species in the genus *Ephemerella,* this female type specimen of *quebecensis* cannot be placed even to species group. Specific characters for the females of this genus have not yet been found.

KEY TO SPECIES

ADULT MALES

1. Third segment of genital forceps long and slender: six times as long as broad, fig. 125 . . . .**33. attenuata**
   Third segment of genital forceps shorter and more stout: never more than four times as long as broad, usually as broad as long, figs. 118, 122, 123, 136, 150 2
2. Third segment of forceps three or four times as long as broad, figs. 118–122 . .3
   Third segment of forceps less than twice as long as broad, sometimes only as long as broad, figs. 123, 124, 126–138 . .7
3. Second segment of forceps relatively short and subangulate, with a sharp and deep mesal constriction, fig. 122 . . . **1. lata**
   Second segment of forceps relatively long and more evenly curved, not having a sharp and deep mesal constriction, figs. 118–121 . . . ,. 4
4. Outer surface of hind femur with numerous small, black dots 5
   Outer surface of hind femur without black dots 6
5. An arcuate, transverse row of four black dots present on each abdominal sternite . . . .**2. tuberculata**
   Abdominal sternites without transverse rows of black dots . . . . . .**3. walkeri**
6. Second segment of forceps only three times as long as basal segment, fig. 121; fore wing only 6–7 mm. long . . . .**4. cornutella**
   Second segment of forceps four times as long as basal segment, fig. 120; fore wing 9 mm. long . . . .**5. cornuta**
7. Penis lobes broadened near apexes and bearing a pair of subapical, lateral tubercles, figs. 123, 124, 126 8
   Penis lobes not broadened near apexes and not bearing a pair of subapical, lateral tubercles; either not expanded near apexes, or, if so, bearing several dorsal or ventral spines, figs. 132, 135–138 12
8. Abdominal tergites 2–7 white (in freshly killed specimens), with a pair of small, brown, lateral dots or streaks on each tergite . . . .**6. frisoni**
   Abdominal tergites tan or brown 9
9. Caudal filaments white, with articulations not darkened . . . .**7. sordida**
   Caudal filaments with at least basal articulations darkened with red or brown 10
10. An arcuate, transverse row of four black dots present on each abdominal sternite . . . .**9. serratoides**
    Abdominal sternites without transverse rows of black dots 11

11. Genital forceps strongly bowed, fig. 128; abdominal sternites light tan, with dark brown, longitudinal streaks near each lateral margin . . . .**8. serrata**
    Genital forceps more nearly straight, fig. 133; abdominal sternites uniformly shaded with tan, lacking lateral, dark brown, longitudinal streaks . . . .**10. deficiens**
12. Penis lobes with a deep, median, apical notch, lateral margins projected posteriorly as relatively narrow processes, as in figs. 130–132, 134–138, 145 13
    Penis lobes with only a very shallow, median, apical notch, lateral margins not produced posteriorly, figs. 139–144, 146–150; penis lobes never bearing stout spines 23
13. Penis lobes entirely without spines, fig. 131; upper face red-brown, with area below antennae bright yellow . . . .**11. needhami**
    Penis lobes provided with spines, as in figs. 130, 132, 134–138 14
14. Penis lobes divided on meson to level of bases of genital forceps; numerous, minute spines borne by mesal surface of each penis lobe, fig. 130 . . . .**12. aurivillii**
    Penis lobes not divided on meson to level of bases of forceps; penis lobes bearing relatively few, large spines, figs. 132, 134–138 15
15. Lateral processes of penis lobes long and slender, extended straight, fig. 134 . . . .**13. septentrionalis**
    Lateral processes of penis lobes short, curved inward at apexes, figs. 132, 135–138 16
16. Second segment of genital forceps not suddenly swollen at apex, figs. 127, 137, 138 17
    Second segment of genital forceps suddenly swollen at apex, figs. 129, 132, 135, 136, 145 19
17. Penis lobes short and broad, with each lobe bearing two or three dorsal spines, three or four ventral spines, and two or three apical spines, fig. 138 . . . .**14. argo**
    Penis lobes narrower and more elongate, with spines arranged differently, figs. 127, 137 18
18. Dorsum of thorax tan and yellow; compound eyes of living insect pinkish tan . . . .**15. ora**
    Dorsum of thorax deep reddish brown; compound eyes of living insect yellow . . . .**16. excrucians**
19. Caudal filaments entirely white, with articulations not darkened . . . .**17. dorothea**
    Caudal filaments with at least basal articulations darkened 20
20. Wings with veins brown . . . .**18. subvaria**
    Wings with veins and crossveins hyaline, or costal longitudinal veins faintly tinged with light yellow 21
21. Penis lobes without apical, submesal spines; each lobe bearing two to three

dorsal spines and seven to nine ventral spines, fig. 129......**19. fratercula**
Penis lobes with at least one pair of apical, submesal spines; each lobe bearing four to six dorsal spines in basal area, figs. 135, 145......................22

22. No ventral spines on penis lobes in basal area, fig. 135............**20. invaria**

One to three ventral spines on each penis lobe in basal area, fig. 145..........
...................................**21. rotunda**

23. Penis lobes showing, from lateral aspect, a large, ventral, subapical enlargement, figs. 140, 141.......**22. prudentalis**
Penis lobes lacking a ventral, subapical enlargement......................24

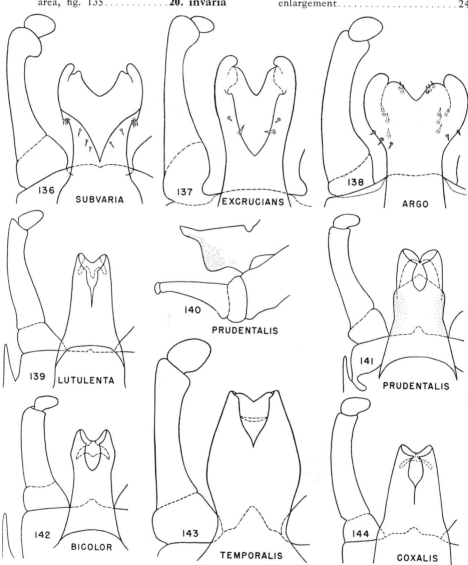

Fig. 136.—*Ephemerella subvaria,* male genitalia.
Fig. 137.—*Ephemerella excrucians,* male genitalia.
Fig. 138.—*Ephemerella argo,* male genitalia.
Fig. 139.—*Ephemerella lutulenta,* male genitalia.
Fig. 140.—*Ephemerella prudentalis,* lateral aspect of male genitalia. (After McDunnough.)
Fig. 141.—*Ephemerella prudentalis,* dorsal aspect of male genitalia. (After McDunnough.)
Fig. 142.—*Ephemerella bicolor,* male genitalia.
Fig. 143.—*Ephemerella temporalis,* male genitalia.
Fig. 144.—*Ephemerella coxalis,* male genitalia. (After McDunnough.)

24. Legs and abdomen having extremely numerous, minute black dots scattered over surfaces..........**23. lutulenta**
Legs and abdomen not having black dots scattered over surfaces............25
25. Penis lobes short, extremely stout, enlarged near bases to form a vase-shaped structure, fig. 143....**24. temporalis**
Penis lobes not vase shaped, figs. 144, 146–150.........................26
26. Caudal filaments white, articulations not darkened................**25. coxalis**
Caudal filaments either entirely dark gray or brown, or light and with darkened articulations....................27
27. Penis lobes with lateral margins evenly, sinuately enlarged toward apexes, and maximum width of apical portion as great as maximum width of basal portion, fig. 146.............**26. simplex**
Penis lobes with lateral margins not evenly and sinuately enlarged toward apexes, and maximum width of apical portion less than maximum width of basal portion, figs. 142, 147–150...28
28. Penis lobes with a subangulate, preapical enlargement on either side, figs. 142, 147..............................29

Penis lobes with a rounded, preapical enlargement on either side, fig. 150, or lobes with lateral margins parallel near apexes, figs. 148, 149..........30
29. Caudal filaments white, with articulations red.......................**28. bicolor**
Caudal filaments gray, with articulations dark gray or black...**29. verisimilis**
30. Penis lobes with rounded, preapical, lateral enlargements, fig. 150......
.....................**30. minimella**
Penis lobes with lateral margins subparallel near apexes, figs. 148, 149..31
31. Penis lobes constricted (or narrower) in apical third, and inner, apical angles of peritreme acute, as in fig. 148; caudal filaments yellow-brown near bases, color blending into white at apexes, and articulations black..**31. funeralis**
Penis lobes constricted in slightly more than basal half, and inner, apical angles of peritreme blunt, fig. 149; caudal filaments light yellow, with articulations red-brown ....**32. aestiva**

MATURE NYMPHS

1. Platelike gills present on abdominal segments 3–7, figs. 165, 166, 172, 173;

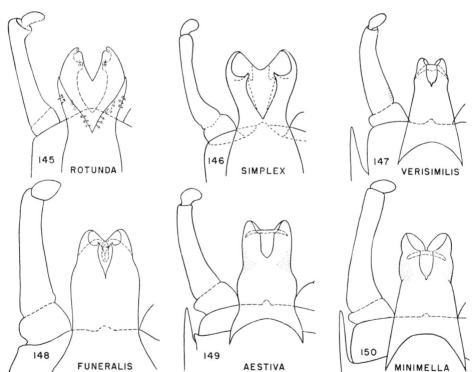

Fig. 145.—*Ephemerella rotunda*, male genitalia.
Fig. 146.—*Ephemerella simplex,* male genitalia.
Fig. 147.—*Ephemerella verisimilis*, male genitalia.

Fig. 148.—*Ephemerella funeralis*, male genitalia.
Fig. 149.—*Ephemerella aestiva*, male genitalia.
Fig. 150.—*Ephemerella minimella*, male genitalia. (After McDunnough.)

filamentous gills absent on first abdominal segment.....................2
Platelike gills present on abdominal segments 4–7, figs. 162–164; filamentous gills present on first segment.......4

Frontal shelf of head without such a notch.............................8
8.   Frontal shelf with a pair of small, triangular, lateral projections, fig. 156.......
.............................**1. lata**

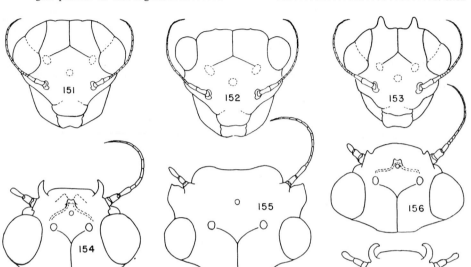

Fig. 151.—*Ephemerella bicolor*, head of mature male nymph, anterior aspect.
Fig. 152.—*Ephemerella bicolor*, head of mature female nymph, anterior aspect.
Fig. 153.—*Ephemerella temporalis*, head of mature nymph, anterior aspect.
Fig. 154.—*Ephemerella cornuta*, head of mature nymph, dorsal aspect.
Fig. 155.—*Ephemerella walkeri*, head of mature nymph, dorsal aspect.
Fig. 156.—*Ephemerella lata*, head of mature nymph, dorsal aspect.
Fig. 157.—*Ephemerella cornutella*, frontal shelf of head of mature nymph, dorsal aspect.

2.   Fore femur not enlarged and toothed on anterior margin, figs. 161, 172, 173..3
Fore femur enlarged and toothed on anterior margin, fig. 160...........6
3.   Apical half of caudal filaments clothed with short spines, fig. 172; maxillary palps reduced or entirely wanting, figs. 167, 169.....................10
Apical half of caudal filaments clothed with long hairs, fig. 173; maxillary palps well developed, fig. 168.....14
4.   Gill on abdominal segment 4 operculate, fig. 164; maxillary palps wanting, fig. 169.................................5
Gill on abdominal segment 4 semioperculate, figs. 162, 163; maxillary palps present, as in fig. 168..............29
5.   Occipital tubercles wanting in male, vestigial in female, figs. 151, 152....22
Occipital tubercles well developed in both sexes, although larger in females than in males, fig. 153.................23

(*WALKERI* Group)

6.   Occipital tubercles present, as in fig. 153.................**2. tuberculata**
Occipital tubercles absent...........7
7.   Frontal shelf of head with a notch beneath each antennal base, fig. 155....
.............................**3. walkeri**

Frontal shelf with a pair of conspicuous, lateral horns, figs. 154, 157.........9
9.   Body of mature nymph 6–7 mm. long; frontal horns so curved inward as to be almost semicircular, fig. 157..........
.............................**4. cornutella**
Body of mature nymph 9–10 mm. long; frontal horns not so markedly incurved, fig. 154..................**5. cornuta**

(*SERRATA* Group)

10.   Maxillary palps entirely wanting, fig. 169.....................**10. deficiens**
Maxillary palps present, although reduced in size, fig. 167............11
11.   Prothorax with a pair of small, dorsal, submedian tubercles near posterior margin....................**8. serrata**
Prothorax lacking dorsal tubercles.....12
12.   Head, thorax, and legs clothed with long hair......................**7. sordida**
Head, thorax, and legs not clothed with long hair.........................13
13.   Caudal filaments black at bases, tips light yellow or white; each abdominal sternite with a pair of dark, lateral streaks and a row of four black dots..
.....................**9. serratoides**
Caudal filaments entirely light yellow or white, fig. 172; each abdominal sternite

with a pair of dark, lateral streaks only
.................................**6. frisoni**

(*INVARIA* Group)

14. Abdominal tergites with a double row of erect, conspicuous tubercles..........
......................**11. needhami**
Abdominal tergites either entirely without paired dorsal tubercles, or with small to minute ones, figs. 165, 166, 173....15

15. Legs markedly long and slender, spider-like............**13. septentrionalis**
Legs not markedly long and slender...16

abdominal tergites 5 and 6 mostly brown.................................18

18. Dorsum of mesothorax freckled with pale dots in addition to the usual 16 relatively large, pale spots...**17. dorothea**
Dorsum of mesothorax uniform deep brown, with relatively large, pale spots only.................**16. excrucians**

19. Lateral spines of apical abdominal tergites well developed, those of eighth segment twice as long as wide at base; hind leg with tibia longer than femur
.........................**12. aurivillii**

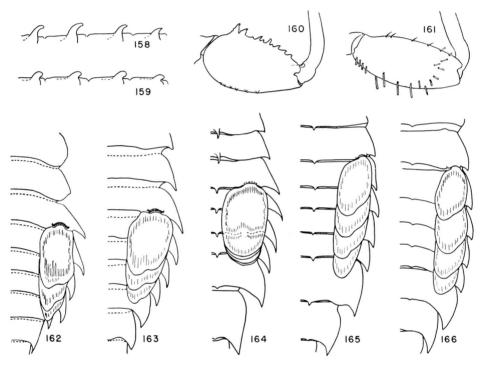

Fig. 158.—*Ephemerella lutulenta*, dorsal tubercles of basal abdominal segments of mature nymph, lateral aspect.
Fig. 159.—*Ephemerella coxalis*, dorsal tubercles of basal abdominal segments of mature nymph, lateral aspect.
Fig. 160.—*Ephemerella cornuta*, fore femur of mature nymph.
Fig. 161.—*Ephemerella needhami*, fore femur of mature nymph.
Fig. 162.—*Ephemerella simplex*, right side of abdomen of mature nymph, dorsal aspect.
Fig. 163.—*Ephemerella lita*, right side of abdomen of mature nymph, dorsal aspect.
Fig. 164.—*Ephemerella temporalis*, right side of abdomen of mature nymph, dorsal aspect.
Fig. 165.—*Ephemerella subvaria*, right side of abdomen of mature nymph, dorsal aspect.
Fig. 166.—*Ephemerella rotunda*, right side of abdomen of mature nymph, dorsal aspect.

16. Paired dorsal tubercles of abdomen entirely lacking......................17
Abdomen with at least rudimentary paired dorsal tubercles...................19

17. Lateral margin of eighth abdominal tergite sinuate, fig. 173; abdominal tergites 5 and 6 almost completely white.......
..............................**14. argo**
Posterior two-thirds of each lateral margin of eighth abdominal tergite straight;

Lateral spines of apical abdominal tergites not so well developed, those of eighth segment as long as broad at base; hind leg with tibia equal to or shorter than femur................20

20. Dorsal abdominal tubercles greatly reduced, almost obsolete...**20. invaria**
Dorsal abdominal tubercles distinct, figs. 165, 166.........................21

21. Dorsal abdominal tubercles relatively

well developed, fig. 165; abdominal tergites lacking pale spots at bases of tubercles..............**18. subvaria**
Dorsal abdominal tubercles more reduced, fig. 166; abdominal tergites having a pale spot at base of each tubercle ....**21. rotunda**

(*BICOLOR* Group)

22. Distance between rows of dorsal abdominal spines suddenly increased on segment 5....................**28. bicolor**
Distance between rows of dorsal abdominal spines uniformly increasing from segments 1 to 7.......**30. minimella**
23. Distance between dorsal abdominal spines on segment 5 less than length of that segment at median line..........24
Distance between dorsal abdominal spines on segment 5 equal to or greater than length of that segment at median line 25
24. Rows of dorsal abdominal tubercles parallel, fig. 164......**24. temporalis**
Rows of dorsal abdominal tubercles converging posteriorly...**22. prudentalis**
25. Posterolateral spine of abdominal segment 3 well developed, its length twice as great as width at base........ 26
Posterolateral spine of abdominal segment 3 reduced, its length only as great as width at base.....................28
26. Posterolateral spine of abdominal segment 9 slightly incurved; lateral margin of segment 9 almost straight...........
......................**31. funeralis**
Posterolateral spine of abdominal segment 9 straight; lateral margin of segment 9 clearly convex...................27
27. Dorsal abdominal tubercles of segments 1–3 long and slender, fig. 158.........
.....................**23. lutulenta**
Dorsal abdominal tubercles of segments 1–3 short and blunt, fig. 159..........
........................**25. coxalis**
28. Distance between rows of dorsal abdominal spines increasing from segments 2 to 7.................**29. verisimilis**
Rows of dorsal abdominal spines almost parallel...................**32. aestiva**
29. Paired dorsal tubercles present on occiput, thorax, and abdomen...**33. attenuata**
Dorsal tubercles wanting on occiput, thorax, and abdomen 30
30. Second and third abdominal tergites with posterolateral angles produced as spinelike projections, fig. 163......**27. lita**
Second and third abdominal tergites with posterolateral angles not produced, fig. 162....................**26. simplex**

WALKERI Group

1. *Ephemerella lata* Morgan

Ephemerella lata Morgan (1911:112).
Ephemerella inflata McDunnough (1926:187).

MALE.—Length of body and of fore wing 6–7 mm. Body generally very dark brown,

almost black, with abdominal tergites slightly lighter in color than thorax. Wings hyaline, with most veins and crossveins hyaline, veins near costal margin of fore wing very light yellow. Genitalia with penis lobes fused almost to tips, penial spines lacking; second forceps segment short and subangulate, with a sharp median constriction, fig. 122; and

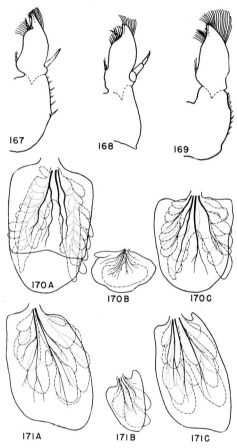

Fig. 167.—*Ephemerella frisoni*, maxilla of mature nymph.
Fig. 168.—*Ephemerella needhami*, maxilla of mature nymph.
Fig. 169.—*Ephemerella deficiens*, maxilla of mature nymph.
Fig. 170A.—*Ephemerella temporalis*, gill of fourth abdominal segment.
Fig. 170B.—*Ephemerella temporalis*, gill of seventh abdominal segment.
Fig. 170C.—*Ephemerella temporalis*, gill of fifth abdominal segment.
Fig. 171A.—*Ephemerella needhami*, gill of third abdominal segment.
Fig. 171B.—*Ephemerella needhami*, gill of seventh abdominal segment.
Fig. 171C.—*Ephemerella needhami*, gill of fifth abdominal segment.

third forceps segment three times as long as wide.

NYMPH.—Head and body without dorsal spines and tubercles. Frontal shelf of head with a pair of small, triangular projections, fig. 156. Fore femur enlarged and toothed on anterior margin.

Known from Maine, Michigan, New York, North Carolina, and Quebec.

2. *Ephemerella tuberculata* Morgan

Ephemerella tuberculata Morgan (1911:112).

MALE.—Length of body and of fore wing 8–9 mm. Body generally very dark red-brown, with middle abdominal tergites usually yellow-brown. Wings hyaline, veins slightly grayish or gray-brown. Genitalia with penis lobes fused almost to tips, penial spines lacking, and inner angles of peritreme acute, fig. 118; second forceps segment evenly bowed from base to apex, a slight inner enlargement near base; third forceps segment four times as long as wide.

NYMPH. — Head with frontal shelf notched below bases of antennae, and with a pair of horns on vertex. Each fore femur enlarged and toothed on anterior margin. Thorax and abdomen with dorsal, paired tubercles.

Known from Maryland, New York, North Carolina, Ontario, Quebec, and Tennessee.

3. *Ephemerella walkeri* Eaton

Baetis fuscata Walker (1853:570).
    Name preoccupied.
Ephemerella walkeri Eaton (1884:129).
    New name.
Ephemerella bispina Needham (1905:43).
Ephemerella fuscata (Walker). McDunnough (1931d:214); Traver (1935a:600).

MALE.—Length of body and of fore wing 7–8 mm. Body generally dark brown, with abdominal tergites slightly lighter toward lateral margins. Wings hyaline, veins slightly stained with yellow. Genitalia with penis lobes fused almost to tips, spines wanting, and inner angles of peritreme blunt, fig. 119; second forceps segment only slightly bowed, with an inner enlargement near base; third forceps segment four times as long as wide.

NYMPH.—Head with a broad frontal shelf having a notch beneath each antennal base, fig. 155; vertex lacking tubercles. Each

fore femur enlarged and toothed at anterior margin. Prothoracic tubercles small and represented by a single pair laterally; dorsal abdominal tubercles minute.

Known from Indiana, New Brunswick, Ontario, and Quebec.

4. *Ephemerella cornutella* McDunnough

Ephemerella cornutella McDunnough (1931b:82; 1931d:211).

MALE.—Length of body and of fore wing 6–7 mm. Body generally very dark brown, abdominal tergites somewhat lighter on mesal area. Wings hyaline, veins slightly tinged with faint yellow; crossveins almost invisible. Genitalia, fig. 121, with penis lobes fused almost to tips, penial spines wanting; second segment of forceps relatively short and straight, an inner enlargement near base; third segment four times as long as wide.

NYMPH.—Frontal shelf of head with a pair of strongly incurved, lateral horns, fig. 157. Fore femur enlarged and bearing teeth on anterior margin. Head, thorax, and abdomen without dorsal tubercles or spines.

Known from Georgia, New Brunswick, New Hampshire, New York, North Carolina, Nova Scotia, Quebec, and West Virginia.

5. *Ephemerella cornuta* Morgan

Ephemerella cornuta Morgan (1911:114).

This species differs from *cornutella* only in that the length of body and of fore wing of male are 9–10 mm. each, second forceps segment of male genitalia is relatively longer and more bowed, fig. 120, and frontal shelf of head of nymph bears a pair of horns which are relatively less incurved, fig. 154.

Known from Connecticut, Maine, New Hampshire, New York, North Carolina, Pennsylvania, Quebec, and Tennessee.

SERRATA Group

6. *Ephemerella frisoni* McDunnough

Ephemerella frisoni McDunnough (1927a:10).

MALE.—Length of body and of fore wing 5.0–6.5 mm. Head dark brown, eyes tan; antennal pedicel tan, flagellum brown. Thoracic notum dark brown, lateral areas, near wing bases, tan; venter dark brown;

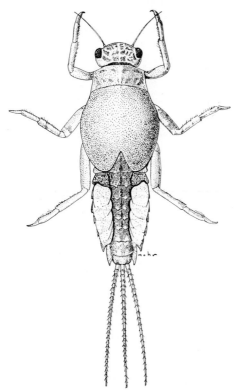

Fig. 172. — *Ephemerella frisoni*, mature nymph, dorsal aspect.

fore femur light brown, fore tibia and tarsus and all of middle and hind legs white; wings hyaline, with brown stain at base of each wing, each fore wing milky in stigmatic area. Abdominal tergites 1–7 and anterior half of tergite 8 white, usually with a fine, median, dorsal, brown line and a pair of small, brown, lateral marks on each tergite; posterior half of tergite 8, and tergites 9 and 10, light brown; abdominal sternum white, usually with a pair of lateral, brown dots on each sternite; caudal filaments entirely white, articulations not darkened. Genitalia, fig. 123, white, with forceps faintly shaded with tan.

NYMPH.—Fig. 172. Length of body 5.0–6.5 mm., of caudal filaments 2.5–3.5 mm. General color light yellowish tan, sometimes almost white, with small and variable brown markings.

Vertex of head, dorsal area of pronotum, and (usually) basal area of front wingpads with irregular, light brown markings; apexes of femora and bases of tibiae usually shaded with light brown; a minute, dark brown

dot present at apicoventral angle of each trochanter. Dorsal part of abdominal tergites fairly uniformly shaded with light brown in area not covered by gills, elongate, dark brown markings along line of inner dorsal margins of gills and on median dorsal line; abdominal venter rather uniformly shaded with light tan, this shading freckled with light yellow or white dots; a minute, dark brown dot present near lateral margin of each abdominal sternite, these dots forming a sublateral row on either side of abdominal venter; basal half of each posterolateral projection of sternites shaded with tan; each caudal filament with a single tan crossband usually present near base.

Head and thorax without tubercles; maxillary palps present, but somewhat degenerated, fig. 167; each tarsal claw with six to eight denticles; posterolateral angles of abdominal segments 4–9 produced, bluntly pointed; posterior margin of second abdominal tergite with a pair of extremely small, submedian tubercles; tergites 3–7 with these tubercles relatively well developed; eighth tergite with tubercles greatly reduced, but discernible, tubercles wanting on ninth tergite; pairs of tubercles converging slightly from tergites 3 to 7, rudimentary tubercles more widely spaced on tergite 8; abdominal segments 1 and 2 without gills, segments 3–7 bearing platelike gills; caudal filaments with relatively few, short setae at each articulation, these setae not longer nor more dense in apical than in basal area of filaments.

Known from Illinois and Missouri. Develops in fairly rapid creeks or small rivers.

Illinois Records.—MUNCIE, Stony Creek: June 8, 1927, T. H. Frison, 1 N; May 22, 1942, Ross & Burks, 1 ♂. OAKWOOD, Salt Fork River: June 6, 1925, T. H. Frison, 1 ♂; June 9, 1926, Frison & Auden, 3 ♂; May 21, 1928, T. H. Frison, 4 N; June 29, 1929, T. H. Frison, 2 N; June 14, 1935, C. O. Mohr, 6 N; May 21, 1936, Mohr & Burks, 3 N; June 11, 1936, C. O. Mohr, 9 N; May 22, 1942, Ross & Burks, 5 N.

7. *Ephemerella sordida* McDunnough

Ephemerella sordida McDunnough
(1925c: 42; 1931d: 205).

MALE.—Length of body 4.5–5.0 mm., of fore wing 5–6 mm. Body very dark brown, almost black, with abdominal sternites somewhat lighter; wings hyaline, all veins only very slightly darker than membrane. Caudal filaments white throughout. Genitalia, fig. 124: penis lobes with a pair of prominent, lateral tubercles, second forceps segment slightly bowéd, and third forceps segment as broad as long.

NYMPH.—Head, thorax, and legs conspicuously hairy; maxillary palp present, but reduced; fore femur not toothed on anterior margin, but bearing a few spicules; abdominal tergites with a double row of submesal, papillate protuberances; caudal filaments bearing only short spines throughout their length.

Known from Ontario and Quebec.

8. Ephemerella serrata Morgan

Ephemerella serrata Morgan (1911:109).

MALE.—Length of body and of fore wing 5–6 mm. Thorax yellow-brown, abdominal tergum red-brown, and abdominal venter yellow-brown, without transverse rows of black dots. Wings hyaline, tinged with brown at bases, and veins faintly shaded. Genitalia, fig. 128, with penis lobes fused almost to tips, and a lateral tubercle present on each penis lobe near tip; second forceps segment bowed toward apex; third forceps segment as broad as long.

NYMPH.—Vertex of head roughened, but without distinct tubercles. Prothorax with a pair of small, submedian, dorsal tubercles near posterior margin; each tarsal claw with three or four denticles. Abdominal tergites with a double row of submedian, wartlike tubercles; caudal filaments bearing only short setae.

Known from Maryland, Massachusetts, New York, North Carolina, Quebec, and West Virginia.

9. Ephemerella serratoides McDunnough

Ephemerella serratoides McDunnough (1931b: 83; 1931d: 207).

MALE.—Length of body and of fore wing 5–6 mm. Thorax yellow-brown, with dark brown shading; abdominal tergites red-brown, with darker brown markings; abdominal sternites lighter in color than tergites, each sternite bearing an arcuate, transverse row of four black dots; wings hyaline, with veins faintly shaded. Genitalia, fig. 126, differ only slightly from those of *serrata*.

NYMPH.—Head and thorax smooth, without tubercles or conspicuous hairs; fore femur bearing a few spicules on posterior margin; each tarsal claw with six or seven denticles. Abdominal tergites with a double row of small, submedian, wartlike protuberances; caudal filaments with only short setae.

Known from Maryland, North Carolina, Quebec, and West Virginia.

10. Ephemerella deficiens Morgan

Ephemerella deficiens Morgan (1911:111).
Ephemerella atrescens McDunnough (1925c:43).

MALE.—Length of body and of fore wing 5–6 mm. Head and body very dark brown, with abdominal sternites uniformly tan. Wings hyaline, with brown staining at wing bases; veins faintly darker than wing membrane. Genitalia, fig. 133: penis lobes fused only two-thirds of the way to tips, and lateral tubercles present on each penis lobe near apex; second forceps segment is relatively straight.

NYMPH.—Head and thorax smooth, without tubercles; maxillary palp entirely wanting, fig. 169; each tarsal claw with eight or nine denticles. Abdominal tergites lacking mid-dorsal tubercles; caudal filaments bearing only short setae.

Known from Georgia, Massachusetts, Michigan, New Brunswick, New Hampshire, New York, North Carolina, Nova Scotia, Ontario, Quebec, and West Virginia.

INVARIA Group

11. Ephemerella needhami McDunnough

Ephemerella excrucians Needham (1905:47), not Walsh. Misidentification.
Ephemerella needhami McDunnough (1925b:171). New name.

MALE.—Length of body and of fore wing 6–8 mm. Head, thorax, and abdominal tergum dark red-brown to almost black; abdominal venter mostly light red-brown.

Head red-brown, face below ocelli yellow; in life, each eye red-tan in upper portion, yellow in lower portion; antennal scape yellow, pedicel and flagellum brown. Thorax dark red-brown to almost black, with yellow spot on sternum between each pair of coxae; fore leg smoky brown, middle and hind legs yellow, with vague brown shading on outer sides of coxa and near apex of femur; wings hyaline, with brown stain at bases, principal veins light yellow. Abdominal tergum dark red-brown, usually with a pair of yellow, submedian spots at anterior margin of each tergite, occasional specimens with a dorsal, median, longitudinal, yellow stripe; sternites light red-brown, with transverse yellow area near anterior margin of each sternite; four minute, black marks on each sternite; genitalia, fig. 131, yellow-brown, with smoky brown shading toward apexes; caudal filaments uniformly gray-brown, articulations not darkened.

FEMALE.—Length of body 7–8 mm., of fore wing 8–9 mm. Color in general lighter than in male, with dark red-brown areas of male being replaced by yellow-brown. Entire sternum of thorax and abdomen yellow, faintly shaded with brown near apex of mesosternum; fore leg yellow, with brown shading near apex of femur; wings faintly brown at bases; caudal filaments very light yellow, basal articulations red.

NYMPH.—Length of body 6–8 mm., of caudal filaments 3.5–4.5 mm. Color extremely variable, ranging from an almost uniformly dark brown form to a form with a light yellow, or white, longitudinal, mesal stripe that extends from vertex to tenth abdominal tergite; lateral margins of pronotum and abdominal segments usually light yellow or white; caudal filaments usually with narrow, brown crossbands throughout their lengths.

Head and thorax without dorsal tubercles; maxillary palps well developed, fig. 168; each tarsal claw with 8 to 10 denticles; each fore- femur with stout, blunt spines along posterior margin and near apex, fig. 161; posterolateral angles of abdominal segments 3–9 produced and spinelike; tergites 2–8 each with a pair of long, submedian tubercles, these forming two almost parallel rows; caudal filaments bearing long setae in apical areas.

Known from Illinois, Indiana, Maine, Michigan, New York, Nova Scotia, and Quebec. Develops in cool, fairly rapid creeks or small rivers.

Illinois Records.—AROMA PARK, Kankakee River: June 11, 1947, B. D. Burks, 4 ♂, 3 ♀. EDDYVILLE, Lusk Creek: April 4, 1946, Burks & Sanderson, 4 N; May 24, 1946, Mohr & Burks, 1 N; May 16–17,

1947, B. D. Burks, 1 ♂, 2 ♀, 1 N. KANKAKEE, Kankakee River: April 30, 1931, T. H. Frison, 45 N; April 23, 1935, Ross & Mohr, 5 N; May 17, 1935, H. H. Ross, 9 N; July

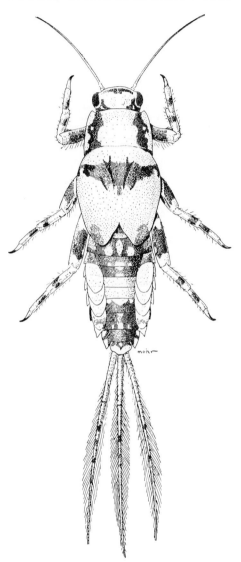

Fig. 173.—*Ephemerella argo*, mature nymph, dorsal aspect.

21, 1935, Ross & Mohr, 1 ♂. MOMENCE, Kankakee River: June 4, 1932, Frison & Mohr, 7 ♂; May 26, 1936, H. H. Ross, 4 ♂; May 17, 1937, Ross & Burks, 24 N; June 1, 1937, B. D. Burks, 1 ♂; May 15, 1938, Ross & Burks, 8 N; May 21, 1940, Mohr

& Burks, 2 N; June 3–4, 1947, B. D. Burks, 6♂, 1♀.

12. *Ephemerella aurivillii* (Bengtsson)

Chitonophora aurivillii Bengtsson
    (1908:243; 1909:8).
Ephemerella aronii Eaton (1908:149).
Ephemerella norda McDunnough (1924d:223).
Chitonophora aurivilliusi Lestage
    (1930a:204). Emended name.
Ephemerella concinnata Traver (1934:219).
    New synonymy.

MALE.—Length of body 10–11 mm., of fore wing 11–12 mm. Thorax red-brown, with lateral areas dark brown; wings hyaline, with stigmatic areas milky and veins brown. Abdominal tergites with anterior third of each segment light brown and posterior two-thirds dark brown; abdominal sternites yellow-brown; caudal filaments uniformly dark brown. Genitalia, fig. 130, distinct from those of all other North American species: penis lobes divided almost to bases, inner margins of lobes provided with numerous minute spines; second forceps segment suddenly enlarged at apex.

NYMPH.—Head and thorax entirely lacking dorsal spines or tubercles; maxillary palps well developed; each fore femur bearing spicules at posterior margin and near dorsal apex; each tarsal claw with 8 to 11 denticles; each abdominal tergite with a pair of light-colored, submesal spots at posterior margin, on each spot a cluster of minute spines and a small, acute tubercle; caudal filaments with long setae near apexes.

Known from Alaska, Alberta, Labrador, Michigan, Montana, Ontario, Pennsylvania, Quebec, and northern Europe. This species is distributed throughout the northern part of the Holarctic region.

13. *Ephemerella septentrionalis* McDunnough

Ephemerella septentrionalis McDunnough
    (1925b:171; 1931d:201).

MALE.—Length of body 8–9 mm., of fore wing 10–11 mm. Dorsum of thorax and abdomen brown; venter yellow-brown; wings and wing veins hyaline. Genitalia, fig. 134, distinct from those of all other eastern North American species: slender lateral projections of penis lobe are virtually straight and directed posteriorly; mesal area of each lobe with two or three stout spines; second forceps segment relatively straight, but enlarged suddenly at apex.

NYMPH.—Head, thorax, and abdomen without dorsal spines or tubercles; legs conspicuously long, slender, and hairy; each tarsal claw with 10 to 12 denticles; caudal filaments bearing long setae in apical regions.

Known from New York, Ontario, and Quebec.

14. *Ephemerella argo* Burks

Ephemerella argo Burks (1949:232).

MALE.—Length of body 8 mm., of fore wing 9 mm. Thorax light yellowish tan, with vague brown markings; abdomen light yellowish tan, tergites with large areas shaded with brown.

Head deep cream colored, vertex with obscure, light gray markings; each antennal scape and pedicel cream colored, flagellum brown; eyes yellowish tan. Legs deep cream colored, apex of fore tibia and second and third fore tarsal segments darkened; wings hyaline, stigmatic areas milky, anterior veins slightly yellow stained, other veins hyaline. Abdominal tergites light yellowish tan, with brown shading: mid-dorsal area of tergites 1–4 almost completely brown, tergites 5 and 6 shaded only near meson, tergites 7–10 completely brown, shaded on mid-dorsal area; venter of abdomen light yellowish, with vague, longitudinal, brown line near each lateral margin of each sternite; genitalia, fig. 138, light yellowish tan; caudal filaments light yellowish tan, articulations brown.

NYMPH.—Length of body 7–9 mm., of caudal filaments 4–5 mm. Body principally yellowish tan, with a prominent, brown color pattern, fig. 173; caudal filaments each with one to three narrow, brown crossbands in middle and apical areas.

Head, thorax, and abdomen without dorsal spines or tubercles; maxillary palps well developed; each tarsal claw with six to nine denticles; posterolateral angles of abdominal segments 3–9 produced, spinelike; caudal filaments bearing long setae in apical areas.

Known from Illinois and Indiana. Probably a large-river species.

Illinois Record.—MOUNT CARMEL, Wabash River: April 2, 1932, Frison & Ross, 2 N.

15. *Ephemerella ora* Burks

Ephemerella ora Burks (1949:235).

MALE.—Length of body 7 mm., of fore wing 8 mm. Head and body yellow, shaded with tan and brown.

Head yellow, shaded with tan between eyes and around ocelli; eyes pinkish tan; each antennal scape and pedicel yellow, flagellum brown. Thorax yellow, with tan shading on dorsomedian area of pronotum, mesonotum mostly tan; legs yellow, with apex of each fore tibia and apex of each of the basal three fore tarsal segments darkened; wings and veins hyaline, with anterior two veins of fore wing stained with faint yellow. Ground color of abdomen yellow, basal tergites with mid-dorsal area of each heavily shaded with brown, apical three tergites pinkish tan; eighth sternite with posterolateral angles pink; genitalia, fig. 127, yellow; caudal filaments light yellow, articulations black.

FEMALE.—Length of body 7.5 mm., of fore wing 9 mm. General color similar to that of male, but lighter; caudal filaments white, articulations black.

Known from Illinois. Apparently a large-river species.

Illinois Record.—MOUNT CARMEL: April 22, 1946, at light, Mohr & Burks, 1 ♂, 1 ♀.

16. *Ephemerella excrucians* Walsh

Ephemerella excrucians Walsh (1862:377).
Ephemerella semiflava McDunnough (1927b:300). New synonymy.

McDunnough (1931d:192) long ago expressed the opinion that *semiflava* would prove to be a synonym of *excrucians*.

MALE.—Length of body 5.5–7.5 mm., of fore wing 7–9 mm. Dorsum of thorax dark red-brown, dorsum of abdomen varying from dark red-brown to almost black; abdominal venter slightly lighter in shade than dorsum.

Head red-brown; eyes yellow in living insect; each antennal pedicel red-brown, flagellum smoky brown. Legs yellow, segments of each fore tarsus darkened at tips. Wings hyaline, veins at costal margin of fore wing slightly yellowish; other veins and all crossveins hyaline; sometimes a faint red-brown staining present at bases of wings. Caudal filaments white, basal articulations light red-brown. Genitalia, fig. 137,

with penis lobes only slightly expanded near apexes, each lobe bearing two or three short, stout dorsal spines in basal position, and sometimes one or two lateral spines on either side; second segment of forceps straight and not expanded at apex.

NYMPH.—(Described from two specimens reared in Michigan by Dr. J. W. Leonard.) Length of body 5–6 mm., caudal filaments 3 mm. General color dark brown, with relatively large, pale markings; caudal filaments white or pale yellow, with faint, darker crossbands throughout.

Head, thorax, and abdomen entirely without dorsal tubercles; maxillary palps present, segmented; each tarsal claw with seven or eight ventral denticles; posterolateral angles of abdominal segments 4–9 produced as small, relatively blunt spines; caudal filaments bearing long, dense setae in apical areas.

Known from Illinois, Michigan, New Brunswick, Oklahoma, Ontario, and Quebec. The only Illinois specimen of this species I have seen is the lectotype in the Museum of Comparative Zoology. Now known to develop in a cold, rapid river.

Illinois Record.—ROCK ISLAND: 12 ♂, 5 ♀, B. D. Walsh (Walsh 1862:377).

17. *Ephemerella dorothea* Needham

Ephemerella dorothea Needham (1908:190).

MALE.—Length of body 5–6 mm., of fore wing 7–8 mm. Head and body pale yellow, almost white. Eyes light red, antennae yellow. Thorax pale yellow, without darker shading; wings and veins hyaline; legs entirely yellow, with each fore tarsus faintly darkened. Abdomen entirely light yellow, without darker areas or shading; caudal filaments white, articulations not darkened. Genitalia, fig. 132, light yellow, with each penis lobe bearing six to eight stout spines in basal position; second forceps segment enlarged at apex.

NYMPH.—Length of body 6–8 mm., of caudal filaments 4–5 mm. General color pale yellow-brown, with small, light tan, freckle-like spots and larger, pale markings; each caudal filament usually with two or three narrow, brown crossbands near tip.

Head, thorax, and abdomen entirely without dorsal tubercles; maxillary palps well developed; each tarsal claw with six to nine denticles; posterolateral angles of abdominal segments 4–9 slightly produced as relatively blunt spines; caudal filaments bearing long, dense setae in apical areas.

Known from Connecticut, Illinois, Indiana, Michigan, New Brunswick, New Hampshire, New York, North Carolina, Pennsylvania, Quebec, South Carolina, Tennessee, Vermont, Virginia, and West Virginia. Develops in fairly large rivers.

Illinois Records.—GOLCONDA: May 13, 1932, Frison, Mohr, & Ross, 4 N. MOUNT CARMEL: Wabash River, April 2, 1932, Frison & Ross, 19 N.

18. *Ephemerella subvaria* McDunnough

Ephemerella subvaria McDunnough (1931*b*: 84; 1931*d*: 194).

MALE.—Length of body 8–9 mm., of fore wing 9–10 mm. Head, thorax, and abdominal dorsum dark brown; abdominal venter light red-brown; wings hyaline, with veins and crossveins light brown. Genitalia, fig. 136, with penis lobes broad and with five or six stout, basal spines on either side; occasionally one or two apical, ventral spines present also; second forceps segment relatively straight, with apex suddenly expanded.

NYMPH.—Head and thorax entirely without dorsal spines or tubercles, but abdominal tergites with relatively well-developed, submedian spines, fig. 165; each tarsal claw with seven to nine denticles; caudal filaments bearing long setae in apical regions.

Known from Michigan, Ontario, Pennsylvania, Quebec, and Wisconsin.

19. *Ephemerella fratercula* McDunnough

Ephemerella fratercula McDunnough (1925*a*: 213).

MALE.—Length of body 7 mm., of fore wing 8 mm. Thorax and dorsum of terminal abdominal segments light brown, basal abdominal segments dark brown on dorsum; venter yellow; wings and wing veins hyaline. Genitalia, fig. 129: each penis lobe with two or three stout, dorsal spines and seven to nine ventral ones; second forceps segment almost straight, suddenly enlarged at apex.

Known from Quebec.

20. *Ephemerella invaria* (Walker)

Baetis invaria Walker (1853: 568).

MALE.—Length of body and of fore wing 8–9 mm. Head light red-brown, eyes reddish orange. Thorax light red-brown; pronotum red-brown, with faint gray shading; thoracic sternum dark red-brown; legs light yellow, almost white, with apex of each fore femur and fore tibia stained with red-brown; wings hyaline, longitudinal veins stained with light brown. Abdominal tergum smoky brown, becoming slightly lighter on apical tergites; sternum uniformly yellow-brown; caudal filaments white, articulations dark. Genitalia, fig. 135, light yellow; each penis lobe bearing five to eight stout, dorsal spines in basal and lateral positions on either side and, in most specimens, a single pair of ventral spines at apex, in some specimens, as many as four apical spines; second forceps segments sharply expanded at apex.

NYMPH.—Length of body 7–9 mm., of caudal filaments 4–5 mm. Color of body varying from dark brown with yellow markings to yellow with small, brown markings; thorax of light-colored specimens usually with two brown crossbands; abdominal tergites 5 and 6, in area between gills, partly or almost completely light colored.

Head and thorax without dorsal tubercles; maxillary palps well developed; each tarsal claw bearing five to seven denticles; abdominal segments 4–9 with well-developed, stout, spinelike projections at posterolateral angles, segment 3 often with rudimentary posterolateral spines; minute to obsolescent dorsal, submedian spines present on abdominal tergites 3–9; caudal filaments bearing long, dense setae in apical areas.

Known from Illinois, Maryland, Michigan, New York, Ontario, Quebec, and Wisconsin. Develops in fairly rapid, moderate-sized rivers.

Illinois Records.—DIXON: Rock River, May 22, 1925, D. H. Thompson, 3 N. OREGON: Rock River, May 15, 1930, Frison & Ross, 1 N. ROCKFORD: Rock River, May 4, 1926, D. H. Thompson, 13 N. ROCKTON: Rock River, May 15, 1926, D. H. Thompson, 1 N.

21. *Ephemerella rotunda* Morgan

Ephemerella rotunda Morgan (1911: 113).
Ephemerella feminina Needham (1924: 309).

MALE.—Length of body and of fore wing 9–11 mm. Thorax light pinkish yellow, with brown markings; abdominal dorsum yellow, with a broad, brown crossband on each segment; venter light yellow or white with, occasionally, faint tan crossbands; wings hyaline, with veins faintly yellow. Genitalia, fig. 145, differ from those of *invaria* in possessing a ventral, subbasal group of two or three spines on each penis lobe.

NYMPH.—Dorsal tubercles on thorax and head absent, but dorsal, submedian abdominal spines present, although small, fig. 166; each tarsal claw with five to eight denticles; caudal filaments with long setae in apical regions.

Known from Michigan, New York, North Carolina, Ontario, Quebec, and Wisconsin.

BICOLOR Group

22. *Ephemerella prudentalis* McDunnough

Ephemerella prudentalis McDunnough (1931a:40).

MALE.—Length of body and of fore wing 7–8 mm. Thorax and basal abdominal tergites red-brown, color gradually changing to yellow on apical tergites; sternites yellow, with black markings; wings hyaline, brown stain at bases; veins hyaline, occasionally tinged with tan. Genitalia, figs. 140, 141, differ from those of all other eastern North American species in having a large, ventral, subapical enlargement on each penis lobe.

NYMPH.—Head with a pair of small occipital tubercles; dorsum of abdominal tergites with two rows of submedian spines converging posteriorly; caudal filaments bearing long setae in apical areas.

Known from Quebec.

23. *Ephemerella lutulenta* Clemens

Ephemerella lutulenta Clemens (1913:335).
Ephemerella lineata Clemens (1913:336).

MALE.—Length of body 8–9 mm., of fore wing 9–10 mm. Thorax dark brown, abdomen yellow-brown, peppered with minute, black dots.

Head yellow-brown, with gray streaks on frons around ocelli and on frontal carina; antennae yellow-brown. Pleura and venter of thorax sprinkled with minute, black dots; wings hyaline, often faintly stained with brown at bases; veins brown; legs yellow-brown, partly or completely covered by a sprinkling of black dots. Both dorsum and venter of abdomen sprinkled with black dots; genitalia, fig. 139, yellow-brown, penis lobes with lateral margins straight and inner peri-

treme angles blunt; second forceps segment slightly bowed toward apex; caudal filaments light yellow-brown near bases, becoming white at apexes, articulations dark brown.

FEMALE.—Length of body and of fore wing same as in male; thorax and abdomen lighter colored than in male, causing sprinkling of black dots to show more clearly; brown on wing veins lighter than in male.

NYMPH.—Length of body 9–12 mm., of caudal filaments 6–8 mm. Uniformly dark yellow-brown, sometimes with small, vague, brown markings on dorsum of thorax; caudal filaments usually uniformly tan, sometimes with faint, brown crossbands.

Occipital tubercles of female minute, those of male obsolete, as in figs. 151, 152; maxillary palps completely absent, as in fig. 169; each tarsal claw with 8 to 10 denticles. Posterolateral angles of abdominal segments 2–9 produced as slender spines, those on segment 2 minute, those on segments 3–8 long and slightly curved inward at apexes; tergites 1–7 each with a pair of long, submedian tubercles, fig. 158, these forming two rows diverging posteriorly; platelike gill of segment 4 operculate, almost or quite covering three posterior pairs of gills; caudal filaments with dense, long setae near apexes.

Known from Illinois, Indiana, Maine, Massachusetts, New Brunswick, New York, North Carolina, Ontario, Quebec, Tennessee, and Wisconsin. Develops in relatively small, shallow lakes.

Illinois Records. — ANTIOCH, Channel Lake: May 16, 1936, Ross & Mohr, 1 N; May 27, 1936, H. H. Ross, 1 ♂ ; May 16, 1938, B. D. Burks, 1 ♀, 1 N.

24. *Ephemerella temporalis* McDunnough

Ephemerella temporalis McDunnough (1924c:74; 1931a:35).

MALE.—Length of body 7–8 mm., of fore wing 8–10 mm. Thorax bright yellow-brown; abdominal dorsum mostly dark brown; sternum light yellow or tan.

Head yellowish tan, eyes light reddish yellow; each antennal scape and pedicel yellowish tan, flagellum gray-brown. Wings hyaline, veins in costal region faintly stained with tan; legs bright yellow, apex of each fore femur and fore tibia shaded with tan. Abdominal tergites 2–7 dark brown, often with a vague, transverse, black mark at posterior margin of each, tergites 8–10 yellow- or red-brown; sternites light yellow or tan, each sternite usually with an arcuate, transverse row of four black dots. Genitalia, fig. 143, bright yellow, penis lobes vase shaped, greatly enlarged at bases, second forceps segment slightly enlarged at apex; caudal filaments yellow to almost white, articulations light reddish brown.

NYMPH.—Length of body 8–10 mm., of caudal filaments 5–7 mm. Body dark brown, flecked with tan dots; abdomen often with a longitudinal, dorsal tan stripe; caudal filaments usually with alternating, broad cross-bands of brown and tan.

Head with well-developed occipital tubercles, fig. 153; maxillary palps wanting. Thorax lacking dorsal tubercles; each tarsal claw bearing 9 to 12 denticles. Postero-lateral angles of abdominal segments 2–9 produced, spinelike, those borne by segments 2 and 3 minute, fig. 164; abdominal tergites 1–4 each with a pair of finger-like, submedian tubercles, tergites 5–7 each with a pair of narrow, acute, submedian spines, and segments 8 and 9 each with a pair of rudimentary spines, these spines and tubercles forming two parallel rows; platelike gills borne by abdominal segment 4 not entirely covering more posterior gills; caudal filaments bearing long, dense setae in apical areas.

Known from Georgia, Illinois, Michigan, New Brunswick, New York, North Carolina, Ontario, Quebec, and Wisconsin. Develops in rather small, shallow lakes.

Illinois Records.—FREEPORT: at light, June 11, 1948, Burks, Stannard, Smith, 3 ♂. GRAYSLAKE: May 26, 1936, H. H. Ross, 1 N. HAVANA, Illinois River: May 21, 1895, C. A. Hart, 1 N; shore of Cook's Island, May 17, 1894, C. A. Hart, 1 N; Quiver Lake, June 1–2, 1894, Smith, 1 N; outlet Quiver Lake, June 1, 1895, C. A. Hart, 1 N. LAKE COUNTY: Cedar Lake, June 19, 1892, Hart & Shiga, 8 N; Fourth Lake, June 16–20, 1892, Hart & Shiga, 14 N; Sand Lake, June 15, 1892, Hart & Shiga, 2 N. WICHERT: June 9, 1948, Burks, Stannard, & Smith, 1 ♂.

25. *Ephemerella coxalis* McDunnough

Ephemerella coxalis McDunnough (1926: 186; 1931a: 37).

MALE.—Length of body and of fore wing 7–8 mm. Thorax and abdomen brown, with large, yellow markings; venter light yellow; wings and wing veins hyaline. Genitalia, fig. 144, quite similar to those of *lutulenta*, with minor differences in details of structure of peritreme opening.

NYMPH.—Head with small occipital tubercles. Thorax without dorsal tubercles; each tarsal claw with six to nine denticles. Abdominal tergites 1–7 bearing two rows of relatively large, submedian tubercles, fig. 159, these two rows diverging posteriorly, caudal filaments with long setae in apical regions.

Known from Georgia, Indiana, North Carolina, Ontario, and Quebec.

26. *Ephemerella simplex* McDunnough

? *Ephemerella unicornis* Needham (1905: 45). *Ephemerella simplex* McDunnough (1925c: 41; 1931d: 208).

MALE.—Length of body and of fore wing 6–7 mm. Head dark brown to black, base of each antennal scape surrounded by a yellowish ring; antennae dark brown; eyes deep red-brown. Thoracic dorsum dark brown to black, with minute, light red-brown markings at sutures; thoracic venter vaguely marked with light brown; fore leg black, fading to gray-yellow toward apex of tarsus; middle and hind legs yellow, coxae brown, apexes of tibiae and tarsal segments shaded with faint brown; all wings hyaline, veins stained with brown, those near costal margin of fore wing darker. Abdominal tergites very dark brown, apical tergites vaguely marked with yellow-brown spots; sternum chiefly dark yellow-brown, slightly lighter on sternites 7 and 8; sternite 9 dark brown to black; genitalia, fig. 146, dark smoky brown; caudal filaments uniformly gray-tan, articulations not darker.

FEMALE.—Length of body 5–7 mm., of fore wing 7–8 mm. In general, same color as male, but with vertex of head and areas of thoracic pleura at wing bases stained with deep yellow or red and thoracic venter with large, dark yellow areas; all legs dusky yellow, with coxae mostly dark brown; wings hyaline, veins near costal margin of each fore wing stained faintly yellow, other veins hyaline; caudal filaments very light yellow, with basal articulations red-brown.

NYMPH.—Length of body 6–8 mm., of caudal filaments 4–6 mm. Body and appendages extremely broad, flat, and hairy.

General color tan to brown, often with vague, dark brown marks on abdominal dorsum. Caudal filaments with two or three narrow, brown crossbands near base.

Head, thorax, and abdomen without dorsal tubercles or spines; maxillary palps well developed; tarsal claws without denticles; posterolateral angles of abdominal segments 4–9 produced, spinelike, fig. 162; platelike gills borne by abdominal segment 4 only partly covering more caudal pairs of gills; caudal filaments bearing short, sparse setae throughout.

Known from Illinois, New Brunswick, New York, North Carolina, Ontario, Quebec, and Tennessee. Develops in fairly rapid, moderate-sized rivers.

Illinois Record.—Rockton: Rock River, June 25, 1947, B. D. Burks, 5 ♂, 7 ♀.

27. Ephemerella lita Burks

Ephemerella lita Burks (1949: 235).

Adult unknown.

Nymph.—Length of body 8 mm., of caudal filaments 5.5 mm. General color light tan, with a few small, brown markings; caudal filaments each with a single, narrow, brown crossband near base.

Head and body flat, conspicuously hairy, without dorsal spines or tubercles; head semiquadrate, with clypeo-genal margin beneath each antennal base slightly incised; maxillary palps well developed; tarsal claws without denticles; posterolateral angles of abdominal segments 2–9 produced, spinelike; platelike gills of abdominal segment 4 semioperculate, only partly covering more caudal pairs of gills, fig. 163; caudal filaments bearing relatively few short setae at each articulation, these setae slightly longer in apical area than in basal area of filaments.

Known from Illinois. Taken in small or moderate-sized, fairly rapid rivers.

Illinois Records.—Dixon: Rock River, May 21–22, 1925, D. H. Thompson, 5 N. Oakwood: Salt Fork River, May 22, 1928, T. H. Frison, 1 N. Rockford: Rock River, June 2, 1927, D. H. Thompson, 1 N.

28. Ephemerella bicolor Clemens

Ephemerella bicolor Clemens (1913: 336).

Male.—Length of body and of fore wing 5–6 mm. Thorax red-brown, abdominal

tergites a lighter red-brown, and venter yellow; wings and veins hyaline. Genitalia, fig. 142, with penis lobes enlarged near apexes.

Nymph.—Head in male lacking occipital tubercles, fig. 151, but in female having very small ones, fig. 152. Each tarsal claw with 8 to 12 denticles. Abdomen with two rows of dorsal, submedian tubercles diverging toward rear, with pair on tergite 5 conspicuously more widely spaced than on anterior tergites; caudal filaments each with long setae in apical area.

Known from Indiana, New Brunswick, New York, Nova Scotia, Ontario, and Quebec.

29. Ephemerella verisimilis McDunnough

Ephemerella virisimilis McDunnough (1930: 57; 1931a: 65).

Male.—Length of body and of fore wing 7–8 mm. Thorax and abdominal tergum dark brown; abdominal venter dark yellow-brown. Wings and veins hyaline. Genitalia, fig. 147, with penis lobes slightly enlarged near apexes.

Nymph.—Head with well-developed occipital tubercles. Tarsal claws each bearing 9 to 12 denticles. Abdomen with two rows of dorsal, submedian tubercles which diverge gradually toward rear; caudal filaments with long, dense setae in apical areas.

Known from Maine, New Brunswick, Ontario, and Quebec.

30. Ephemerella minimella McDunnough

Ephemerella minimella McDunnough (1931a: 63).

Male.—Length of body and of fore wing 6 mm. Thorax and basal abdominal tergites mostly very dark brown, with apical two abdominal tergites lighter brown; sternum tan, gradually fading to white on apical two abdominal sternites. Genitalia, fig. 150: penis lobes with a rounded, preapical enlargement on either side, inner peritreme angles acute, and second forceps segment relatively straight.

Nymph.—Occipital tubercles wanting in male and vestigial in female; dorsal, submedian abdominal spines forming two rows evenly diverging posteriorly; caudal filaments each with long setae in apical region.

Known from Quebec.

31. *Ephemerella funeralis* McDunnough

Ephemerella funeralis McDunnough
(1925a: 210; 1931a: 39).

MALE.—Length of body and of fore wing 6–8 mm. Thorax and abdominal tergites generally red-brown, with sternites lighter red-brown. Wings hyaline, with all veins faintly yellow, and those veins near costal margin of deeper hue. Genitalia, fig. 148: lateral margins of penis lobes nearly straight and parallel near tips, inner peritreme angles acute; second forceps segment almost straight.

NYMPH.—Small occipital tubercles present. Each tarsal claw with 7 to 10 denticles. Abdomen with two rows of dorsal, submedian tubercles diverging posteriorly; caudal filaments with long, dense setae in apical areas.

Known from Georgia, Indiana, New York, Ohio, Ontario, Quebec, South Carolina, Virginia, and West Virginia.

32. *Ephemerella aestiva* McDunnough

Ephemerella aestiva McDunnough (1931a: 64).

MALE.—Length of body and of fore wing 6–7 mm. Thorax and abdominal tergites generally very dark brown; basal and apical abdominal sternites shaded with light brown, with median ones white. Genitalia, fig. 149: penis lobes each with lateral margin near apex nearly straight and parallel, inner peritreme angle blunt; second forceps segment relatively straight.

NYMPH.—Head with occipital tubercles. Abdomen with two almost parallel rows of dorsal, submedian spines; caudal filaments bearing long setae in apical regions.

Known from Quebec.

33. *Ephemerella attenuata* McDunnough

Ephemerella attenuata McDunnough
(1925c: 42; 1931d: 209).

MALE.—Length of body and of fore wing 6 mm. Thorax and basal abdominal tergites very dark brown, apical tergites lighter brown; basal sternites light brown, apical three sternites almost white; wings and veins hyaline. Genitalia, fig. 125, distinct from those of all other North American species: penis lobes fused almost to tips and with a subapical, angulate projection on either side; second forceps segment slightly bowed, and third six times as long as broad.

NYMPH.—Occipital, thoracic, and dorsal abdominal tubercles present; each tarsal claw with 8 to 10 denticles; first pair of platelike gills semioperculate, as in fig. 162; caudal filaments bearing long setae in apical regions.

Known from Maryland, Ontario, and Quebec.

BAETISCIDAE

This family includes only the genus *Baetisca,* which was placed in the subfamily Baetiscinae of the family Baetidae by Traver (1935a: 555) and in the family Baetiscidae of the superfamily Heptagenioidea by Ulmer (1933: 209). Whereas I agree with Ulmer that *Baetisca* represents a group sufficiently distinct to be properly considered a family rather than a subfamily, I do not agree that it has heptageniid affinities. The wing venation and the number of clearly differentiated segments in the hind tarsus in the adults plainly show that this group has no near affinities with the heptageniid type. It is one of the mayflies, such as *Prosopistoma,* which has no known, closely related forms in the Recent fauna.

14. *BAETISCA* Walsh

Baetisca Walsh (1862: 378).

The compound eyes in the adult males of *Baetisca* are large, almost contiguous on the meson. The eyes project posteriorly so as almost completely to cover the pronotum. Each of these eyes is composed of a ventral portion made up of small facets and a much larger dorsal portion of large facets. The division between these two portions is not clearly marked, although in life the lower portion is slightly darker than the upper. The compound eyes in the females are widely separated, and each eye, in life, has a vertical, anterior, colored stripe near the mesal margin. In the males, the fore leg is about as long as the body; the five-segmented fore tarsus is more than twice as long as the fore tibia, and the fore tibia and fore femur are nearly equal in length. In adults of both sexes, a pair of slender and acutely pointed prosternal projections arise between the fore coxae. Each of the middle and hind tarsi in the males and each tarsus in the females has four clearly differentiated segments. The thorax is quite thickset.

The wings of some adults are washed with red or orange, but those of most are hyaline. In the fore wing, there are quite numerous, weak crossveins, fig. 30, numerous, short marginal intercalaries, and the outer wing margin is always slightly scalloped. The median intercalary vein and M_2 extend almost to the wing base. There are no cubital intercalaries. Vein 1A extends to the outer wing margin, and a series of irregular, weak intercalary veins extends from 1A to the anal wing margin. The hind wing is almost circular in outline, fig. 176, has a broad costal projection near the wing base, and numerous weak crossveins and numerous marginal intercalaries; vein M is forked near the center of the wing.

The abdomen is stocky in the basal half, but the segments are markedly more slender and elongate from the sixth segment posteriorly. The male genitalia, fig. 174, very similar throughout the genus, are composed of a pair of three-segmented forceps and a pair of subconical penis lobes which are fused on the median line almost to the tips. The apical margin of the terminal abdominal segment in the females has a pair of submedian, triangular projections with a V-shaped notch on the meson between them. The median caudal filament is vestigial in both the male and female adults.

In the subimagoes, the wings are heavily shaded with dark brown or black; white spots surround the crossveins and often two vague, white bands extend obliquely across each wing.

The nymphs, the first one of which was described by Walsh (1864), are among the most unique and distinct of all mayfly naiads, fig. 181. In these nymphs, the integument is more heavily armored than in any other Nearctic species. The head is small and hypognathous; a pair of small frontal horns is usually present between the bases of the antennae; and the genae are produced above the bases of the mandibles as a pair of spines or small, flat ledges. These projections of the head vary in size and shape among the different species. The distal margins of the labium and labrum meet to close completely the mouth opening anteriorly; the buccal cavity is closed laterally by the mandibles. The labial and maxillary palps each have three segments; the second segment of the labial palp has an apicolateral projection which forms, with the third seg-

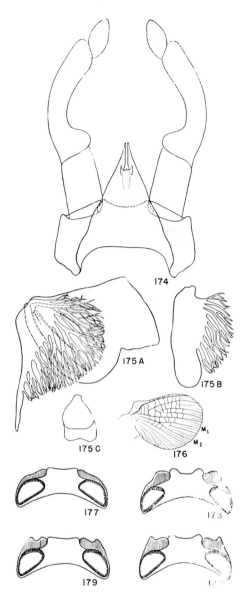

Fig. 174.—*Baetisca obesa*, male genitalia.
Fig. 175A.—*Baetisca bajkovi*, gill of first abdominal segment.
Fig. 175B.—*Baetisca bajkovi*, gill of fourth abdominal segment.
Fig. 175C.—*Baetisca bajkovi*, gill of sixth abdominal segment.
Fig. 176.—*Baetisca obesa*, hind wing.
Fig. 177.—*Baetisca lacustris*, head of mature nymph, dorsal aspect.
Fig. 178.—*Baetisca bajkovi*, head of mature nymph, dorsal aspect.
Fig. 179.—*Baetisca laurentina*, head of mature nymph, dorsal aspect.
Fig. 180.—*Baetisca obesa*, head of mature nymph, dorsal aspect.

ment, a forceps. Each mandible bears, at the apex, two long incisors, a large, tufted lacinia, and a broad, molar surface provided with numerous, lamellate teeth. The hypopharynx is large—about one-half as wide as the labrum—and has a pair of broad, thin parapsides.

The entire thoracic notum is fused and modified to form a carapace, which covers the thorax and the first five abdominal tergites, as well as the anterior part of the sixth abdominal tergite. The legs are relatively short and stout, with long, slender, edentate claws which are almost as long as the tibiae. Each of the abdominal segments 1–6 bears a pair of platelike, fissured gills, fig. 175. These gills normally are completely concealed by the thoracic carapace. Abdominal tergite 6 has a median, truncated, pyramidal lamina against which the apex of the carapace fits. Each of abdominal segments 6–9 has a pair of blunt, posterolateral projections. Tergite 9 has also a median dorsal spine on the posterior margin. Tergite 10 has a median notch on the posterior margin; this tergite is completely surrounded posteriorly and laterally by the incised ninth tergite. Sternite 9 is produced posteriorly and has a median notch on the posterior margin in both males and females. There are three relatively short caudal filaments, each of which bears a dense fringe of setae on both the inner and outer margins.

Baetisca was revised by Traver (1931*c*: 45), and the nymphs were keyed out by McDunnough (1932*b*: 213). Both nymphs and adults were again treated by Traver (1935*a*: 558), and, since the appearance of Traver's and McDunnough's comprehensive papers, three additional Nearctic species have been described, viz., *bajkovi* Neave, *rogersi* Berner, and *thomsenae* Traver.

Specific characters for the females of this genus have not yet been found.

KEY TO SPECIES

ADULT MALES

1. Wings partly or almost entirely washed with pink............**1. rubescens**
   Wings hyaline...........................2
2. All longitudinal veins of fore wing brown; caudal filaments white, with articulations brown..........**2. laurentina**
   Longitudinal veins of fore wing posterior to R_1 hyaline; veins C, Sc, and R_1 lightly shaded with yellow, and bases stained with brown; caudal filaments

usually entirely white, basal articulations sometimes brown..............3
3. Abdominal venter almost completely shaded with brown........**3. obesa**
   Abdominal venter white or very pale yellow.............................4
4. Middle abdominal tergites light brown; fore tibia white, shaded brown at apex; fore wing 10 mm. long......**4. bajkovi**
   Middle abdominal tergites dark red-brown; entire fore tibia tan; fore wing 8 mm. long............**5. lacustris**

NYMPHS

1. Mesonotum with both lateral and dorsal spines.............................2
   Mesonotum with lateral spines only, as in fig. 181.......................3
2. Frontal tubercles of head reduced, virtually wanting, fig. 179....**2. laurentina**
   Frontal tubercles of head relatively well developed, fig. 180..........**3. obesa**
3. Frontal projections of head reduced; genal shelf small, fig. 177............
   ....**5. lacustris**
   Frontal projections of head relatively well developed; genal shelf well developed, fig. 178..................**4. bajkovi**

1. *Baetisca rubescens* (Provancher)

Cloe unicolor Provancher (1876:267), not Hagen. Misidentification.
Cloe rubescens Provancher (1878:127, 144). New name.

MALE.—Length of fore wing 8–9 mm. Base and costal area of fore wing and most of hind wing flushed with a pink stain; thoracic notum dark red-brown, abdominal tergites red-brown, and abdominal sternites lighter reddish or yellowish brown; genital forceps and caudal filaments white or pale yellow.

The nymph is unknown.

The species is known from Quebec.

2. *Baetisca laurentina* McDunnough

Baetisca laurentina McDunnough (1932*b*:214).

MALE.—Length of fore wing 9–10 mm. Compound eyes in life yellow in upper portion, darker yellow, with brown flecks, in lower portion; head and antennae yellow-brown. Mesonotum dark chestnut brown, darker brown at apex of scutellum; mesopleura light brown, sternum dark brown. All legs yellow-brown, with each fore leg slightly darker than others; wings hyaline, with all longitudinal veins of fore wing light brown and crossveins hyaline. Abdominal tergites dark brown; sternites light brown to almost white; genital forceps tan or red-brown; caudal filaments light yellow or tan, articulations dark brown.

NYMPH.—Length of body 8–10 mm. Frontal tubercles of head virtually wanting, only faintly indicated; each gena slightly produced above base of mandible as a small, subtriangular shelf, fig. 179; dorsal and lateral spines of mesonotal shield long and relatively slender; mesonotal shield relatively long and slender, with a maximum width, not including lateral spines, two-thirds as great as maximum length.

Known from Illinois, Michigan, New Brunswick, Ontario, and Quebec. Develops in cool, fairly rapid streams.

Illinois Record.—AROMA PARK: Kankakee River, June 4, 1947, B. D. Burks, 1 ♂.

3. Baetisca obesa (Say)

Baetis obesa Say (1839:43).

MALE.—Length of fore wing 9–11 mm. Compound eyes tan, lower portion slightly darker; head and antennae yellow-brown. Mesonotum red-brown, darker at apex of scutellum; thoracic pleura yellow-brown; sternum yellow-brown, becoming lighter toward posterior margin. Each fore leg light yellow, apex of femur, of tibia, and of each tarsal segment darkened with yellow-brown; middle and hind legs almost white, brown shading at apex of each tarsal segment; wings hyaline, veins C, Sc, and R_1 of fore wing brown-shaded at bases, light yellow distad. Abdominal tergites dark brown, becoming chestnut brown on posterior tergites; sternum light brown, slightly darker on apical three sternites; genital forceps, fig. 174, and caudal filaments usually white or very faintly stained with tan; basal articulations of caudal filaments sometimes brown.

NYMPH.—Length of body 8–10 mm. Frontal tubercles of head well developed, projecting as a pair of rounded protuberances; each gena produced above base of mandible as a triangular ledge, fig. 180; dorsal and lateral spines on mesonotal shield relatively short and stout; mesonotal shield relatively long and slender, with a maximum width, not including lateral spines, two-thirds as great as maximum length.

Known from Illinois, Indiana, Manitoba, Michigan, New Hampshire, New York, and Wisconsin. Develops in cool, fairly rapid streams.

Illinois Records.—HAVANA: Illinois River, April 18, 1894, C. A. Hart, 1 N. MOMENCE: Kankakee River, June 1, 1937, B. D. Burks, 1 ♂. RICHMOND: at light, June 4, 1938, Ross & Burks, 1 ♂. ROCK ISLAND: 20 ♂, 14 ♀ (Walsh, 1862: 378).

4. Baetisca bajkovi Neave

Baetisca bajkovi Neave (1934: 166).
    Description of nymph.
Baetisca bajkovi Neave. Daggy (1945: 388).
    Description of adult ♂.

This species differs from *lacustris* only by minor and possibly intergrading characters both in the nymph and the adult. It may, thus, eventually be necessary to place *bajkovi* as a synonym of *lacustris*. Until actual intergrades have been found, it is best, however, to continue to separate the two. All specimens referred at present to *bajkovi* came from streams, while *lacustris* has been reported as having come only from lakes.

MALE.—Length of fore wing 10 mm. Compound eyes yellowish tan, with lower portion of each slightly darker; head and

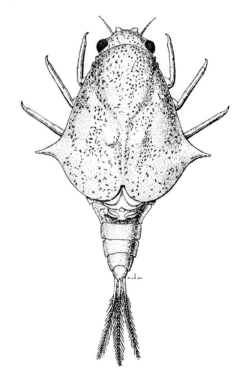

Fig. 181.—*Baetisca bajkovi*, mature nymph, dorsal aspect.

antennae tan. Mesonotum brown, shaded with darker brown at apex of scutellum; thoracic pleura light brown; sternum tan, becoming white on metasternum. Each fore leg white, with faint brown shading at apexes of femur, tibia, and tarsal segments; middle and hind legs white, each with faint brown shading at apexes of tarsal segments; wings hyaline, veins C, Sc, and R_1 brown at bases, light yellow distad. Abdominal tergum chestnut brown, the sternum white; the genital forceps and the caudal filaments white.

NYMPH.—Length of body 7–9 mm. Frontal tubercles of head projecting as a pair of rounded protuberances; each gena produced above base of mandible as a broad, somewhat rounded ledge, fig. 178; long, relatively stout, lateral spines present on mesonotal shield, fig. 181; dorsal mesonotal spines wanting; maximum width of mesonotal shield, not including lateral spines, three-fourths to four-fifths as great as maximum length of shield.

Known from Illinois, Indiana, Manitoba, and Minnesota. Develops in fairly rapid creeks and moderate-sized rivers. The early instar nymphs occur in the swift current, while the late instar nymphs migrate to the comparatively still eddies along the banks.

Illinois Records.—BYRON: Rock River, May 20, 1927, 1 N. DIXON: Rock River, May 22, 1925, D. H. Thompson, 6 N. EAST DUBUQUE: Mississippi River, May 9, 1941, Mohr & Burks, 1 N. GOLCONDA: Big Grand Pierre Creek, May 13, 1932, Frison & Ross, 3 N. HAVANA: Spoon River, April 22, 1898, C. A. Hart, 1 N; White Oak Creek, June 8, 1940, Ross, Riegel, & Burks, 4 N. KANKAKEE: Kankakee River, May 22, 1912, 1 N. MOMENCE: Kankakee River, May 16, 1940, B. D. Burks, 2 N; May 21, 1940, Mohr & Burks, 2 N; June 1, 1940, B. D. Burks, 1 ♂, 1 N. OAKWOOD: Salt Fork River, April 18, 1948, Burks & Stannard, 2 N. OREGON: Rock River, May 25, 1927, D. H. Thompson, 2 N. PECATONICA: Pecatonica River, June 3, 1926, D. H. Thompson, 1 N; May 31, 1927, D. H. Thompson, 2 N. PROPHETSTOWN: Rock River, May 4, 1940, B. D. Burks, 1 N. QUINCY: May 17, 1940, Mohr & Burks, 1 N. ROCK ISLAND: Rock River, April 18, 1931, T. H. Frison, 1 N. ST. JOSEPH: Salt Fork River, June 11, 1940, Thompson & Burks, 19 N.

5. *Baetisca lacustris* McDunnough

Baetisca lacustris McDunnough (1932b: 214).

MALE.—Length of fore wing 8 mm. Wings entirely hyaline, with veins C, Sc, and R_1 of fore wing only very faintly tinged with yellow; thoracic notum chestnut brown; abdominal tergites red-brown; sternites white or very pale yellow; genital forceps pale yellow; caudal filaments white.

NYMPH.—Frontal tubercles of head greatly reduced and genae with small, shelf-like projections, fig. 177; mesonotal shield with long and sharp lateral spines; dorsal mesonotal spines wanting.

Known from Manitoba, Ohio, and Ontario.

OLIGONEURIIDAE

The inclusion here of this family is necessitated by the occurrence in Illinois and Indiana of a curious nymph, fig. 184, which is certainly referable to the Oligoneuriidae. A remarkable adult, also belonging to this family, but not congeneric with our nymph, has been taken in Saskatchewan and Utah.

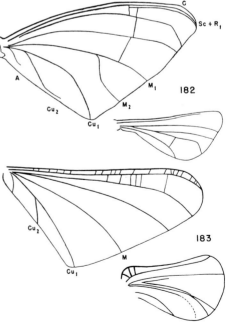

Fig. 182. — *Lachlania saskatchewanensis*, wings. (After Ide.)
Fig. 183. — *Oligoneuria anomala*, wings. (After Eaton.)

In Ulmer's classification (1933: 207), the family Oligoneuriidae includes seven genera, all, up to quite recently, considered to be Palearctic, Neotropical, or Ethiopian in distribution. The members of this family, in the imago stage, are characterized by having the fore wing with vein Sc absent or fused with R, the number of longitudinal veins ranging from four to seven, and the few remaining crossveins restricted to the anterior two to five interspaces; the hind wing has no crossveins, or, at most, very few crossveins, all restricted to the anterior part of the wing. The median caudal filament may be well developed or vestigial. In the nymphs of the Oligoneuriidae, the fore femur and fore tibia have each a dense fringe of long setae on the inner side; the gills of the first abdominal segment are large and situated on the venter, the gills of the following six segments are small, flat, and slender, and are dorsal in position. Each cercus bears a fringe of long setae on the mesal side only; the median caudal filament is either well developed or vestigial.

15. *OLIGONEURIA* Pictet

Oligoneuria Pictet (1845:290, pl. 47).

In the fore wing of the known adults of this genus, vein M is unbranched, veins Cu_1 and Cu_2 diverge near the center, and Rs diverges from R near the base of the wing, fig. 183. The median caudal filament is well developed.

Oligoneuria ammophila Spieth

Oligoneuria ammophila Spieth (1937:139; 1938a:1).

This species is known only from the nymph. It may be that, when the adult is found, the generic assignment will have to be changed. A complete description of the nymph is given by Spieth (1937:139).

Length of mature specimens 10–12 mm. This nymph, fig. 184, bears gill tufts on the maxillae. All the abdominal gills are single: each gill of pair on segment 1, fig. 184, large, finely dissected, and located on the ventral side in such a position as to project anteriorly between the hind coxae; gills of segments 2–7 small, simple, obovate, with acute apexes, and dorsal in position.

The nymph of *ammophila* holds its legs in a most unusual position: normally, the fore

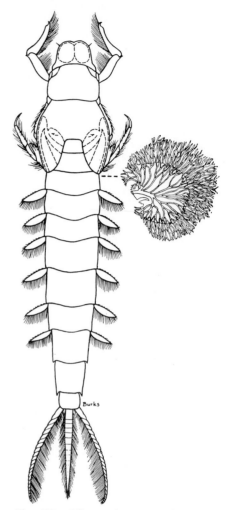

Fig. 184.—*Oligoneuria ammophila*, nymph, dorsal aspect. Small figure at right shows detail of gill of first abdominal segment, borne on venter.

legs are held close to the mouth opening, beneath the head; the middle legs extend almost straight posteriorly, beneath the thorax and basal abdominal segments; the hind legs extend laterally from the metathorax, at right angles to the longitudinal axis of the body. In this position, the middle legs extend much farther posteriorly than do the hind legs. Such an arrangement of the legs probably fits the nymph to maintain its footing in the rather loose sands in which it has been found to live. As is shown in fig. 184, only parts of the coxae and trochanters of the middle legs are visible from the dorsal aspect.

Known from Illinois and Indiana. Lives in shallow, rapid streams with sandy bottoms.

Illinois Records. — HILLSDALE: Rock River at mouth of Canoe Creek, dredging sandy bottom 15 yards from bank, July 30, 1925, R. E. Richardson, 1 ♂ N; Rock River at foot of Lephardt's Island, dredging clean sandy bottom 15 yards from bank, July 30, 1925, R. E. Richardson, 1 ♂ N.

16. *LACHLANIA* Hagen

Lachlania Hagen (1868:372).

This genus is represented in North America only by *saskatchewanensis* Ide (1941: 154), described from a single adult female. Nymphs and adults of *Lachlania* sp. are recorded by Edmunds (1948b:43) as occurring in Utah.

In *Lachlania* adults, the fore wing, fig. 182, has only three or four crossveins; vein M branches a little beyond mid-length, and veins Cu_1 and Cu_2 fork close to the wing base. The legs are greatly atrophied and probably nonfunctional. The median caudal filament is vestigial.

LEPTOPHLEBIIDAE

This family corresponds exactly to the subfamily Leptophlebiinae of the family Baetidae in Traver's classification (1935a: 504), and to the family Leptophlebiidae of Ulmer's classification (1933:201).

In this family, each compound eye in the adult males is composed of a large upper portion of comparatively large facets, and a small lower portion of smaller and darker-colored facets. These two portions of the eye are distinctly separated, but the upper portion is not set on a well-developed stalk, as in the Baetinae, figs. 255–257. The compound eye in the females is of the same size as the lower portion of the eye in the males. The fore tarsus in the males has five segments; all tarsi in the females and the middle and hind tarsi in the males have four clearly differentiated segments. In all Nearctic genera, the two claws borne by each tarsus are dissimilar, with one claw hooked and one claw lobed.

In all Nearctic genera of Leptophlebiidae, the adults have two pairs of wings; in some exotic leptophlebiids, the adults lack hind wings. The fore wings in Nearctic forms

have numerous crossveins, figs. 185–192, and all the principal longitudinal veins are preserved complete, except that the basal part of vein M_2 is obsolete in some genera. Vein R_5 often is sharply bent posteriorly near its base, that is, close to its point of divergence from R_4, as in figs. 185–187. Veins Cu_1 and Cu_2 are separated at their bases; Cu_1 is straight, but Cu_2 at about mid-length is sharply bent toward the anal wing margin. There are two or four cubital intercalary veins. The hind wing may or may not have a costal projection. There is considerable variation within this family in the venation of the hind wing.

The male genitalia are composed of a pair of elongate penis lobes, which are partly or completely fused on the mesal margins, and a pair of four-segmented forceps. The basal segment of each arm of the forceps is often almost completely fused with the second segment; the second segment is elongate and straight; the third and fourth segments are small and semiquadrate or triangular. The terminal abdominal sternite in the females usually has a median cleft or indentation on the posterior margin. There are always three well-developed caudal filaments; the median one may be somewhat shorter and more slender than the cerci.

The nymphs, figs. 199, 212, are typically slender and somewhat flattened forms. They commonly inhabit still water or water with reduced current, such as that in eddies along banks of streams or rivers. The tarsal claws have minute, ventral denticles. Abdominal segments 1–7 bear gills; these gills are slender-lamelliform to filamentous, figs. 193–196. The nymphs always have three long caudal filaments, all of which have relatively inconspicuous setae uniformly distributed over the filaments.

KEY TO GENERA

ADULTS

1. Hind wing without a costal angulation, and with costal margin slightly concave near mid-length, figs. 185–187......2
   Hind wing with a well-marked costal angulation, figs. 188–192......3
2. Fore wing with Cu_1 and Cu_2 closer together than Cu_1 and M in subbasal region where these veins are subparallel, figs. 185, 186......**17. Leptophlebia**
   Fore wing with Cu_1 and Cu_2 separated in subbasal region by a space equal to that separating Cu_1 and M, fig. 187...
   ....**18. Paraleptophlebia**

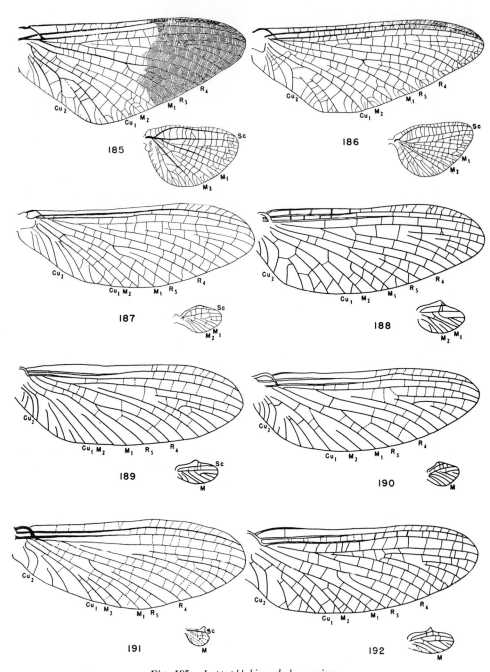

Fig. 185.—*Leptophlebia nebulosa*, wings.
Fig. 186.—*Leptophlebia cupida*, wings.
Fig. 187.—*Paraleptophlebia praepedita*, wings.
Fig. 188.—*Thraulodes speciosus*, wings. (After Traver.)
Fig. 189.—*Habrophlebia vibrans*, wings. (After Traver.)
Fig. 190.—*Choroterpes basalis*, wings. (After Traver.)
Fig. 191.—*Habrophlebiodes americana*, wings. (After Traver.)
Fig. 192.—*Traverella presidiana*, wings. (After Traver.)

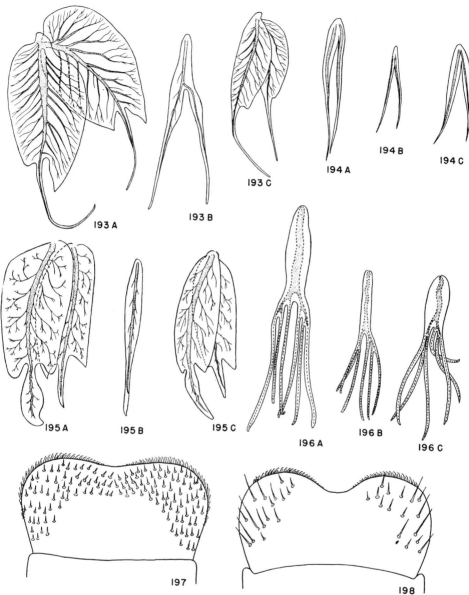

Fig. 193*A.*—*Leptophlebia* sp., gill of fifth abdominal segment.
Fig. 193*B.*—*Leptophlebia* sp., gill of first abdominal segment.
Fig. 193*C.*—*Leptophlebia* sp., gill of seventh abdominal segment.
Fig. 194*A.*—*Paraleptophlebia praepedita*, gill of fifth abdominal segment.
Fig. 194*B.*—*Paraleptophlebia praepedita*, gill of first abdominal segment.
Fig. 194*C.*—*Paraleptophlebia praepedita*, gill of seventh abdominal segment.
Fig. 195*A.*—*Choroterpes basalis*, gill of fifth abdominal segment.
Fig. 195*B.*—*Choroterpes basalis*, gill of first abdominal segment.
Fig. 195*C.*—*Choroterpes basalis*, gill of seventh abdominal segment.
Fig. 196*A.*—*Habrophlebia vibrans*, gill of fifth abdominal segment.
Fig. 196*B.*—*Habrophlebia vibrans*, gill of first abdominal segment.
Fig. 196*C.*—*Habrophlebia vibrans*, gill of seventh abdominal segment.
Fig. 197.—*Paraleptophlebia praepedita*, nymphal labrum, dorsal aspect.
Fig. 198.—*Habrophlebiodes americana*, nymphal labrum, dorsal aspect.

3. Vein M of hind wing forked, fig. 188....
   ....**19. Thraulodes**
   Vein M of hind wing simple, figs. 189–
   192..................................4
4. Vein Sc of hind wing extending nearly or
   quite to apex of wing, fig. 189........
   ....**20. Habrophlebia**
   Vein Sc of hind wing ending near costal
   angulation, figs. 190–192............5
5. Costal angulation of hind wing small,
   rounded at apex, fig. 190; penis lobes
   simple, without appendages, fig. 214..
   ....**21. Choroterpes**
   Costal angulation of hind wing prominent,
   almost or quite acute at apex, figs. 191,
   192; penis lobes bearing decurrent appendages, figs. 215, 216............6
6. Male forceps base divided into two triangular lobes; decurrent appendages of
   penis lobes projecting anterolaterally,
   fig. 215; costal projection of hind wing
   finger-like, stubby at apex, fig. 191....
   ....**22. Habrophlebiodes**
   Male forceps base entire, not divided into
   triangular lobes; decurrent appendages
   of penis lobes projecting toward meson,
   and a pair of slender appendages arising from forceps base, fig. 216; costal
   angulation of hind wing acutely pointed
   at apex, fig. 192.......**23. Traverella**

Mature Nymphs

1. Gills of first abdominal segment similar
   in type to gills borne by more posterior
   segments, as in figs. 194, 196........2
   Gills of first abdominal segment of a different type from gills borne by more
   posterior segments, as in figs. 193, 195;
   each gill of first pair filamentous, each
   gill of following pairs double and lamellate..............................6
2. Abdominal segments 2–9 with posterolateral spines.........**19. Thraulodes**
   Abdominal segments 8 and 9 only bearing
   posterolateral spines.................3
3. Each abdominal gill lamelliform, the
   margins of each finely dissected to form
   numerous, long filaments; gills on
   segments 1–5 bilamellate; gills becoming progressively smaller from anterior
   to posterior abdominal segments......
   ....**23. Traverella**
   Each abdominal gill not lamelliform, the
   margins not finely dissected........4
4. Gills on abdominal segments 2–7 each
   consisting of two clusters of slender
   filaments borne on a single, narrow
   stalk, fig. 196......**20. Habrophlebia**
   Gills on abdominal segments 2–7 bifid to
   bases, each part a very slender lamella, fig. 194.....................5
5. Apical margin of labrum only slightly indented on meson, fig. 197...........
   ....**18. Paraleptophlebia**
   Apical margin of labrum deeply indented
   on meson, fig. 198.................
   ....**22. Habrophlebiodes**
6. Each gill of pair borne by first abdominal
   segment a single filament; apical extensions of gills on segments 2–7 some-

what spatulate at apexes, fig. 195....
.....................**21. Choroterpes**
Each gill of pair borne by first abdominal
segment bifid at apex; apical extensions
of gills on segments 2–7 slender, acute
at apexes, fig. 193..................
..................**12. Leptophlebia**

17. *LEPTOPHLEBIA* Westwood

Leptophlebia Westwood (1840:31).
Blasturus Eaton (1881:193).
Euphyurus Bengtsson (1917:177).

In accordance with the researches of Ide
(1935a: 123), the American species formerly
placed in *Blasturus* are now placed in *Leptophlebia*. It may be noted that Banks (1900:
245) published this synonymy much earlier.

The members of this genus are medium to
large mayflies with predominantly dark
yellow-brown bodies. The fore tarsus in the
males varies from one to one and two-thirds
times as long as the fore tibia. The wings,
figs. 185, 196, are clear or partly stained
with brown, and all veins and most crossveins are brown. In the fore wing, the
posterior branch of the outer fork (vein R_5)
is sharply bent posteriorly near the base,
vein M_2 diverges from M_1 in the subbasal
region, the basal costal crossveins are weak
or wanting, and there are two long, cubital
intercalary veins. The hind wing has no
costal angulation, and vein M is forked near
the base of the wing.

The male genitalia, fig. 200, are quite
uniform in structure throughout the genus.
The genital forceps arise from a medially
fissured base and have four or five segments,
of which the apical two are minute. The
penis lobes are fused on the meson at the
bases only; each penis lobe bears a stout,
mesal, decurrent appendage. The position of
the apexes of these appendages determines
whether they look like a "scarf" or a
"hood"; actually the structure of these appendages is very similar throughout the
genus. The terminal abdominal sternite in
the females has a triangular, median notch
on the posterior margin. The three caudal
filaments may be equal in length and thickness or the median filament may be slightly
shorter and weaker than the cerci.

The stout-bodied nymphs, fig. 199, are
vigorous swimmers. The thorax and abdomen are slightly flattened on the dorsum,
but the head is held almost in a hypognathous
position. Each of the maxillary and labial

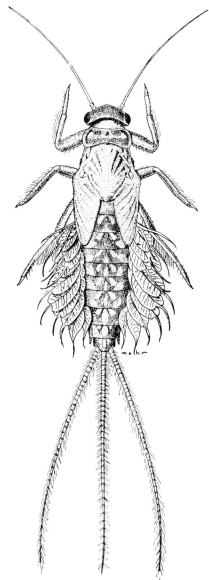

Fig. 199.—*Leptophlebia* sp., mature nymph, dorsal aspect.

palps has three segments. Each antenna is longer than the head and thorax combined. Each tarsal claw has two rows of ventral denticles, a long row from base to near the tip on the outer side, and another, shorter row near the tip on the inner side. Gills of the first abdominal segment are bifid and filamentous, fig. 193B; the gills on segments 2–7 are double and lamelliform, figs. 193A and C, each lamella having a terminal, filamentous extension. Each of the three caudal

filaments is slightly longer than the head and body combined.

In this genus, specific characters for the females and nymphs have not yet been found.

KEY TO SPECIES

ADULT MALES

1. Abdominal tergites 2–7 white, with dark spiracular dots..........**1. johnsoni**
   Abdominal tergites 2–7 partly or almost completely covered by dark brown shading on a tan background........2
2. Fore wing with apical one-fourth to two-fifths shaded with brown, fig. 185; occasionally a spot at outer apical angle of wing hyaline..........**2. nebulosa**
   Fore wing completely hyaline except for faint, brown shading in stigmatic area, fig. 186..................**3. cupida**

1. *Leptophlebia johnsoni* McDunnough

Leptophlebia johnsoni McDunnough (1924c:73).
Blasturus gracilis Traver (1932a:133).

MALE.—Length of body and of fore wing 8–9 mm. Head very dark, glossy brown, with eyes slightly lighter brown. Thorax dark brown to black on dorsum and light brown to tan on pleura and sternum; wings hyaline, with veins and most crossveins dark brown, and each fore wing with a light brown cloud covering stigmatic and outer apical areas. Abdominal segments 2–7 white, with dark brown, or black, spiracular dots and tan ganglionic marks; segments 8–10 dark brown; genital forceps white, penis lobes yellow; caudal filaments tan, with articulations dark brown.

Known from Connecticut, Massachusetts, New Hampshire, New York, North Carolina, Ontario, and Quebec.

2. *Leptophlebia nebulosa* (Walker)

Palingenia nebulosa Walker (1853:554).
Potamanthus odonatus Walsh (1862:372).

No authentic Walsh material of *odonatus* is known to be in existence, but the characters given in Walsh's original description of the species certainly indicate that *odonatus* is a synonym of *nebulosa*. When Spieth (1940:327) examined the type of *nebulosa* in the British Museum, he found no reason to alter the concept of the species as currently identified.

MALE.—Length of body and of fore wing 10–12 mm. Head dark brown, eyes slightly lighter brown. Entire thorax very dark brown, almost black, each fore leg brown, middle and hind coxae brown, rest of middle and hind legs tan; wings hyaline, veins and most crossveins tan or brown, crossveins toward posterior margin in either wing often hyaline, stigmatic crossveins of each fore wing extremely numerous, anastomosed, a brown cloud covering outer, apical one-fourth to two-fifths of fore wing, fig. 185, but sometimes with extreme outer, apical angle hyaline, making wing appear to have a broad subapical, brown crossband. Dorsum of abdomen almost or entirely dark brown, sometimes becoming tan along lateral margins of tergites, and often with a pair of submesal, short, lunate, tan marks present at anterior margin of each tergite; sternites 1 and 2 light brown, sternites 3–8 tan or yellow with, sometimes, vague brown shading, sternites 8 and 9 brown; genitalia, fig. 200, with forceps tan and penes brown; caudal filaments brown, articulations darker brown.

FEMALE.—Length of body 10–12 mm., of fore wing 12–14 mm. Color much as in male, but somewhat lighter. Head with yellow shading on each side, near eyes. Thorax with yellow areas on pleura, with sternum mostly yellow-brown; wings without brown shading. Dorsum of abdomen as in male, sternum entirely tan or yellow-brown, apical two sternites always a little lighter than others; caudal filaments tan or light brown, articulations darker brown.

Known from northeastern and midwestern states and southeastern Canada. Develops in ponds or in the still eddies along the banks of streams.

Illinois Records.—CHESTERVILLE: April 15–May 1, 1936, Ross & Mohr, 9 ♂. DANVILLE: May 9, 1926, T. H. Frison, 1 ♂. GEORGETOWN: April 14, 1930, Frison & Ross, 4 ♂, 2 ♀. HAVANA: April 15, 1898, Hart, 1 ♂; April 18, 1894, Hart, 1 ♂; April 21, 1898, Hart, 1 ♂; April 22, 1898, Hart, 11 ♂, 7 ♀; April 24 & 25, 1898, Hart, 16 ♂, 10 ♀; April 28 & 29, 1898, Hart, 1 ♂, 2 ♀. HOMER: April 27, 1907, 1 ♂. MAHOMET: April 16, 1925, T. H. Frison, 4 ♂, 1 ♀; April 23, 1925, T. H. Frison, 12 ♂, 3 ♀. ST. JOSEPH: Salt Fork River, May 3, 1914, 1 ♂, 2 ♀. URBANA: April 25, 1949, J. E. Porter, 1 ♂. WATSON: April 15–May 5, 1936, Ross & Mohr, 10 ♂. WAUCONDA: April 30, 1942, Ross & Burks, 2 ♂, 1 ♀. WHITE HEATH: April 16, 1932,

Ross & Riegel, 1 ♂; April 22, 1917, 1 ♂, 1 ♀; April 28, 1916, 1 ♂; Sangamon River, May 5, 1940, H. H. Ross, 2 ♂, 4 ♀; May 10, 1938, H. H. Ross, 1 ♂.

3. Leptophlebia cupida (Say)

Ephemera cupida Say (1823: 163).
Ephemera hebes Walker (1853: 538).
Palingenia concinna Walker (1853: 553).
Palingenia pallipes Walker (1853: 553).
Baetis ignava Hagen (1861: 47).

MALE.—Length of body 9–11 mm., of fore wing 10–12 mm. Head very dark brown, almost black, eyes slightly lighter brown. Thorax uniformly very dark brown to black. Each fore leg dark brown, middle and hind legs slightly lighter brown; wings hyaline, with brown staining at bases and light brown shading in stigmatic area of each fore wing, fig. 186; stigmatic crossveins numerous, only slightly or not at all anastomosed. Dorsum of abdomen dark brown, with tan crossline at posterior margin and tan stripe along either lateral margin of each tergite, often a large, irregular, median, tan mark on each tergite; sternites 2–7 tan or very light brown, sternites 1, 8, and 9 washed with brown; genital forceps at base tan or yellow, becoming brown at apexes, penes brown; caudal filaments gray-brown, articulations darker brown.

FEMALE.—Length of body 9–11 mm., of fore wing 10–12 mm. Generally slightly lighter in color than the male. Head dark brown only between and posterior to ocelli, otherwise tan. Thorax dorsally dark brown, pleura marked with tan, sternum tan on meson; wings usually not darkened at bases, sometimes faintly so; each fore leg dark brown, middle and hind legs yellow-brown or tan. Dorsum of abdomen dark brown, each tergite yellow or tan at posterior and lateral margins, entire sternum yellow or tan; caudal filaments as in male.

Known from the northeastern and midwestern states and southeastern Canada. Develops in ponds or in the still eddies along the banks of streams.

Illinois Records.—HEROD: March 24, 1939, Ross & Burks, 1 ♂; April 4, 1948, 3 ♂; April 8–9, 1947, B. D. Burks, 2 ♂, 3 ♀. KICKAPOO STATE PARK: May 4, 1947, Ross & Stannard, 1 ♂. ROCK ISLAND: 11 ♂, 3 ♀ (Walsh 1862: 372). RUDEMENT: Blackman Creek, April 7, 1947, B. D. Burks, 1 ♀.

URBANA: University Grounds, April 8, 1889, Marten, 5 ♂ ; April 15, 1898, Hart, 10 ♂ . WATSON: April 9–24, 1936, Ross & Mohr, 19 ♂ ; April 11, 1932, Ross & Mohr, 2 ♂ .

18. *PARALEPTOPHLEBIA* Lestage

Paraleptophlebia Lestage (1917:340) ;
    Ulmer (1920a:113, 116).

The genus *Paraleptophlebia* was originally distinguished from *Leptophlebia* in the nymphs only. Lestage's description of the genus appeared in a study of the nymphs of Palearctic mayflies. Ulmer (1920c:113; 1933: 202), however, gave characteristics for the separation of the adults of the two genera. Traver (1934:189; 1935a:510) transferred most of the American species of *Leptophlebia* to *Paraleptophlebia*.

The adults of *Paraleptophlebia* are small, extremely frail mayflies. They have clear or faintly tinted wings and most of them have predominantly dark brown bodies; the males of some species have the middle abdominal segments. mostly white. The fore wing, fig. 187, has the posterior branch of the outer fork (vein R_5) sharply bent posteriorly near the base, vein M_2 diverges from M_1 near the wing base, and the cubital intercalary veins are detached at their bases. The hind wing is broad and lacks a costal angulation, and vein M is forked near the base of the wing. The fore leg in the males has the tarsus one and one-half to two times as long as the fore tibia. The terminal abdominal sternite in the females is deeply cleft on the meson. The three caudal filaments are equal in length.

The male genitalia in *Paraleptophlebia*, figs. 201–211, consist of a pair of slender penis lobes, which are more or less fused on the meson, and a pair of four-segmented forceps arising from a medianly cleft base. Each penis lobe bears one or two appendages. The shape and arrangement of these appendages, as well as the shape of the penis lobes themselves, provide excellent characters for the recognition of species. Some species also show striking differences in the form of the forceps. The male genitalia may be cleared, stained, and mounted on microscope slides for study, but the operation must be done with great care, as the minute, fragile genital structures of the males of this genus are easily broken or distorted. The genitalia on many dry specimens are suitable for study without special preparation, and the same is true for specimens in alcohol.

The agile nymphs, fig. 212, are small and rather flat dorsoventrally. In life, they hold their heads in a semihypognathous position. They are not strong swimmers, and are most often to be seen crawling among debris and gravel on the bottoms of shallow pools or eddies. They move with a characteristic, snakelike motion.

In most species, the mouth-parts are similar to those to be found in most heptageniid nymphs. There are three segments in each labial palp and three in each maxillary palp. Also, in most species, each mandible is short and terminates in two or three toothed incisors. In a few western species, however, the body of the mandible is greatly elongated and tusklike, projecting anteriorly far past the labrum. In the species with tusked mandibles, the left mandible retains two incisors while the right has but one. Each antenna is as long as the head and thorax combined. The tarsal claws are single, slender, and long, with numerous ventral denticles. Each of abdominal segments 1–7 bears a pair of bifid, filamentous gills, all of which are of the same type, fig. 194. The posterolateral angles of abdominal segments 8 and 9, or 9 only, are produced as slender spines. The three caudal filaments are of about equal length.

Nymphs of this genus develop principally in shallow, fairly rapid streams of small or moderate size. These streams usually have bottoms of coarse gravel.

The key to nymphs given below is based on a study of nymphal material in the Cornell University collection and the material reared in Illinois. Gordon (1933: 116) and Traver (1935a:514) have published keys for the nymphs of *Paraleptophlebia*, and Ide (1930b:207) has published descriptions and figures of the nymphs of several northeastern species.

Specific characters for the separation of females of this genus have not yet been found.

KEY TO SPECIES

ADULT MALES

1. Forceps with a large, dorsal enlargement near base, visible in lateral aspect, figs. 201, 203.....................2
   Forceps without a large, dorsal enlargement near base.....................3

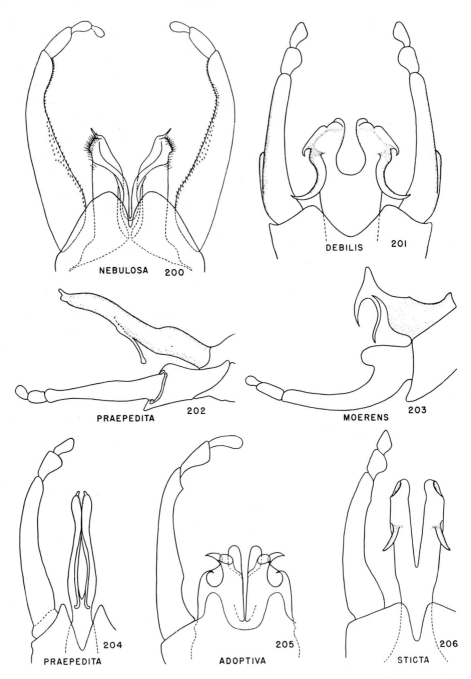

Fig. 200.—*Leptophlebia nebulosa*, male genitalia.
Fig. 201.—*Paraleptophlebia debilis*, male genitalia.
Fig. 202.—*Paraleptophlebia praepedita*, male genitalia, lateral aspect.
Fig. 203.—*Paraleptophlebia moerens*, male genitalia, lateral aspect. (After McDunnough.)
Fig. 204.—*Paraleptophlebia praepedita*, male genitalia.
Fig. 205.—*Paraleptophlebia adoptiva*, male genitalia. (After McDunnough.)
Fig. 206.—*Paraleptophlebia sticta*, male genitalia.

2. All longitudinal veins of fore wing tan, with C, Sc, and R slightly darker; abdominal tergites 2–7 each with a pair of posterolateral, dark brown spots and a dark brown crossband at posterior margin....................**1. debilis**
   Veins C, Sc, and R of fore wing a faint tan, other longitudinal veins hyaline; abdominal tergites 2–7 each with a dark brown crossband at posterior margin, but lacking posterolateral spots......
   ....**2. moerens**

3. Penis lobes long, straight, and slender, figs. 202, 204, 206.................4

   Penis lobes relatively short and broad, figs. 205, 207–211..................5

4. Each penis lobe with a terminal papilla, figs. 202, 204..........**3. praepedita**
   Each penis lobe with a lateral depression at tip; papilla wanting, fig. 206......
   ....**4. sticta**

5. Abdominal tergites 2–6 almost entirely dark brown......................6
   Abdominal tergites 2–6 almost or completely white......................7

6. Penis lobes without decurrent appendages, each lobe with a beaklike, lateral projection, fig. 205......**5. adoptiva**

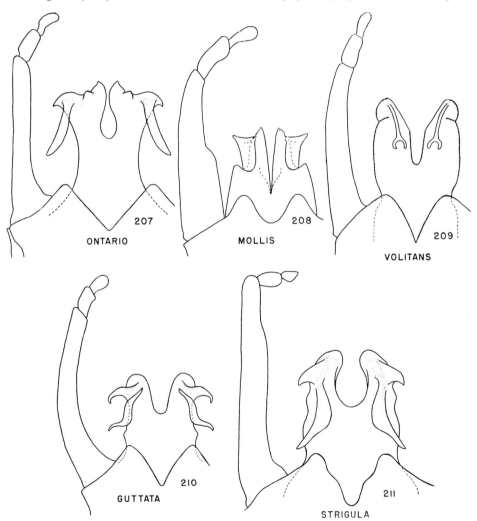

Fig. 207.—*Paraleptophlebia ontario,* male genitalia.
Fig. 208.—*Paraleptophlebia mollis,* male genitalia. (After McDunnough.)
Fig. 209.—*Paraleptophlebia volitans,* male genitalia.
Fig. 210.—*Paraleptophlebia guttata,* male genitalia.
Fig. 211.—*Paraleptophlebia strigula,* male genitalia.

Penis lobes each with a large, decurrent appendage, fig. 207.......**6. ontario**

7. Penis lobes without decurrent appendages, fig. 208..................**7. mollis**
Penis lobes with decurrent appendages, figs. 209–211......................8

8. Penis lobes without apicolateral projections; decurrent appendages slender, bifid at apexes, fig. 209.....**8. volitans**
Penis lobes with apicolateral projections; decurrent appendages not bifid at apexes...........................9

9. Abdominal tergites 2–7 white, with spiracular dots; mesal apical angles of penis lobes divergent, fig. 210........
..........................**9. guttata**
Abdominal tergites 2–7 white, with spiracular dots and a large, brown spot near each posterolateral angle; mesal apical angles of penis lobes convergent, fig. 211................**10. strigula**

MATURE NYMPHS

1. Gills borne by abdominal segments 3–5 dividing into two branches at a point at least one-third the distance from base to apex of gill; tracheae of these gills with numerous, prominent, lateral branches............................2
Gills borne by abdominal segments 3–5 divided into two branches at a point not more than one-sixth the distance from base to apex of gill; tracheae of these gills with only a few, minute, lateral branches....................3

2. Anterior margin of labrum slightly indented on meson; each gill of abdominal segments 3–5 divided at a point about one-third distance from base to apex of gill.....................**7. mollis**
Anterior margin of labrum not indented on meson; each gill of abdominal segments 3–5 divided at a point about one-half distance from base to apex of gill
.....................**5. adoptiva**

3. Abdominal venter with a pair of longitudinal, sublateral brown bands......
.....................**3. praepedita**
Abdominal venter without sublateral bands..............................4

4. Only abdominal segment 9 with posterolateral angles produced as spines....5
Both abdominal segments 8 and 9 with posterolateral angles produced as spines
..................................6

5. Abdominal segments 2–6 each with a black streak along either lateral margin
.....................**10. strigula**
Abdominal segments 2–6 with only a small, black spot near base of each gill
.....................**9. guttata**

6. Gills borne by abdominal segments 3–5 with long, sparse, marginal hair......
.....................**8. volitans**
Gills without marginal hair..........7

7. Tibiae light yellow, with brown band in middle and at base of each.........
.....................**1. debilis**
Tibiae light yellow or tan, with brown shading at bases only.............8

8. Combined length of second and third segments of maxillary palp one and one-half times as great as length of first segment...................**6. ontario**
Combined length of second and third segments of maxillary palp equal to length of first segment....**2. moerens**

1. *Paraleptophlebia debilis* (Walker)

Baetis debilis Walker (1853:569).
Leptophlebia mollis Needham (1908:189), not Eaton. Misidentification.
Leptophlebia separata Ulmer (1920a:27; 1921:255).

MALE.—Length of body and of fore wing 8–9 mm. Head, thorax, and apex of abdomen dark brown, abdominal segments 2–7 white, with transverse, brown lines at posterior margins of tergites; wings hyaline, with tan longitudinal veins and colorless crossveins; genital forceps, fig. 201, a faint tan, with penis lobes slightly darker; caudal filaments white.

Known from the northern states and the southern part of Canada.

2. *Paraleptophlebia moerens* (McDunnough)

Leptophlebia moerens McDunnough (1924b:94).

MALE.—Length of body and of fore wing 5.5–6.5 mm. Head dark brown, almost black; eyes in life brown; antennae dark brown, each becoming hyaline at apex of flagellum. Thorax dark brown; fore leg light brown, with femur and tibia shaded with dark brown; middle and hind legs light brown, with coxae dark brown, femora and tibiae shaded with dark brown, and tarsi white; wings hyaline, with faint brown staining at base of fore wing, veins and crossveins colorless. First abdominal segment dark brown; second tergite shaded with light brown, with dark brown crossband at posterior margin; tergites 3–6 white, with dark brown crossband at posterior margin of each; apical three segments chestnut brown; sternites 2–6 white, with large, median, orange-tan spot on each; forceps base brown, forceps, fig. 203, tan, penis lobes yellow-brown; caudal filaments white.

FEMALE.—Length of body and of fore wing 7 mm. Head and thorax red-brown, lighter than in male; legs light brown, with dark brown shading at base of each tibia; wings very faintly stained with tan, and longitudinal veins of fore wing stained a faint yellow-brown. Abdomen uniformly red-brown, each segment with posterior margin slightly darkened; terminal abdominal sternite with a relatively shallow, broad, rounded, median excavation on posterior margin; caudal filaments a faint yellowish tan.

NYMPH.—Length of body 7–8 mm. Head light yellow-brown, with lateral areas near eyes shaded with red-brown. Thorax light brown, yellowish laterally; legs light yellow-

genitalia, figs. 202, 204, light brown; caudal filaments uniformly tan.

FEMALE.—Length of body 5–6 mm., of fore wing 5.5–6.5 mm. Coloration similar to that of male, except that red-brown replaces dark brown, and tan replaces light brown; all crossveins of fore wing, except those in anal and cubital areas, tan; cubital and anal crossveins of fore wing, and all crossveins in hind wing, hyaline; posterior margin of seventh abdominal sternite produced posteriorly to form a long, pointed

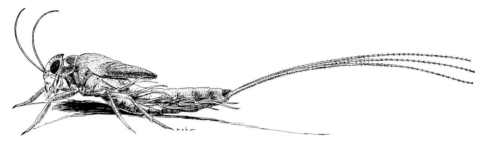

Fig. 212.—*Paraleptophlebia praepedita*, mature nymph, lateral aspect.

brown, with brown shading near apexes of femora. Abdomen light brown, each tergite with two pairs of vaguely defined, light yellowish spots; gills hyaline, tracheae gray, with only a few, minute, lateral branches; posterolateral angles of tergites 8 and 9 produced as spines; caudal filaments light yellow-brown.

Known from the northeastern and mid-western states and eastern Canadian provinces.

Illinois Record.—HEROD: Branch Big Grand Pierre Creek, May 2, 1946, Burks & Sanderson, 1 ♂.

3. *Paraleptophlebia praepedita* (Eaton)

Leptophlebia praepedita Eaton (1884:99).

MALE.—Length of body and of fore wing each 4.0–5.5 mm. Head very dark brown, almost black; eyes in life dark red-brown; antennae brown, each becoming hyaline at tip of flagellum. Thorax very dark brown, with yellow-brown markings on pleura; legs usually uniformly light brown, femora sometimes slightly darkened; wings hyaline, each fore wing slightly brown-stained at base, longitudinal veins tan, crossveins colorless. Abdomen usually uniformly brown, middle segments sometimes slightly lighter brown on dorsal meson and at anterior margins;

ovipositor; posterolateral angles of terminal sternite acuminate, median emargination of posterior margin deep and triangular; caudal filaments tan.

NYMPH.—Fig. 212. Length of body 5–7 mm. Head and body tan, vaguely marked with light yellow on dorsum of thorax; legs yellow, shaded with tan near apexes of femora, in middle of tibiae, and near bases of tarsi; abdominal tergum tan, marked with light yellow on meson and near posterolateral angles of each tergite; abdominal venter yellow, with longitudinal, tan bar parallel with and near to each lateral margin; gills hyaline, central tracheal stripes purplish gray; posterolateral angles of abdominal tergites 8 and 9 produced as spines; caudal filaments uniformly tan.

Known from the northeastern and mid-western states and eastern Canada. This is the commonest species of *Paraleptophlebia* in Illinois.

Illinois Records.—DIONA: June 7, 1941, Ross & Mohr, 1 ♂. EDDYVILLE: Lusk Creek, May 16–17, 1947, B. D. Burks, 1 ♂. FOX RIDGE STATE PARK: May 13–17, 1938, Ross & Burks, 6 ♂, 1 ♀, 1 N; May 25, 1942, Ross & Riegel, 1 ♂, 2 N. HEROD: May 2–9, 1942, Burks & Mohr, 4 ♂, 7 ♀, 16 N; May 13, 1939, Burks & Riegel, 2 ♂; May 27, 1942, B. D. Burks, 71 ♂, 8 ♀; May 2, 1946, Mohr

& Burks, 1 ♂. MONTICELLO: June 6, 1947, Jack Warner, 3 ♂. MUNCIE: May 24, 1914, 1 ♂. OAKWOOD: May 22, 1942, Ross & Burks, 6 ♂; May 28, 1948, Burks & Evers, 1 ♂; May 29, 1936, Ross & Mohr, 2 ♂. RICHMOND: June 14, 1938, Mohr & Burks, 1 ♂. ROSECRANS: Des Plaines River, June 8, 1938, B. D. Burks, 2 ♂; May 22–27, 1938, Ross & Burks, 10 ♂, 7 ♀, 8 N. URBANA: May 20, 1914, 1 ♂. WADSWORTH: June 3, 1943, Ross & Sanderson, 32 ♂, 13 ♀.

4. *Paraleptophlebia sticta* new species

This species is most closely related, in type of genitalia, to *praepedita,* but differs in that the penis lobes are much shorter and thicker, and lack terminal papillae; the decurrent appendages of the penis lobes also are shorter and stouter in *sticta* than they are in *praepedita.* The two species differ in color, as abdominal tergites 2–6 in *praepedita* usually are uniformly brown, but sometimes have lighter brown areas on dorsal meson and at anterior margins; these tergites in *sticta* are white, faintly tinged with tan, and have dark brown, posterolateral markings.

MALE.—Length of body 5 mm., of fore wing 6 mm. Head dark chestnut brown, antennae yellow-brown, each becoming hyaline toward apex of flagellum. Thorax dark chestnut-brown; wings hyaline, longitudinal veins near costal margin of each fore wing tinged with tan, other veins and all crossveins hyaline, stigmatic crossveins not anastomosed, 9 to 11 in number, slanted; legs pale tan, with coxae brown. First abdominal segment brown; tergites 2–6 white, with faint tan tinge, a dark brown spot near posterolateral angle of each tergite and a transverse line on posterior margin, a longitudinal, dark brown line in spiracular region, and a dark brown circle at each spiracle; posterior two-thirds of tergite 7 and all of tergites 8–10 brown of a lighter shade than thorax; sternites 2–7 white, tinged with tan, unmarked; sternite 8 light tan, with transverse, brown line at anterior margin; sternite 9 tan, with dark brown shading at lateral and anterior margins; forceps base, fig. 206, light tan, with deep, median fissure, forceps white, faintly tinged with tan, penis lobes tan; caudal filaments white, articulations not darkened.

Holotype, male.—Watson, Illinois, April

21, 1932, Ross & Mohr. Specimen in alcohol, genitalia on a microscope slide.

Paratype. — INDIANA. — SPENCER: McCormick's Creek, April 28, 1941, W. E. Ricker, 1 ♂. Specimen in alcohol.

5. *Paraleptophlebia adoptiva* (McDunnough)

Leptophlebia adoptiva McDunnough (1929: 169).

MALE.—Length of body and of fore wing 7–8 mm. Head and thorax very dark brown, abdomen a lighter brown; wings hyaline, longitudinal veins brown; genital forceps, fig. 205, yellow-brown and penis lobes brown; caudal filaments uniformly light brown.

Known from Michigan, New York, Ontario, and Quebec.

6. *Paraletophlebia ontario* (McDunnough)

Leptophlebia ontario McDunnough (1927b: 299).

MALE.—Length of body and of fore wing 5–6 mm. Head dark brown; eyes in life red-brown; antennae tan, each with flagellum yellow. Thorax dark brown; legs light yellow-brown, with coxae darkened, each fore femur and fore tibia stained with brown, and all tarsi quite light, almost white; wings hyaline, with faint, brown staining at base of fore wing, veins C, Sc, and R_1 of fore wing pale brown, all other veins and all crossveins hyaline. Abdominal tergum dark brown, tergites 2–7 paler brown on meson and across anterior margin of each; first abdominal sternite brown, other sternites pale brown, almost white; genital forceps, fig. 207, tan, penis lobes yellow-brown; caudal filaments mostly white, faintly brown-stained at bases.

FEMALE.—Length of body 5–6 mm., of fore wing 5.5–6.5 mm. Head and body a uniform dark yellow-brown, with posterior margins of abdominal segments shaded darker; legs light yellow; all longitudinal veins of fore wing tan; ovipositor only very slightly produced; terminal abdominal sternite with a deep, triangular, median notch on posterior margin; caudal filaments white.

NYMPH.—Length of body 4.5–6.0 mm. Head and body a rich tan, with vague, darker shading on abdominal tergites; abdominal sternites a faint tan; antennae white; legs white, with light brown shading at apexes of femora and bases of tibiae; abdominal gills hyaline, with tracheae purple and showing a few minute, lateral branches; posterolateral angles of abdominal tergites 8 and 9 produced as spines; caudal filaments tan.

Known from Illinois, New York, Ohio, and Ontario.

Illinois Records.—ALTO PASS: branch of Clear Creek, May 23, 1946, Mohr & Burks, 1 ♂. EDDYVILLE: Belle Smith Spring, June 7, 1946, Mohr & Burks, 1 ♂. MUNCIE: June 26, 1948, L. J. Stannard, 1 ♂. WOLF LAKE, Hutchins Creek: May 14–25, 1940, Mohr & Burks, 16 ♂, 19 ♀, 22 N; May 31, 1940, B. D. Burks, 7 ♂, 12 ♀; May 14–29, 1946, Mohr & Burks, 4 ♂, 1 ♀, 4 N.

7. *Paraleptophlebia mollis* (Eaton)

Cloe mollis Hagen (1861:53). Nomen nudum.
Leptophlebia mollis Eaton (1871:88).

MALE.—Length of body and of fore wing 8 mm. Head, thorax, and apex of abdomen dark brown; abdominal segments 2–7 entirely white, without markings; wings hyaline, with veins and crossveins colorless; genital forceps, fig. 208, white, penis lobes tan; caudal filaments white.

Known from the northeastern states and southeastern Canadian provinces.

8. *Paraleptophlebia volitans* (McDunnough)

Leptophlebia volitans McDunnough (1924b:95).

MALE.—Length of body and of fore wing 5–6 mm. Head, thorax, and apex of abdomen dark brown; abdominal segment 2 shaded with brown, segments 3–7 white, with brown markings at posterior margins of tergites and in spiracular areas; wings hyaline, with veins and crossveins colorless; genital forceps, fig. 209, white, and penis lobes tan; caudal filaments white.

Known from the Appalachian region and southeastern Canada.

9. *Paraleptophlebia guttata* (McDunnough)

Leptophlebia guttata McDunnough (1924b:95).

MALE.—Length of body and of fore wing 5–6 mm. Head, thorax, and apex of ab-

domen very dark brown; abdominal segments 2–7 white, with a pair of brown spiracular dots on each tergite; wings hyaline, with veins and crossveins colorless; genital forceps, fig. 210, white, penis lobes light yellow-brown; caudal filaments white.

Known from the northeastern states and eastern Canadian provinces.

10. *Paraleptophlebia strigula* (McDunnough)

Leptophlebia strigula McDunnough (1932b:209).

MALE.—Length of body and of fore wing 7 mm. Head, thorax, and apex of abdomen brown; abdominal segments 2–7 white, with fairly extensive, brown shading near posterolateral angles of each tergite, a longitudinal, dark brown line present in each spiracular area; wings hyaline, with crossveins colorless and longitudinal veins stained with tan; genital forceps, fig. 211, white, penis lobes tan; caudal filaments white.

Known from Ontario and Pennsylvania.

19. *THRAULODES* Ulmer

Thraulodes Ulmer (1920a:33).

This genus includes but two Nearctic species, although Ulmer (1920c:116) assigned nine Neotropical and Mexican species to it.

The adults of *Thraulodes* are of medium size, with mostly tan or yellow-brown bodies and clear wings. The wing veins are strong and usually brown; the crossveins, as well as the longitudinal veins, fig. 188, are quite well developed. In the fore wing, the posterior branch of the outer fork (R_5) is not strongly bent posteriorly near its origin, and vein M is branched at about mid-length. There are two long, cubital intercalary veins which are joined to the veins Cu_1 and Cu_2 by strong crossveins. The hind wing has a costal projection, and vein Sc ends just distad of the costal angulation. The crossveins of the hind wing are restricted to the area near the costal angulation.

The fore tarsus in the males is slightly shorter than the fore tibia. The male genital forceps arise from an undivided base and have the two apical segments of each arm relatively minute; the penis lobes are divided to the bases and bear apical appendages.

The nymphs have three-segmented labial and maxillary palps, with the maxillary palp forceps-like at the apex; the head is flattened; all the gills are double and lanceolate, diminishing in size from abdominal segment 1 to 7; the posterolateral angles of abdominal segments 2–9 are produced as slender spines; the three caudal filaments are of the same length. These characters are drawn from Peruvian specimens in the Cornell University collection determined as of this genus by Needham & Murphy.

Thraulodes speciosus Traver (1934:201), described from Texas, and *arizonicus* McDunnough (1942:117), from Arizona, are the only known Nearctic species.

20. *HABROPHLEBIA* Eaton

Habrophlebia Eaton (1881:195).

The Nearctic species of *Habrophlebia* consist of small, slender mayflies with clear wings. The fore tarsus in the males is only two-thirds as long as the fore tibia; the tibiae of all legs in both sexes are conspicuously long and slender, always much longer than the femora. All longitudinal veins of both wings, fig. 189, are hyaline or faintly tinted at the bases only; the crossveins are all but invisible. Each fore wing has two long and two alternating, short cubital intercalaries, which are free at the bases; vein R_5 is slightly bent posteriorly near the base; vein M_2 diverges from M_1 near the base of the wing, but the basal part of M_2 is obsolete. The hind wing has a costal angulation, and vein Sc ends at or near the apex of the wing. The genital forceps arise from a medianly, deeply fissured base; the length of the apical two forceps segments combined is almost as great as the length of the long basal segment. The penis lobes are divided nearly to the base, fig. 213; each lobe bears a long, bladelike, apical appendage. The terminal abdominal sternite in the females is deeply cleft on the meson of the posterior margin. The median caudal filament is longer than the cerci.

In the nymphs, the body is slender and only slightly flattened; the head is subtriangular and is held in a somewhat prognathous position. The antennae are slender and each about as long as the head and thorax combined. The maxillary palp has three segments, and the labial has three. The tarsal claws are single and relatively short, and bear a single ventral row of denticles. Each of the gills, fig. 196, is single at the base, then quickly branches into two rami, each of which is then subdivided into three or more slender terminal filaments. The median caudal filament is longer than the cerci.

This genus includes two Nearctic species, neither of which has yet been taken in Illinois. One of the two is, however, widely distributed and might eventually be found to occur here.

Habrophlebia vibrans Needham

Habrophlebia vibrans Needham (1908:192).
Habrophlebia jocosa Banks (1914:614).

MALE.—Head and thorax dark brown; legs predominantly white; wings hyaline,

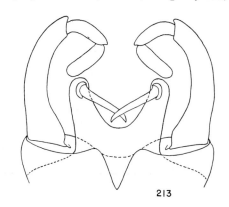

213

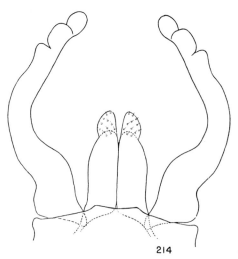

214

Fig. 213.—*Habrophlebia vibrans*, male genitalia.
Fig. 214.—*Choroterpes basalis*, male genitalia.

with a small, brown-stained area at base of each fore wing. Basal and apical abdominal tergites dark brown, with middle tergites brown at posterolateral areas only, the central and anterior areas white; anterior and posterior abdominal sternites tan, and middle sternites white; genitalia, fig. 213, tan; caudal filaments predominantly white.

Known from the eastern states and southeastern Canadian provinces.

21. *CHOROTERPES* Eaton

Choroterpes Eaton (1881:194).

The members of the genus *Choroterpes* are small, extremely delicate mayflies. The males have predominantly dark brown or black bodies, with the abdominal venter of each considerably lighter in color. The females are rather uniformly light brown. The compound eyes of the males are semi-turbinate, as there is in each eye a very slight development of a stalk to set off the upper portion from the lower. The wings in both sexes are stained with brown or red-brown at the bases; otherwise, the wings are clear, with the crossveins and posterior longitudinal veins hyaline. The veins near the costal margin of each fore wing are faintly stained with brown. The posterior branch of the outer fork (vein R_5) in the fore wing is not bent posteriorly near the base, fig. 190; vein M_2 diverges from M_1 near the center of the wing, but the basal part of M_2 is obsolescent; there are four cubital intercalary veins, the anterior pair long and the posterior pair short. Each hind wing has a small costal angulation, vein Sc ends just distad of this angulation, and vein M is unbranched.

In the male genitalia, fig. 214, there is a pair of long penis lobes which are fused on the meson only at the bases; these penis lobes lack appendages. Each arm of the forceps, arising from an undivided base, is four segmented, with the suture setting off the basal segment from the second segment extremely obscure. In the females, the terminal abdominal sternite is only slightly or not at all indented on the meson of the posterior margin.

In the nymphs, the head and body are flattened dorsoventrally. The antennae are relatively short—only about as long as the head is wide. There are three segments in each of the labial and maxillary palps. The head itself is quadrate and strongly prognathous. Each tarsal claw is single, relatively short, thick at the base, and provided with a single row of ventral denticles near the tip. The first abdominal segment bears a pair of single, filamentous gills, fig. 195*B*; segments 2–7 have each a pair of bifid, lamelliform gills, each lamina having a spatulate terminal extension, figs. 195*A*, 195*C*, which varies in shape among the different species. The median caudal filament is longer than the cerci.

Choroterpes basalis (Banks)

Leptophlebia basalis Banks (1900:248).

MALE.—Head and thorax very dark brown; wing bases heavily stained with brown. Abdominal dorsum dark brown, variegated with white, and abdominal venter white; penis lobes, fig. 214, brown; genital forceps and caudal filaments white.

Widely distributed throughout the eastern states and southeastern Canada, it should be taken in Illinois eventually.

22. *HABROPHLEBIODES* Ulmer

Habrophlebiodes Ulmer (1920a:39).

The members of the genus *Habrophlebiodes* are small, extremely delicate, brown mayflies. In the males, the fore tarsus is as long as the fore tibia. The wings are clear or faintly stained with yellow; the longitudinal veins are well marked, but the crossveins are all but invisible. The posterior branch of the outer fork (vein R_5) in the fore wing is bent posteriorly near the base, fig. 191; vein M_2 is obsolete at the base, but apparently diverges from M_1 near the center of the wing; there are two cubital intercalary veins. Each hind wing has a finger-like costal projection, and vein Sc ends just distad of the projection.

The male genital forceps arms arise from a deeply cleft base, fig. 215; the apical two segments are minute and subtriangular. The penis lobes are fused to the tips; each lobe bears a decurrent, lateral appendage. In the females, the well-developed ovipositor is formed by a prolongation of the seventh sternite and underlies the eighth. The ninth sternite is deeply cleft on the meson of the apical margin. The median caudal filament is longer than the cerci.

The nymphs are very similar to those of

Paraleptophlebia, but differ in the shape of the labrum, figs. 197, 198; in *Habrophlebiodes,* the anterior margin has a deeper median cleft. The body is depressed; the head is semiflattened and held in a nearly hypognathous position. Each maxillary palp has three segments, and the labial three. Each antenna is as long as the head and thorax combined. The claws are long and slender and each bears a single ventral row of denticles. The abdominal gills are all of the same type. Each gill has a slender stem which subdivides to produce two long, slender, lanceolate filaments. The caudal filaments are longer than the body.

This genus includes three species, one of which occurs in Illinois.

Habrophlebiodes americana (Banks)

Habrophlebia americana Banks (1903:235).
Choroterpes betteni Needham (1908:194).

After a study of Banks' types in the Museum of Comparative Zoology and the remains of Needham's types at Cornell University, I agree with McDunnough (1925a: 210) that *betteni* is a synonym of *americana.* It has two closely related species: *brunneipennis* Berner (1946:61), a southern species with amber-tinted wings, and *annulata* Traver (1934:199), a species described from Oklahoma that has the posterior margins of the abdominal segments edged with black.

MALE.—Length of body and of fore wing 4.5–5.5 mm. Head dark red-brown, antennae tan. Thorax dark brown; fore coxa and femur brown, the latter darker at apex, fore tibia tan, brown at apex, fore tarsus light tan; middle and hind legs mostly light tan or yellowish, with coxae brown, and femur of hind leg darkened with brown in middle and at apex; wings hyaline, longitudinal veins faintly brown-stained, crossveins hyaline. Abdomen mostly dark brown, with pale yellowish markings of variable extent at anterior margins of middle sternites and tergites; genital forceps, fig. 215, light tan to brown, penis lobes light yellow-brown; caudal filaments white or light tan, articulations darkened.

FEMALE.—Length of body and of fore wing 5.5–6.5 mm. Head, thorax, and abdomen uniformly rich red-brown, but considerably lighter than in male; legs and wings marked as in male. Ovipositor projecting

slightly past base of ninth sternite. Caudal filaments tan, darkened at articulations.

NYMPH.—Length of body 5–6 mm. Head and dorsum of body chestnut brown, venter of thorax and basal abdominal sternites white or light yellow, shading to tan on posterior abdominal sternites; abdominal gills hyaline, tracheae black; caudal filaments uniformly light brown.

Known from the eastern and midwestern states and eastern Canadian provinces. Occurs in the shallow, comparatively still eddies along the banks of streams.

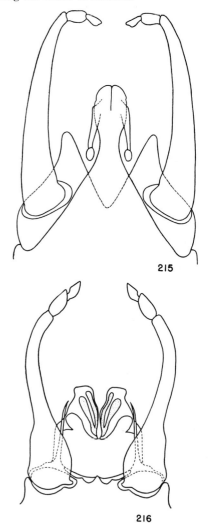

215

216

Fig. 215.—*Habrophlebiodes americana,* male genitalia.

Fig. 216.—*Traverella albertana,* male genitalia.

Illinois Records.—HEROD: May 23, 1946, Ross & Mohr, 1 ♂, 1 ♀; June 20, 1940, Mohr & Riegel, 1 ♂, 2 N; July 16, 1947, L. J. Stannard, 1 ♂.

23. *TRAVERELLA* Edmunds

Traverella Edmunds (1948c: 141).

Mayflies of this genus are medium sized and rather heavy bodied. They are strikingly marked with contrasting snow-white and very dark brown areas. The upper portion of each compound eye in the males is extremely broad and is set off from the lower portion by a rudimentary stalk. Two prominently projecting lobes beneath the frontal shelf of the adult head probably are the rudimentary maxillae. The fore tarsus in the males is shorter than the fore tibia. The wings in adults of both sexes are slightly cloudy and tinged with gray-brown so that they look somewhat like subimago wings. The veins and crossveins, fig. 192, are well marked and usually dark brown. The posterior branch of the outer fork (vein R_5) in the fore wing is bent rearward near the base; vein M_2 diverges from M_1 near the center of the wing; there are two long, cubital intercalary veins; vein Cu_2, near the middle, is obliquely bent toward the anal wing margin. The hind wing has a long, acute costal angulation, and vein Sc ends at the distal angle of this projection.

In the male genitalia, the base of the forceps is undivided but bears a median tooth, fig. 216, and a pair of slender lateral processes which lie beside the penis lobes. The penultimate segment of each arm of the forceps is small and semiquadrate; the apical segment is minute and subtriangular. Each penis lobe has one long, slender appendage. In the females, the terminal abdominal sternite projects rearward past the bases of the caudal filaments, and its posterior margin has a deep, median indentation; the eighth sternite has a median, membranous structure which is eversible and undoubtedly serves as an ovipositor. The three caudal filaments are subequal in length in both sexes.

In the nymphs, the body is flattened and the head is quadrate and hypognathous. The labrum is flat and greatly developed, being fully one-third as long as the head. Each of the three-segmented maxillary palps is held in such a position that it projects laterally from beneath the head, the second segment resting parallel to the lateral margin of the head. The apical palpal segment bears a dense brush of long setae. Each labial palp has three segments. Each antenna is almost twice as long as the head. The tarsal claws are relatively short, with a single row of denticles on the ventral side of each. The gills on abdominal segments 1–7 are all of the same type, decreasing in size from the first segment to the seventh. Each gill is bifid, and each element is lamelliform, with the margins finely dissected. Each of the three caudal filaments is as long as the body.

This genus includes only two known Nearctic species: *albertana* (McDunnough) (1931b: 82), occurring in Utah, Saskatchewan, and Alberta, and *presidiana* (Traver) (1934: 199), described from Texas.

BAETIDAE

The family Baetidae, as here defined, corresponds to a combination of the Baetidae of the Baetoidea and the Siphlonuridae of the Heptagenioidea in Ulmer's classification (1933), and the name is used in a considerably more restricted sense than by Traver (1935a: 427).

The eyes in the males are large, figs. 241, 255–257, and, in many species, each eye is divided into two distinct sections: a lower portion composed of relatively small facets and an upper portion composed of larger facets. In eyes that are divided, the upper portion of each eye is set on a platform which completely separates it from the lower portion, fig. 257. The wing venation in the various members of this family varies from a type approaching that of the fossil Permian mayfly, *Protereisma,* to a much reduced type in which many longitudinal veins and crossveins have been eliminated. Parts of veins may also be atrophied in the wings with reduced venation. This partial atrophy of veins is usually evident toward the bases of the wings. The hind wing in the various genera may be either well developed, or reduced in size and venation, or wanting entirely. The hind tarsus in the adults of both sexes have three or four clearly differentiated segments.

The male penis lobes vary from a well-developed type with relatively complex structure, as in figs. 242–246, to a greatly reduced, almost structureless type, as in figs. 267–269, in the most simplified genera, such as *Cloeon* and *Baetis*. The median caudal filament is vestigial.

All the nymphs are streamlined, rather fishlike forms, and typically vigorous swimmers. The head is not flattened dorsoventrally, as in the Heptageniidae, and the compound eyes are lateral. Usually, the abdominal gills are single and more or less platelike, but, when they are double, the lower element of the pair is not composed of a mass of fibrillae, except in the genus *Isonychia*. In various genera, the median caudal filament may be either well developed or vestigial.

KEY TO SUBFAMILIES

Adults

1. Vein M₂ of fore wing detached at base from stem of M, figs. 31, 220–222; hind wing greatly reduced or wanting entirely; hind tarsus with only three clearly defined segments, fig. 15......
   ....**Baetinae,** p. 113
   Vein M₂ of fore wing not detached at base from stem of M, figs. 217–219; hind wing well developed; hind tarsus with four clearly defined segments, figs. 18, 20................................2
2. Gill remnants present at base of rudimentary maxilla and at base of fore-coxa...........**Isonychiinae,** p. 108
   Gill remnants absent.................
   ....**Siphlonurinae,** p. 98

Mature Nymphs

1. Each abdominal gill composed of a plate-like dorsal element and a ventral fibrillar tuft, fig. 225; fore coxa and maxilla with gill tufts...**Isonychiinae,** p. 108
   All abdominal gills platelike; gills usually single, but, when double, both elements of each gill platelike, figs. 223, 224, 226–228; fore coxa and maxilla without gills....................2
2. Posterolateral angles of each apical abdominal tergite prolonged as thin, flat spines, figs. 240B, 247, 254; labrum with anterior margin entire or with a broad, median, V-shaped notch, fig. 229...........**Siphlonurinae,** p. 98
   Posterolateral angles of apical abdominal tergites not prolonged as spines, figs. 266, 298; labrum with a median, square notch, fig. 231.......**Baetinae,** p. 113

SIPHLONURINAE

The subfamily Siphlonurinae, as here defined, corresponds very closely to Ulmer's family Siphlonuridae (1933:209).

In the Siphlonurinae, each compound eye in the adult males is made up of an upper portion composed of large facets and a lower

portion of smaller facets, but the two portions of the eye are not distinctly separated, fig. 241. The fore tarsus in adult males is always much longer than the fore tibia. Gill remnants are wanting on the head and thorax of adults of both sexes. The fore wing in this subfamily is readily distinguished from the fore wing in all other mayflies, in that the cubital intercalary veins form a series of parallel, often sinuate but usually not branched veins extending from vein Cu₁ to the anal margin of the wing, and in that vein Cu₁ is straight throughout its length, fig. 219. In the hind wing, vein M is either not forked or forked in the basal half of its length. In many species, the wings are wholly or partly shaded with brown, yellow, or tan, with prominently colored veins and crossveins. In the adults, the median caudal filament is always vestigial.

The vigorous, fishlike nymphs, figs. 240B, 247, of the members of this subfamily are strong and rapid swimmers, almost always living in rapidly flowing water. The tarsal claws of the nymphs are long and slender, but are always shorter than the tibiae, fig. 26. Gills are borne only by the abdomen, and these gills are platelike and usually single, but when they are double both parts of the individual gill are platelike. There are always three well-developed caudal filaments. Each cercus has long setae on the inner side only. The nymphs of *Siphlonurus* have been shown to be at least in part predaceous (Morgan 1913:386). The structure of the mouth-parts of the nymphs of another genus, *Parameletus*, indicates that it also may be predaceous.

KEY TO GENERA

Adults

1. Abdominal segments 5–9 with broad, flat, lateral expansions, fig. 233; median ventral spine present on mesosternum and metasternum....**24. Siphlonisca**
   Abdominal segments without broad, lateral expansions; no median ventral spines present on thorax..................2
2. Hind wing with an acute costal angulation, and vein M forked near base, fig. 237.
   ....**25. Ameletus**
   Hind wing with a blunt, or with no, costal angulation, and vein M either not forked or forked well distad of the base, figs. 218, 219.....................3
3. Vein M of hind wing simple, not forked, fig. 218..........**26. Parameletus**
   Vein M of hind wing forked, fig. 219....
   ....**27. Siphlonurus**

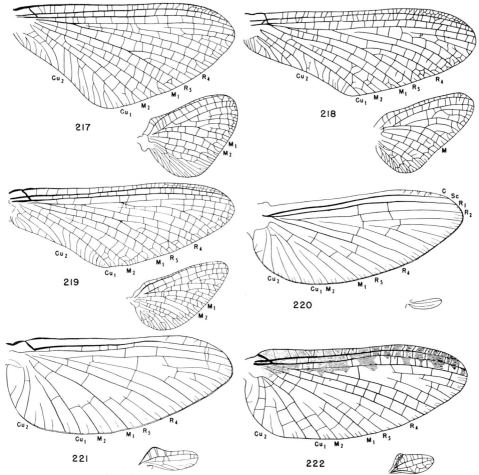

Fig. 217.—*Isonychia rufa*, wings.
Fig. 218.—*Parameletus midas*, wings. (After Traver.)

Fig. 219.—*Siphlonurus quebecensis*, wings.
Fig. 220.—*Baetis propinquus*, wings.
Fig. 221.—*Callibaetis fluctuans*, wings.

Fig. 222.—*Callibaetis ferrugineus*, wings.

MATURE NYMPHS

1. A stout, median, ventral spine on meso– and metasternum.....**24. Siphlonisca**
   No median ventral spines on thorax...2
2. A conspicuous, transverse pecten of spines present on margin of each maxilla, fig. 240*A*..........**25. Ameletus**
   No pecten of spines present on maxilla..3
3. Apical segment of labial palp and an apposed, thumblike projection of penultimate palp segment forming a forceps, fig. 230............**26. Parameletus**
   Labial palp not forceps-like at apex.....
   ....**27. Siphlonurus**

24. *SIPHLONISCA* Needham

Siphlonisca Needham (1909:71).

This strikingly distinct genus is at once recognizable because of the wide, flat lateral extensions on the margins of abdominal segments 5–9 of the adults, fig. 233, and on all abdominal segments of the nymphs, fig. 234. A midventral spine is present on the mesosternum and metasternum in both the nymphs and the adults. The abdominal gills of the nymphs are single and platelike, with the margins slightly irregular. The median caudal filament is well developed in the nymphs, but is vestigial in the adults.

Siphlonisca most closely resembles *Oniscigaster* McLachlan (1873:108; 1874:139), described from New Zealand.

Siphlonisca aerodromia Needham (1909: 71), known from New York, is the only described species. The male genitalia of this species are shown in fig. 232.

25. *AMELETUS* Eaton

Ameletus Eaton (1885: 210).

In the members of this genus, each compound eye, in both the males and females, has a slightly oblique, contrastingly colored band extending across the outer surface, fig. 235. This band is visible only in freshly killed or living specimens, as the color pattern of the eyes quickly disappears after

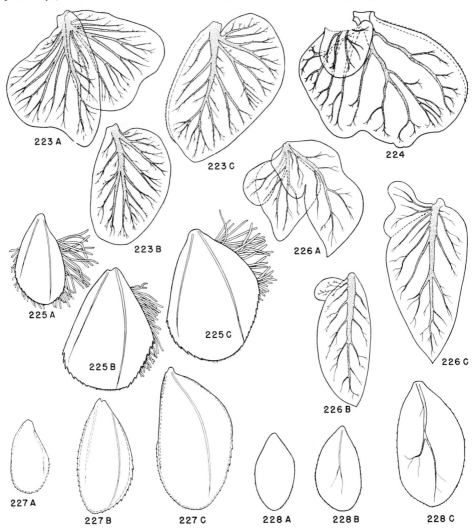

Fig. 223A.—*Siphlonurus marshalli*, gill of first abdominal segment.
Fig. 223B.—*Siphlonurus marshalli*, gill of seventh abdominal segment.
Fig. 223C.—*Siphlonurus marshalli*, gill of fifth abdominal segment.
Fig. 224.—*Siphlonurus alternatus*, gill of fifth abdominal segment.
Fig. 225A.—*Isonychia* sp., gill of first abdominal segment.
Fig. 225B.—*Isonychia* sp., gill of seventh abdominal segment.
Fig. 225C.—*Isonychia* sp., gill of fifth abdominal segment.
Fig. 226A.—*Callibaetis skokianus*, gill of first abdominal segment.
Fig. 226B.—*Callibaetis skokianus*, gill of seventh abdominal segment.
Fig. 226C.—*Callibaetis skokianus*, gill of fifth abdominal segment.
Fig. 227A.—*Ameletus lineatus*, gill of first abdominal segment.
Fig. 227B.—*Ameletus lineatus*, gill of seventh abdominal segment.
Fig. 227C.—*Ameletus lineatus*, gill of fourth abdominal segment.
Fig. 228A.—*Baetis brunneicolor*, gill of first abdominal segment.
Fig. 228B.—*Baetis brunneicolor*, gill of seventh abdominal segment.
Fig. 228C.—*Baetis brunneicolor*, gill of fourth abdominal segment.

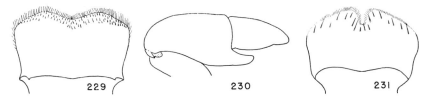

Fig. 229.—*Siphlonurus marshalli,* labrum of mature nymph, dorsal aspect.
Fig. 230.—*Parameletus columbiae,* libial palp of mature nymph.
Fig. 231.—*Baetis vagans,* labrum of mature nymph, dorsal aspect.

death, regardless of the method of preservation of specimens. The fore wing is typical for the subfamily; the hind wing has an acute costal projection, and vein M is forked near the base of the wing, fig. 237. The male penis lobes are always separated to the base, and the forceps base is deeply excavated on the meson, as in fig. 236. The apical abdominal sternite of the females has a median notch on the posterior margin, figs. 238, 239. There are two well-developed caudal filaments.

The streamlined, vigorously swimming nymphs, fig. 240*B,* are distinguished from all other known mayfly nymphs by the pecten of spines borne by each maxilla, fig. 240*A.* The legs are relatively short, and the tarsal claws are uniformly single, nondenticulate, slender, and much shorter than the tibiae. The gills are single and platelike, each gill having a single, stout, rodlike stiffener near each dorsal margin and a weaker but otherwise similar rod near each ventral margin, fig. 227. There are three well-developed caudal filaments; the cerci bear long, dense setae on the inner sides only.

The genus *Ameletus* includes 26 Nearctic species, 22 of which occur in the western states. Two of the remaining four species have been taken only in Quebec and Nova Scotia. The other two species occur in the Midwest and both normally are parthenogenetic. Males are unknown for one species and only two male specimens of the other species are known to have been collected.

KEY TO SPECIES

ADULT FEMALES

Venter of abdomen without ganglionic markings; terminal abdominal sternite as in fig. 239; lateral margins of apical sternite slightly incised..............**1. lineatus**
Venter of abdomen with brown ganglionic markings; terminal abdominal sternite as in fig. 238; lateral margins of apical sternite straight......................**2. ludens**

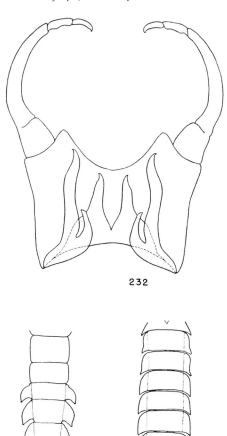

Fig. 232.—*Siphlonisca aerodromia,* male genitalia.
Fig. 233.—*Siphlonisca aerodromia,* abdomen of adult male, ventral aspect.
Fig. 234.—*Siphlonisca aerodromia,* abdomen of mature male nymph, ventral aspect.

MATURE NYMPHS

Brown crossbar at posterior margin of terminal abdominal sternite joining three longitudinal, brown stripes on abdominal venter.... ....**1. lineatus**
Entire terminal abdominal sternite of abdomen shaded with brown, this shading sometimes also extending over one or two sternites anterior to terminal one..........**2. ludens**

1. *Ameletus lineatus* Traver

Ameletus lineatus Traver (1932a:194).

FEMALE.—Length of body 11 mm., of fore wing 12 mm. Head yellow-brown, vertex with a longitudinal, median, dark brown line; eyes light brown, with a yellow, dark brown-bordered stripe on outer surface, fig. 235; antennae smoky brown. Thorax dark yellow-brown, with yellow markings at sutures and on pleura around coxal bases; wings hyaline, veins and crossveins brown; legs yellow-brown, with tarsi somewhat darkened. Abdomen yellow-brown, without well-marked color pattern, although slightly darkened at posterior margins of each tergite; sternite 6 with a median, dark brown mark; lateral margins of terminal abdominal sternite slightly incised near apex; caudal filaments light, articulations dark, brown.

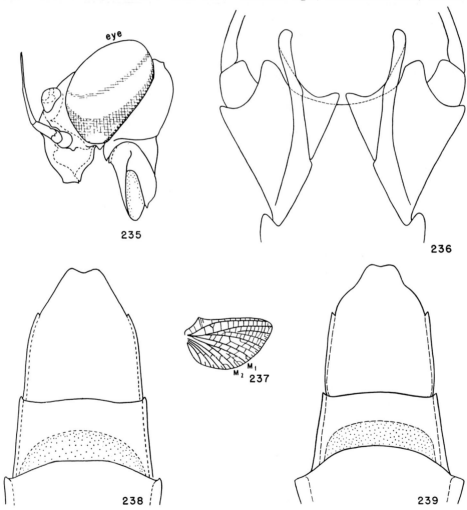

Fig. 235.—*Ameletus lineatus*, head of adult female, lateral aspect.

Fig. 236.—*Ameletus shepherdi*, male genitalia.

Fig. 237.—*Ameletus lineatus*, hind wing.

Fig. 238.—*Ameletus ludens*, terminal abdominal sternites of adult female.

Fig. 239.—*Ameletus lineatus*, terminal abdominal sternites of adult female.

NYMPH.—Length of body 11–13 mm., of caudal filaments 5–6 mm. Head and body cream colored, with vague, light brown markings. Tarsi ordinarily with a light brown band at base and a dark brown band at apex of each. Abdominal venter with three longitudinal, brown stripes, one median, another near each lateral margin; a brown crossband at posterior margin of sternite 9 joining the three longitudinal stripes; caudal filaments with an extremely broad, brown crossband in middle and a narrow, brown crossband at apex of each filament.

This species, reported from North Carolina and Illinois, is known only from female adults and female nymphs. The nymphs are found among debris and emergent vegetation along the banks of swift, cool streams.

Illinois Records.—CORA: April 24, 1939, Burks & Riegel, 1 ♀ N. HEROD: Gibbons Creek, March 14, 1946, Ross & Burks, 1 ♀ N; Herod Spring, March 14, 1946, Ross & Burks, 1 ♀ N; April 4, 1946, Burks & Sanderson, 2 ♀ N; Gibbons Creek, April 7–10, 1947, B. D. Burks, 2 ♀, numerous ♀ N; May 15, 1941, Mohr & Burks, 4 ♀ N. RUDEMENT, Blackman Creek: April 2, 1932, Frison & Ross, 4 ♀ N; April 4, 1946, Burks & Sanderson, 3 ♀ N; April 4–8, 1947, B. D. Burks, numerous ♀ N.

2. Ameletus ludens Needham

Ameletus ludens Needham (1905:36).

This species differs from *lineatus* only in having brown ganglionic markings on the abdominal sternites of the adult female and in having a differently shaped apical abdominal sternite, fig. 238. The nymph of *ludens,* fig. 240, differs from that of *lineatus* in having the entire ninth abdominal sternite brown, and the longitudinal, brown stripes

on the abdominal venter relatively wider. The only two adult male specimens of *ludens* known to have been collected were described by Needham (1924:308).

Although *ludens* has been taken in the neighboring state of Indiana as well as in New York and West Virginia, it has not yet been collected in Illinois.

26. PARAMELETUS Bengtsson

Parameletus Bengtsson (1908:242).
Potameis Bengtsson (1909:13).
Sparrea Petersen (1909:554).
Siphlonuroides McDunnough (1923:48).
Palmenia Aro *in* Lestage (1924a:35).

In *Parameletus,* the fore wing, fig. 218, is rather narrow and elongate, the stigmatic crossveins are anastomosed, and the outer wing margin has fairly numerous, short and irregular intercalary veins. The hind wing usually has a low, broadly rounded costal projection, and vein M is always simple and unbranched. In the nymphs, the labial palps are forceps-like, fig. 230; the abdomen is somewhat flattened dorsoventrally, as in *Leptophlebia,* and the gills are broad and single, with the tracheation dense and pinnately branched.

There are no Illinois species of *Parameletus; croesus* (McDunnough) and *midas* (McDunnough) (1923:48–9) were described from Ontario; *columbiae* McDunnough (1938:31) occurs in the western states and British Columbia.

27. SIPHLONURUS Eaton

Siphlonurus Eaton (1868:89).
Siphlurus Eaton (1871:37, 125).
    Emendation, unnecessarily proposed.
Siphlurella Bengtsson (1909:11).

The genus *Siphlonurus* includes a fairly large number of species of large, strikingly

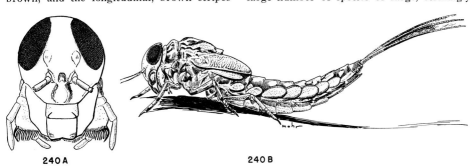

240A   240B

Fig. 240*A.*—*Ameletus ludens,* head of mature nymph, anterior aspect.
Fig. 240*B.*—*Ameletus ludens,* mature nymph, lateral aspect.

marked mayflies. Each compound eye in living males and females has contrastingly colored stripes extending across the outer surface. The head and thorax usually are mostly dark brown, while the abdomen always has a conspicuous color pattern of contrasting light and dark areas which are somewhat annular in arrangement. The fore wing, fig. 219, is long and relatively narrow, with membrane usually hyaline and veins dark. There are numerous, irregular, marginal intercalary veins, and the stigmatic crossveins are usually anastomosed. The cubital intercalary veins are typical for this subfamily. The hind wing has a broadly rounded, inconspicuous costal angulation, and vein M is forked at a point midway between the base and the outer margin of the wing.

The male genitalia in the various species show the most strikingly distinct structural differences to be found in any Nearctic mayfly genus, figs. 242–246. The median caudal filament is represented by a minute vestige in the adults of either sex.

The vigorous, streamlined nymphs, fig. 247, typically inhabit quiet pools along the edges of streams. They also occur commonly in shallow pools filled by seepage water on rock ledges, as well as in small, shallow pools fed only intermittently with fresh water. They are not rheophilus, except in the early instars of some species. The mature nymphs are found invariably in quiet water. As Morgan (1913: 386) has shown, *Siphlonurus* nymphs are in part predaceous.

In the *Siphlonurus* nymphs, the compound eyes are lateral, and the head is hypognathous. The tarsal claws are slender and pointed, and considerably shorter than the tibiae. Abdominal segments 1–7 bear platelike gills, the first two pairs of which invariably are double, while the more posterior pairs are single, except in *alternatus*. The posterolateral angles of the abdominal tergites are produced and spinelike. There are three well-developed caudal filaments; each cercus bears a fringe of long setae on the mesal side only.

This genus includes 18 Nearctic species, 4 of which are known to occur in Illinois; 1 other species may be taken here eventually.

Reliable characteristics for the separation of the females of these species have not yet been found.

KEY TO SPECIES

ADULT MALES

1. Membrane of hind wing completely shaded with brown; male genitalia relatively simple, fig. 245 . . . .**5. marshalli**
   Membrane of hind wing hyaline; male genitalia relatively complex, figs. 242–244, 246 2
2. Each abdominal sternite with a pair of dark brown, lateral, triangular marks, these spots connected on meson by a large V-shaped mark with its apex at median point of anterior margin of sternite; male genitalia with a prominent, serrated bulge on median margin of each inner process, fig. 242 . . . .**3. quebecensis**
   Abdominal sternites not with lateral, triangular spots connected on meson by V-shaped marks; male genitalia with no serrated bulge on mesal margin of each inner process 3
3. Each abdominal sternite with a median, anterior, brown spot, a pair of oblique, lateral marks, and a pair of submedian dots; male genitalia with a pair of broad, dorsal flaps, fig. 246 . . . .**2. alternatus**
   Abdominal sternites not with such markings; male genitalia without broad, dorsal flaps 4
4. Abdominal venter with an interrupted, longitudinal, median stripe; inner processes of male genitalia finger-like at apexes, fig. 243 . . . .**1. rapidus**
   Abdominal venter unmarked or with only faint, lateral, triangular marks on anterior sternites; inner processes of male genitalia nipple-like at apexes, fig. 244 . . . .**4. typicus**

MATURE NYMPHS

1. Each gill borne by abdominal segments 3–6 with a dorsal, recurved flap, fig. 224 . . . .**2. alternatus**
   Each gill borne by abdominal segments 3–6 simple, without a dorsal, recurved flap, fig. 223C 2
2. Abdominal sternites 4–8 each with a pair of broad, longitudinal, lateral, brown bands and four submedian dots . . . .**5. marshalli**
   Abdominal sternites almost entirely brown, light yellow only on median, triangular area at posterior margin and on small area at anterolateral angles of each sternite . . . . . .**3. quebecensis**

1. *Siphlonurus rapidus* McDunnough

Siphlonurus rapidus McDunnough (1924c: 75).

MALE.—Membrane of fore wing faintly stained with tan, almost hyaline, costal and subcostal crossveins well developed, and

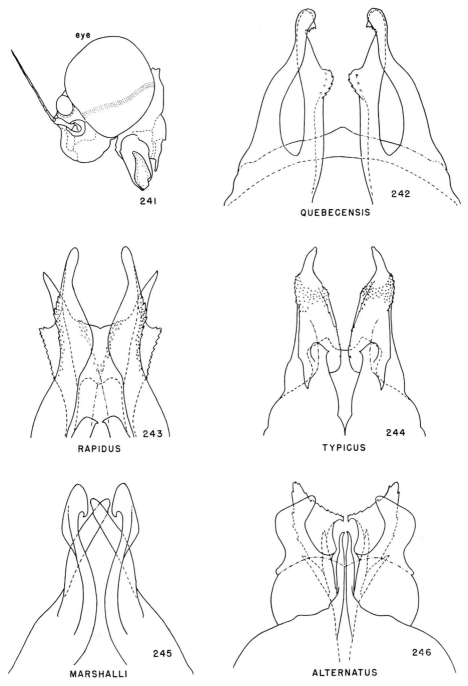

Fig. 241.—*Siphlonurus alternatus*, head of adult male, lateral aspect.

Fig. 242. — *Siphlonurus quebecensis*, male genitalia.

Fig. 243.—*Siphlonurus rapidus*, male genitalia.

Fig. 244.—*Siphlonurus typicus*, male genitalia.

Fig. 245.—*Siphlonurus marshalli*, male genitalia.

Fig. 246.—*Siphlonurus alternatus*, male genitalia.

stigmal crossveins anastomosed; dorsum of body brown, with yellow markings, abdominal venter almost entirely light yellow or white, with a discontinuous, longitudinal, brown stripe on meson. Male genitalia, area milky. Ground color of abdomen light tan to almost white, with brown shading; each tergite with a broad, transverse, shaded area at posterior margin, a large, triangular spot near each posterolateral angle, and a

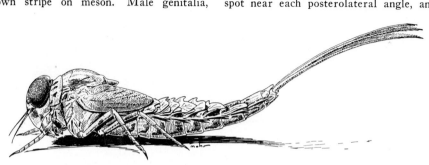

Fig. 247.—*Siphlonurus alternatus*, mature nymph, lateral aspect.

fig. 243, with inner processes elongate, finger-like at apexes, and grossly serrate on outer margins; outer processes acute at apexes, with lateral margins flaring and finely serrate; both inner and outer processes with numerous, minute spines.

The nymph is unknown.

This species is known from Connecticut, Massachusetts, Michigan, New Hampshire, New York, and Quebec.

2. Siphlonurus alternatus (Say)

Baetis alternata Say (1824:304).
Baetis annulata Walker (1853:567).
Baetis femorata Provancher (1876:267),
    not Say. Misidentification.
Siphlurus alternans Provancher (1878:127).
    Misspelling.

MALE.—Length of body 11–13 mm., of fore wing 10–12 mm. Head light yellow, with brown shading at bases of ocelli, on vertex along mesal margin of each compound eye and on meson, and across frontal shelf below antennae; compound eye gray, with two light and two alternating, dark stripes on outer surface of lower portion; antennae tan, each scape slightly darkened at base and apex. Thorax tan, with light yellow to almost white markings on pleura and sternum. Legs light yellow, each with brown annulations at base and near apex of femur, at base and apex of tibia, and at apex of each tarsal segment, all annulations narrow except one near apex of femur; wings hyaline, veins and crossveins dark brown, costal and subcostal crossveins weak, stigmal crossveins anastomosed and stigmal

broad, longitudinal stripe on meson; each sternite with a pair of oblique, lateral marks, a median spot on anterior margin, and a pair of submesal dots in center; genitalia, fig. 246, light tan; caudal filaments tan at bases, becoming white at apexes, articulations dark brown.

FEMALE.—Length of body 11–13 mm., of fore wing 12–14 mm. Color pattern as in male, but background slightly lighter and brown shading less intense; caudal filaments almost entirely white, only faintly tan-shaded near bases, with articulations chocolate brown. Apical margin of terminal abdominal sternite produced, nipple-like on meson.

NYMPH.—Fig. 247. Length of body 11–13 mm., of caudal filaments 6–7 mm. Color patterns of thorax, legs, and abdomen very similar to those of the adult, but markings of abdominal venter somewhat broader than in the adult; each gill on abdominal segments 1 and 2 composed of two equal-sized plates, gills on the following segments composed of a large plate with a much smaller, recurved, dorsal plate, fig. 224; posterolateral, spine-like prolongations of tergites darkened at tips; each caudal filament with a broad, brown crossband near middle and a narrower crossband at tip.

Known from Illinois, Indiana, Michigan, New York, Nova Scotia, Ontario, Quebec, and Wisconsin.

Illinois Records.—FREEPORT: at light, June 10–11, 1948, Burks, Stannard, & Smith, 1 ♂, 3 ♀. ROCKFORD: Long slough of Rock River, May 13, 1927, D. H. Thompson, 1 N. (Walsh 1862:369 records this species from near Chicago on the Des Plaines River; Coal Valley Creek, Rock Island County; and Rock Island.)

3. *Siphlonurus quebecensis* (Provancher)

Baetis canadensis Provancher (1876: 267), not Walker. Misidentification.
Siphlurus quebecensis Provancher (1878: 127). New name.
Siphlurus annulatus Provancher (1878: 144), not Walker. Erroneous citation.
Siphlurus triangularis Clemens (1915a: 250).

MALE.—Length of body 9–12 mm., of fore wing 10–14 mm. Head yellow, with dark brown to black shading at bases of ocelli, on vertex at mesal margins of compound eyes, and across frontal shelf below antennae; compound eyes gray, the lower portion with two light bands alternated with two dark bands extending across outer surface; antennae light brown. Thorax brown, with light yellow or white marks on pleura and sternum; wings hyaline, veins and crossveins dark brown, costal and subcostal crossveins weak, stigmatic crossveins anastomosed and stigmatic areas milky; legs light brown, with dark brown marks at apexes of femora, tibiae, and tarsal segments. Ground color of abdomen light yellow-tan, dark brown shading covering all but anterior quarter of each tergite; triangular, shaded area at lateral margin of each sternite, these triangles connected on meson by a V-shaped mark with its apex on median point of anterior margin of each sternite; genitalia, fig. 242, brown; cerci light tan at bases, becoming almost white at apexes, articulations dark brown.

FEMALE.—Length of body 10–12 mm., of fore wing 12–14 mm. Coloration identical with that of male, except that caudal filaments are slightly lighter in color; apical margin of terminal abdominal sternite with a small, rounded, median notch.

NYMPH.—Length of body 10–12 mm., of caudal filaments 5–6 mm. Body light tan, with brown shading; legs tan, each with brown annulation near apex of femur, at base of tibia, and at base and apex of tarsus, these annulations not always completely encircling leg. Gills on abdominal segments 1 and 2 double, others single; spinelike, posterolateral angles of abdominal tergites dark brown at apexes; abdominal sternites with dark markings broad, abdominal color pat-

tern otherwise identical with that of adult; caudal filaments each with a broad, brown crossband near apex.

Known from Connecticut, Illinois, Maine, Michigan, New York, North Carolina, Ontario, Quebec, South Carolina, and Wisconsin.

Illinois Record.—SOUTH BELOIT: Rock River, May 31, 1927, D. H. Thompson, 3 N.

4. *Siphlonurus typicus* (Eaton)

Siphlurus typicus Eaton (1885: 222).
Siphlonurus berenice McDunnough (1923: 49).
Siphlonurus novangliae McDunnough (1924c: 75).

Spieth (1941a: 93) studied Eaton's type and established the above synonymy.

MALE.—Length of body 9–10 mm., of fore wing 10–11 mm. Head chiefly yellowish tan, light yellow on face below ocelli and shaded with dark brown at bases of ocelli; antennae very light tan, almost white; eyes gray. Thorax light yellow-brown, with white spots on pleura and on sternum; membrane of wings faintly stained with tan, veins and crossveins rich red-brown, costal and subcostal crossveins well developed, stigmatic crossveins very little anastomosed; legs yellow, darkened with brown at apexes of tibiae and tarsal segments. Abdominal ground color pale yellow to white, with light red-brown shading: each tergite with transverse, broad, shaded area at posterior margin, triangular, shaded area at each posterolateral angle, and broad, median, longitudinal spot; sternum virtually or quite unmarked, at most with faint, transverse, sinuate brown markings on anterior sternites; genitalia, fig. 244, yellow-tan; caudal filaments tan at bases, fading to white at apexes, articulations red-brown.

Nymph unknown.

Siphlonurus typicus is known from Connecticut, Illinois, Maine, Massachusetts, New Hampshire, New York, Pennsylvania, and Quebec.

Illinois Record.—ALTO PASS: April 30, 1942, Mohr & Burks, 1 ♂.

5. *Siphlonurus marshalli* Traver

Siphlonurus marshalli Traver (1934: 236).

MALE.—Length of body and of fore wing 10–13 mm. Head chiefly dark brown, light yellow on face below ocelli and on frontal shelf, dark brown shading present around bases of ocelli; antennae tan, base of each flagellum shaded with dark brown; eyes tan. The thorax dark brown, with light yellow markings on pleura and sternum. Membrane of each fore wing hyaline, with dark brown shading around crossveins, this shading most extensive in discal area of wing; membrane of hind wing almost or quite completely stained dark brown; veins and crossveins of fore wing strong, stigmal crossveins anastomosed; each fore leg dark yellow-brown, with tarsus slightly lighter, middle and hind legs yellow, faintly shaded with red-brown on coxae, near apexes of femora, and at apexes of tarsal segments. Abdominal dorsum dark brown at base and on apical tergites, intervening tergites lighter brown; basal and apical sternites brown, intermediate sternites almost white, with vague tan mark near lateral margin of each sternite; genitalia, fig. 245, yellowish tan; caudal filaments uniformly gray-brown near bases, fading to almost white at apexes, articulations in apical part of each filament faintly stained with brown.

FEMALE.—Length of body 11–14 mm., of fore wing 13–15 mm. Color pattern much as in male, except generally slightly lighter. Each fore leg faintly darker than middle and hind legs; fore wing as in male, hind wing with membrane not quite entirely shaded with brown, small hyaline areas present in center of most cells. Middle abdominal segments only slightly lighter than anterior and posterior ones; apical margin of terminal abdominal sternite produced posteriorly, evenly rounded from side to side, or margin very slightly irregular on meson; caudal filaments uniformly tan throughout, or occasionally becoming a little lighter in shade toward apexes.

NYMPH.—Length of body 11–16 mm., of caudal filaments 6–8 mm. Color pattern of thorax much as in adult, with coloration of adult wings distinctly visible; legs showing only a faint indication of darker shaded areas present in adult legs. Dorsum of abdomen light, with a pair of submedian dots at anterior margin of each tergite; postero-lateral, spinelike projections of tergites usually not darkened at tips; each sternite with a pair of broad, longitudinal, brown bands near lateral margins and four submedian, brown dots; gills borne by abdominal seg-

ments 1 and 2 double, others single, fig. 223, caudal filaments each with a broad, dark brown crossband at mid-length.

Known from Arkansas and Illinois.

Illinois Records.—ALTO PASS: April 30, 1942, Mohr & Burks, 1 ♂, 4 N; Jan. 25, 1947, Burks, Stannard, & Riegel, 1 N. DIXON SPRINGS: March 13, 1946, Ross & Burks, 7 N; April 4–6, 1946, Burks & Sanderson, 2 ♂, 1 ♀, 5 N. FOUNTAIN BLUFF: May 15, 1932, Ross & Mohr, 1 N. GIANT CITY STATE PARK: April 2–21, 1942, Ross & Burks, 5 ♂, 4 ♀, 23 N; May 16–29, 1946, Burks & Sanderson, 3 ♂, 1 ♀, 9 N. GOREVILLE, Fern Cliff: March 24, 1939, Ross & Burks, 1 N; April 4–23, 1942, Ross & Burks, 2 ♂, 1 ♀, 5 N. HEROD: June 1–3, 1939, Burks & Riegel, 2 ♀, 2 N.

ISONYCHIINAE new subfamily

The subfamily Isonychiinae is here erected for the reception of a single North American genus, *Isonychia,* which has long been the cause of radical disagreement among mayfly workers. This genus has been considered to have both heptageniid and baetid relationships. Ide (1930b:227) and Spieth (1933:329) included it in the family Heptageniidae; Ulmer (1933:210) placed it in his superfamily Heptagenioidea; and Traver (1935a:477) placed it in the Baetidae. *Isonychia* is, in my opinion, clearly baetid in its family relationships and shows some similarity to *Siphlonurus.* However, both nymphal and adult characteristics are too greatly at variance with those of *Siphlonurus* to permit the two genera to be placed in the same subfamily.

The adults of *Isonychia* have gill remnants persisting at the base of each vestigial maxilla and at the base of each fore coxa. The fore tarsus in the males is approximately as long as the fore tibia. In the fore wing, fig. 217, the cubital intercalaries are a series of short, sinuate, and forked veins which extend from Cu_1 to the anal margin of the wing, much as in *Hexagenia* and *Potamanthus.* In the hind wing, vein M is forked very near the outer margin of the wing. The male genitalia, figs. 248–253, are of a type quite different from those of all other baetid mayflies. The nymphs bear tufted, filamentous maxillary and fore coxal gills, and unique abdominal gills, each of which is composed of an upper, platelike member and a lower, filamentous tuft, fig. 225. These characteristics may be contrasted with those of the members of the Siphlonurinae, as given on page 98 above.

28. ISONYCHIA Eaton

Isonychia Eaton (1871:134).
Chirotonetes Eaton (1881:21). New name, unnecessarily proposed.
Jolia Eaton (1881:192).
Chirotenetes Needham (1905:25).
    Misspelling.

In Isonychia, the adult males have large compound eyes that are contiguous on the dorsal meson; each of these eyes is composed of an upper portion of large facets and an indistinctly separated lower portion of smaller facets. The outer surface of each eye is crossed by a pair of oblique, contrastingly colored stripes. Each compound eye in the adult females is approximately one-half the size of that of the males, and usually it has a single, broad, light-colored stripe extending across the outer surface.

In both sexes of all the species of Isonychia occurring in eastern North America, the fore leg is mostly or entirely red-brown, and the middle and hind legs are light yellow or white. The fore tarsus in the males is approximately equal in length to the fore tibia, and is only one and one-third to one and one-half times as long as the fore tarsus in the females. The hind tarsus in both sexes has four clearly differentiated segments. The wings, fig. 217, are relatively broader and shorter than in the members of the subfamily Siphlonurinae. The fore wing has the stigmal crossveins sometimes anastomosed, and the cubital intercalaries extend from vein Cu_1 to the anal wing margin as a series of sinuate, branched veins. The hind wing lacks an angulate or acute, basal costal projection, and vein M is forked very near the outer wing margin. Rudimentary gills persist on the fore coxa in the adults of both sexes. There is a large, blunt, median projection on the mesosternum, between the mid-coxae.

The male genitalia are composed of a pair of short penis lobes and a pair of four-segmented forceps, the first segment of each arm being very obscurely set off. The forceps base is medially excavated to form a more or less U-shaped cavity, and the characteristically short, stubby penes lie over or within this cavity, figs. 248–253. In this genus, there are four types of penis lobes: the sicca, fig. 252, the bicolor, fig. 253, the diversa, fig. 249, and the sayi, fig. 248. The terminal abdominal sternite in the adult females is usually emarginate on the meson of the posterior margin. The median caudal filament in both sexes is represented by a minute vestige.

The nymphs, fig. 254, are streamlined, vigorously swimming forms which invariably develop in the rapidly flowing water of creeks and smaller rivers. They are known to be in part predaceous (Morgan 1913:386; Clemens 1917:23); their food is principally vegetable detritus and algae, especially diatoms. Each maxilla and fore coxa bears tufts of filamentous gills. Each tarsal claw is single, acutely pointed, and short, being only one-fourth to one-third as long as the tibia. Each fore leg has a dense comb of long, stout setae on the inner margin, fig. 254. This comb of setae is used in gathering food. Each fore tibia bears an apical spur which is nearly one-half as long as the fore tarsus. Each abdominal segment has small, flat, lateral expansions at the lateral margins; the posterolateral angles of these expansions are produced as large, spinelike projections on segments 8 and 9. Abdominal segments 1–7 bear each a pair of gills, the individual gill being composed of a dorsal, platelike member and a ventral, filamentous tuft, fig. 225. There are three well-developed caudal filaments, and each cercus has a dense fringe of setae on the inner side only.

Reliable characteristics for the separation to species of nymphs and females of this genus have not yet been found.

KEY TO SPECIES

ADULT MALES

1. Forceps base with only a shallow, median excavation so that penis lobes are almost completely hidden when viewed from ventral side; penis lobes with acute lateral teeth, fig. 248........**1. sayi**
    Forceps base with a deep caudal excavation so that penis lobes are exposed when viewed from ventral side; penis lobes without lateral teeth, figs. 249–253....................................2
2. Penis lobes relatively long, mushroom shaped at apexes, fig. 249...**2. diversa**
    Penis lobes not mushroom shaped at apexes, figs. 250–253...............3
3. Fore tibia white, shaded with brown at base and at apex............**3. arida**

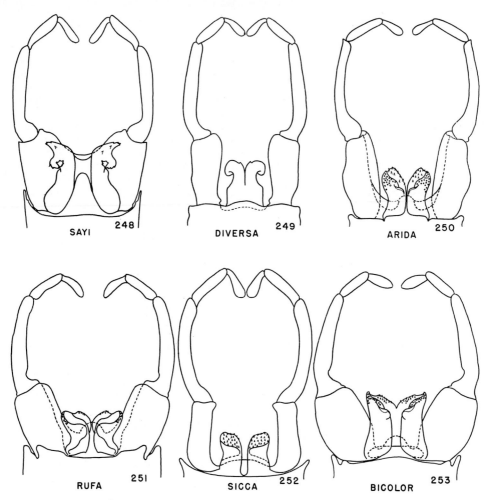

Fig. 248.—*Isonychia sayi,* male genitalia.
Fig. 249.—*Isonychia diversa,* male genitalia.
Fig. 250.—*Isonychia arida,* male genitalia.

Fig. 251.—*Isonychia rufa,* male genitalia.
Fig. 252.—*Isonychia sicca,* male genitalia.
Fig. 253.—*Isonychia bicolor,* male genitalia.

Fore tibia entirely red-brown, sometimes shaded with darker brown at apex..4
4. Crossveins of disc of fore wing brown; penis lobes relatively short and narrow, fig. 252.....................**4. sicca**
Crossveins of disc of fore wing hyaline; penis lobes relatively longer and broader, figs. 251–253...................5
5. Fore wing with stigmatic crossveins relatively numerous and anastomosed, fig. 217.....................**5. rufa**
Fore wing with stigmatic crossveins relatively few, not anastomosed..........
.......................**6. bicolor**

1. Isonychia sayi new species

Baetis arida? Say. Walsh (1862:370).
Baetis arida Walsh, not Say. Hagen (1863:170) ; Walsh (1863:191).

Siphlurus aridus Walsh, not Say. Eaton (1871:129).
Chirotonetes aridus Walsh, not Say. Eaton (1885:206).
Isonychia arida Walsh, not Say. McDunnough (1931c:159) ; Traver (1935a:485).

The name *sayi* is proposed for the species Walsh identified as *Baetis arida* Say. As is explained on page 111 below, Say's species is another form, long unrecognized. I have seen some of Walsh's original material, as well as McDunnough's and Traver's, and have found that the concept of *arida* Walsh, not Say, has not changed since Walsh's time.

MALE.—Length of body 9–12 mm., of fore wing 10–13 mm. Head light red-brown, antennae white, tinged with brown. Thorax dark red-brown, lighter on pleura; each fore leg brown, with apex of femur darker; wings hyaline, veins faintly tinged with tan, crossveins hyaline. Abdomen dark brown, each tergite and sternite with a large, light yellow-brown spot at either anterolateral angle; each tergite also with a smaller, median, light spot on anterior margin; minute, longitudinal, black lines in spiracular region; genitalia, fig. 248, yellow-brown; caudal filaments light yellow or white.

FEMALE.—Length of body 10–13 mm., of fore wing 12–15 mm. Head yellow-brown. Thorax light red-brown; wings hyaline, veins and crossveins brown. Abdomen dark red-brown, with large, conspicuous, light tan markings: light spot at either anterolateral angle of each tergite, tenth tergite usually entirely light, and entire lateral third of each sternite light; black spot near each spiracle; posterior margin of apical abdominal sternite entire.

Holotype, male.—Rock Island, Illinois, Walsh, 1863. Specimen dry, on pin; genitalia on a microscope slide.

Allotype, female.—Same data as for holotype. Specimen dry, on pin.

The holotype and allotype are in the collection of the Museum of Comparative Zoology and the paratypes listed below are in the collection of the Illinois Natural History Survey.

Known from Indiana, Illinois, and Kansas.

Paratypes.—ILLINOIS.—DIXON: June 27, 1935, DeLong & Ross, 1 ♂. GULFPORT: Crystal Lake, June 10, 1939, J. S. Ayars, 1 ♂. OQUAWKA: June 13, 1932, H. L. Dozier, 2 ♀. PROPHETSTOWN: Rock River, July 24–25, 1947, Burks & Sanderson, 2 ♀. ROCKFORD: at light, June 29, 1938, B. D. Burks, 1 ♀. ROCK ISLAND: B. D. Walsh, 1 ♂, 2 ♀.

2. *Isonychia diversa* Traver

Isonychia diversa Traver (1934: 244).

This species is included here because it represents one of the four types of male genitalia to be found in the genus. Head and body dark red-brown; wings hyaline, with hyaline veins and crossveins; genitalia, fig. 249, and caudal filaments white.

Known from Tennessee.

3. *Isonychia arida* (Say)

Baetis arida Say (1839: 42).

The original description of this species mentions only one character that is specific rather than generic. This is "anterior tibiae whitish, obscure at base and tip." Unfortunately, the species identified as *arida* by Walsh (1862: 370) has the anterior tibiae completely brown. Hagen (1863: 191) noticed this discrepancy, but concluded that the normal range of variation in the species would include forms with brown and with white tibiae. All workers since Walsh's time have followed his determination of *arida*, although to my knowledge a form of it with white tibiae has never been found. McDunnough (1931c: 159) stated that there was considerable doubt in his mind that Walsh's determination of *arida* was correct, but that "there seems nothing to be gained by altering his determination and changing the generally accepted idea of the species." I certainly would have followed the same course were it not for the fact that we have an Illinois specimen with white fore tibiae that matches Say's description of *arida* in all particulars. This specimen is quite different from the species determined by Walsh as *arida* and must either be determined as *arida* Say or be described as a new species of extremely doubtful validity. After full consideration of the problem, I have decided that the former course is preferable. Accordingly, the species identified as *arida* by Walsh is renamed *sayi* on page 110 and the name *arida* is here applied to this Illinois specimen, which fully agrees with Say's description.

MALE.—Length of body 9 mm., of fore wing 11 mm. Head light tan, antennae tan, becoming white at apexes of flagella. Thorax red-brown; wings hyaline, veins and crossveins hyaline, stigmal crossveins relatively few, not anastomosed. In fore leg, femur brown, lighter at base; tibia white, darkened with brown at base and tip; tarsus white, segments slightly shaded with brown at apexes. Abdomen red-brown, a transverse stripe of black shading at posterior margin of each tergite; genitalia, fig. 250, white; caudal filaments white.

This species is known from Illinois and Indiana.

Illinois Record.—MOMENCE: at light, August 16, 1938, Ross & Burks, 1 ♂.

4. *Isonychia sicca* (Walsh)

Baetis sicca Walsh (1862:371).

The lectotype male of this species is in the Museum of Comparative Zoology.

Male.—Length of body 9–11 mm., of fore wing 10–12 mm. Head dark red-brown, antennae brown, fading to light yellow at tips of flagella. Thorax dark red-brown; wings hyaline, veins and crossveins brown; each fore leg with femur and tibia red-brown, tarsus yellow, with apex of each

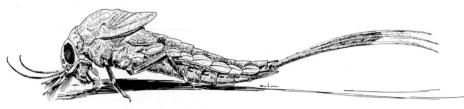

Fig. 254.—*Isonychia sicca*, mature nymph, lateral aspect.

tarsal segment darkened. Abdomen dark red-brown or very dark purplish red, venter slightly lighter than dorsum; genitalia, fig. 252, light brown or yellow-brown; caudal filaments light yellow to white, with articulations near bases darkened.

Female.—Length of body 10–12 mm., of fore wing 12–14 mm. Head tan, shaded with red-brown. Thorax yellow-brown, with sternum darker brown; wings as in male. Abdomen dull, dark red-brown or purplish red, becoming lighter, yellow brown on apical three segments; caudal filaments light yellow, articulations not darkened.

Known from Illinois, Iowa, Nebraska, and Ontario.

Illinois Records.—East Dubuque: at light, July 21, 1927, 5♀. Foster: Mississippi River, July 4, 1939, B. G. Berger, 1♂. Fulton: July 20, 1927, Frison & Glasgow, 1♂, 11♀. Harrisburg: at light, Aug. 16, 1937, Ross & Ritcher, 1♂, 1♀. Havana: 1♀; June 23, 1894, F. Smith, 1♀; June 24, 1894, 1♀; June 25, 1894, 1♀. Homer: June 30, 1925, T. H. Frison, 1♀. Mount Carmel: June 30, 1906, 3♀. Oregon: July 9, 1925, T. H. Frison, 1♂. Quincy: Mississippi River, July 6, 1939, Mohr & Riegel, 1♂; Aug. 13, 1889, C. A. Hart, 1♀. Rock Island: 3♂, 2♀ (Walsh 1862:371). Savanna: July 19, 1892, Forbes, Shiga, Hart, & McElfresh, 3♂, 8♀; July 22, 1892, Hart & Forbes, 4♂; July 27, 1892, McElfresh, Shiga, Forbes, & Hart, 1♀; July

29, 1892, Forbes & Shiga, 1♀. Wilmington: at light, Aug. 6, 1947, Burks & Sanderson, 1♂.

5. *Isonychia rufa* McDunnough

Isonychia rufa McDunnough (1931c:162).

Male.—Length of body and of fore wing 10–12 mm. Head light brown, antennae brown, shading to white at apexes of flagella. Thorax bright red-brown, sternum dull red-brown or beige; wings hyaline, veins faintly yellow, crossveins hyaline, stigmatic crossveins anastomosed; fore femur and tibia red-brown, and tarsus yellow, with brown shading at apexes of segments. Abdomen bright red-brown on dorsum, venter lighter, chestnut brown; transverse, black line at posterior margin of each abdominal tergite, this black line often interrupted on meson; genitalia, fig. 251, yellow-brown; caudal filaments light yellow or tan, articulations at base darker.

Female.—Length of body 10–12 mm., of fore wing 12–14 mm. Head and thorax yellow-brown, the latter often also with reddish tinge; veins and crossveins of wings tan or light yellow, stigmatic crossveins anastomosed. Abdomen red-brown on dorsum, pinkish yellow on venter; apical abdominal segment yellow; caudal filaments yellow to white.

Known from Illinois, Iowa, Kansas, Nebraska, and Ohio.

Illinois Records.—Apple River Canyon State Park: July 3, 1946, Burks & Sanderson, 1♂. Aurora: July 9, 1925, T. H. Frison, 2♀; July 17, 1927, Frison & Glasgow, 7♀. Dixon: May 31, 1914, 1♂. Freeport: at light, Aug. 4, 1948, 3♀. Kankakee: July 9, 1948, Ross & Burks, 1♀; Aug. 2, 1938, Burks & Boesel, 1♂. Monmouth: at light, June 23, 1948, L. J. Stannard, 1♂. Oakwood: July 14, 1939, Burks & Riegel, 1♂. Onarga: at light, July 9, 1948, Ross & Burks, 1♀. Oregon: July 4, 1946, Burks

& Sanderson, 5♂, 2♀. PEORIA: July 13, 1940, F. F. Hasbrouck, 1♂. QUINCY: June 2, 1939, Burks & Riegel, 1♂ ; June 25, 1940, Mohr & Riegel, 1♂. ST. CHARLES: at light, July 8, 1948, Ross & Burks, 2♂. SAVANNA: July 22, 1892, Hart & Forbes, 1♂ ; July 20, 1927, T. H. Frison, 1♀. WEST CHICAGO: July 9, 1948, Ross & Burks, 1♀. WHITE HEATH: Sangamon River, Aug. 2, 1939, Ross & Riegel, 2♂. WILMINGTON: at light, Aug. 6, 1947, Burks & Sanderson, 7♂, 4♀.

6. *Isonychia bicolor* (Walker)

Palingenia bicolor Walker (1853:552).
Chirotenetes albomanicata Needham (1905:31).

MALE.—Length of body and of fore wing 10–12 mm. Head brown; scape and pedicel of each antenna light brown, flagellum tan at base, becoming yellow toward apex. Thorax dark red-brown, almost black; fore leg same color, with tarsus white, segments shaded with brown at apexes; wings hyaline, veins and crossveins colorless except at costal margin, where they are tan. Abdomen very dark red-brown, the apical segment lighter brown, a narrow, transverse, black band at posterior margin of each tergite; genitalia, fig. 253, tan; caudal filaments light yellow or white, a few basal articulations brown.

FEMALE.—Length of body and of fore wing 12–16 mm. Head yellow or tan, shaded with brown. Thorax yellow-brown, darker on venter; wings with veins and crossveins light brown. Abdomen bright red-brown, with transverse, black-shaded stripe at posterior margin of each tergite; black, longitudinal line and spot at each spiracle; caudal filaments white.

Known from the northeastern and midwestern states and the eastern Canadian provinces.

Illinois Records. — EDDYVILLE: Lusk Creek, May 15–23, 1946, Mohr & Burks, 3♂. KANKAKEE: May 31, 1938, Burks & Mohr, 1♂ ; June 5, 1932, Frison & Mohr, 1♂ ; June 15, 1938, Ross & Burks, 2♂ ; June 17, 1939, B. D. Burks, 1♂, 1♀ ; July 10, 1925, T. H. Frison, 5♂ ; July 18, 1925, T. H. Frison, 1♂ ; Aug. 1, 1933, Ross & Mohr, 1♂ ; Aug. 2–4, 1938, Burks & Boesel, 9♂, 7♀ ; Aug. 16, 1938, Ross & Burks, 1♂, 3♀. MOMENCE: June 15, 1938, Ross & Burks, 1♂, 2♀. OAKWOOD: June 6, 1925,

T. H. Frison, 1♂ ; June 9, 1926, Frison & Auden, 1♂ ; July 24, 1939, B. D. Burks, 3♂, 2♀. POPLAR BLUFF: June 20, 1943, T. H. Frison, 1♂. ROCKFORD: June 13, 1931, Frison & Mohr, 1♂. ROCK ISLAND: June 7, 1937, Burks & Riegel, 1♂ ; June, 1933, C. O. Mohr, 2♂. WILMINGTON: at light, Aug. 6, 1947, Burks & Sanderson, 2♂, 3♀.

BAETINAE

The most simplified of North American mayflies belong to the subfamily Baetinae. The wing venation is always reduced, both through complete loss of some veins and through partial atrophy of the veins that persist. The basal part of the outer branches of vein Rs and the base of vein M_2 of the fore wing are always atrophied, as in figs. 31, 220–222. The hind wing, figs. 220–222, 270–284, is greatly reduced in size and venation, or wanting entirely. The homologies of the longitudinal veins that persist in the hind wing have not been conclusively determined. When three longitudinal veins persist, however, they perhaps represent the remnants of Sc, R, and M. The middle and hind tarsi, in both sexes, have only three clearly differentiated segments, fig. 15. The male genitalia, figs. 260, 267–269, 289–297, are greatly reduced, the penis lobes being virtually amorphous, membranous, internal structures. There is, between the bases of the forceps, a flaplike penis cover which, in some species of the genus *Baetis,* is obscure. Each arm of the forceps has four segments; the separation between the second and third segments often is so obscure that each arm appears to have only three segments. The adults uniformly have the median caudal filament vestigial in both sexes.

In the members of this subfamily, antigeny is more pronounced than in other mayflies. The compound eyes in the male adults are greatly enlarged and divided, each eye consisting of two distinctly separated portions, figs. 255–257. The eyes in the females are relatively small and simple. This hypertrophy of the eyes in the males has led to the development of a marked difference between the two sexes in the shape of the head. This difference in head shape can be seen in the nymphs as well as in the adults. The nymphs in even the early instars show this difference in head shape. Because of this, the male nymphs in all stages of development may appear quite unlike the female nymphs.

The nymphs, figs. 266, 298, are streamlined and fishlike in body form, each with a labrum having a square notch on the meson of the anterior margin, fig. 231, one or two pairs of wingpads, slender, denticulate, and single tarsal claws, figs. 264, 265, usually single and platelike gills, and two or three well-delevoped caudal filaments.

This entire subfamily is very difficult to treat taxonomically, as really good structural characters for the separation of species have not yet been found. The various species are at present distinguished almost entirely on differences in the color patterns which, unfortunately, in most species are subject to considerable variation. The differentiation of species throughout this subfamily is, thus, made on a rather insecure basis. It often is not possible to separate the females of this subfamily to genus.

KEY TO GENERA

Adult Males

1. Fore wing and hind wing with relatively numerous crossveins, figs. 221, 222.. ....**29. Callibaetis**
   Fore wing with relatively few crossveins, fig. 220; hind wing with very few crossveins or with none, figs. 270–284, or hind wing wanting entirely.........2
2. Hind wing present, although often greatly reduced...........................3
   Hind wing absent.....................5
3. Marginal intercalary veins of fore wing single, as in fig. 221.**30. Centroptilum**
   Marginal intercalary veins of fore wing in pairs, figs. 31, 220, 222..........4
4. Hind wing greatly reduced and either without venation or with traces only of a single longitudinal vein............. ....**31. Heterocloeon**
   Hind wing relatively well developed, with two or three longitudinal veins, figs. 270–284....................**32. Baetis**
5. Marginal intercalary veins of fore wing in pairs, as in fig. 220................ ....**33. Pseudocloeon**
   Marginal intercalary veins of fore wing single, as in fig. 221................6
6. Second forceps segment of male with a prominent, angular projection on mesal margin, fig. 299........**34. Neocloeo.i**
   Second forceps segment of male simple, without a mesal projection, fig. 300... ....**35. Cloeon**

Mature Nymphs

1. Gills single, platelike on all abdominal segments, fig. 228.................2
   Gills double on at least some abdominal segments, or each gill a thin, somewhat

irregular sheet with a recurved, dorsal or ventral flap, as in fig. 226........6
2. Hind wingpad absent................3
   Hind wingpad present................4
3. Maxillary palp with two segments; median caudal filament usually vestigial...... ....**33. Pseudocloeon**
   Maxillary palp with three segments; median caudal filament well developed ....**34. Neocloeon**
4. Median caudal filament as well developed as the cerci......**30. Centroptilum**
   Median caudal filament reduced or vestigial.............................5
5. Suture between second and third segments of labial palp partly or completely obliterated; second segment without an apicomesal projection, fig. 261, median caudal filament vestigial ....**31. Heterocloeon**
   Suture between second and third segments of labial palp well marked; second segment with an apicomesal projection, figs. 258, 259, 262, 263, median caudal filament reduced or vestigial........ ....**32. Baetis**
6. Hind wingpad absent.......**35. Cloeon**
   Hind wingpad present................7
7. Maxillary palp with three segments; each abdominal gill with an inconspicuous dorsal flap........**30. Centroptilum**
   Maxillary palp with two segments; each gill borne by abdominal segments 1 and 2 with two well-developed laminae, each gill borne by more posterior segments with only a relatively small, recurved ventral flap, fig. 226.......... ....**29. Callibaetis**

29. *CALLIBAETIS* Eaton

Callibaetis Eaton (1881:196).

In *Callibaetis,* the upper portion of each compound eye in the males is stalked, but this stalk is relatively low, fig. 257. The width of the vertex separating the compound eyes in the females is about twice as great as the length of one eye. In the fore wing in the males, the basal costal crossveins are weak or wanting, but these crossveins are well developed in the fore wing in the females. The fore wing in the males usually is not pigmented, but in the females it is, at least in the costal and subcostal interspaces in Nearctic species. Two general types of arrangements of the crossveins of the fore wing are to be found in this genus: in one type, there are relatively few crossveins, with none very near the posterior wing margin, and the crossveins form a single irregular row across the wing, fig. 221; in the other type, there are relatively numerous crossveins, some of them located near the posterior wing margin, and the

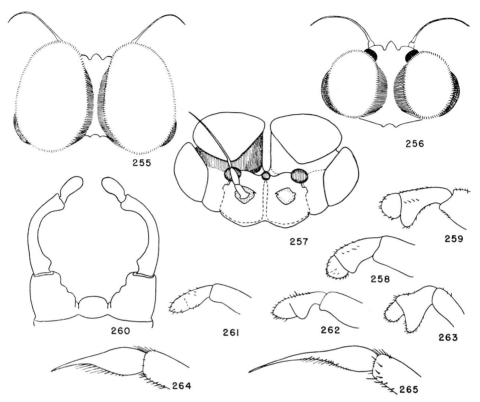

Fig. 255.—*Baetis intercalaris*, head of adult male, dorsal aspect.
Fig. 256.—*Baetis flavistriga,* head of adult male, dorsal aspect.
Fig. 257.—*Callibaetis fluctuans,* head of adult male, anterior aspect.
Fig. 258.—*Baetis intercalaris,* labial palp of mature nymph.
Fig. 259.—*Baetis brunneicolor,* labial palp of mature nymph.
Fig. 260.—*Callibaetis skokianus,* male genitalia.
Fig. 261.—*Heterocloeon curiosum,* labial palp of mature nymph. (After Ide.)
Fig. 262.—*Baetis vagans,* labial palp of mature nymph.
Fig. 263.—*Baetis frondalis,* labial palp of mature nymph. (After Ide.)
Fig. 264.—*Callibaetis fluctuans,* claw of middle leg of mature male nymph.
Fig. 265.—*Callibaetis ferrugineus,* claw of middle leg of mature male nymph.

crossveins form two or more quite irregular rows across the wing, fig. 222. The hind wing is well developed, with abundant crossveins.

The bodies in most species of this genus are thickly sprinkled with minute, brown dots set in small depressions. Similar punctate dots also are often found on the legs. The male genitalia consist of a pair of four-segmented forceps, with a rounded or conic penis cover located between the bases of the forceps. Each basal forceps segment is short and wide, the second segment is narrow, tapering, and indistinctly separated from the third segment, the latter is long, slender, and bowed, while the fourth segment is short, being only about twice as long as wide.

In *Callibaetis* subimagoes, the wings are dark gray, with the paths of the veins and crossveins white.

The nymphs, fig. 226, are streamlined forms which swim with a rapid, darting motion. They live in still water, usually in permanent ponds. The nymphal maxillary palps have two and the labial palps have three segments. The tarsal claws are long and slender, and provided with a row of minute ventral denticles, figs. 264, 265. The abdominal gills are sheetlike and slightly undulated, with a dense net of pinnately branching tracheae. The first and second pairs of gills are always double, with the ventral member often bearing a secondary, recurved flap. The gills on segments 3–6 are single, each having a well-developed, recurved ventral flap. These ventral gill flaps decrease in size from front to rear, so that the recurved flap borne by each gill of segment 3 is almost as large as the gill itself, while the recurved flap of each gill of segment 7 is so small as to be easily overlooked, fig. 226. There are three equally long caudal filaments; the cerci are fringed with long setae on the mesal side only.

Many of the species of this genus are said to be ovoviviparous.

This genus includes about 20 Nearctic species, 3 of which occur in Illinois.

Characteristics for the separation to species of the females of this genus in both adult and nymphal stages have not yet been found.

KEY TO SPECIES

ADULT MALES

1. Crossveins in fore wing in area posterior to vein R₁ relatively few in number, forming a single row across disc of wing, and with none located very near outer wing margin, fig. 221....**1. fluctuans**
   Crossveins in fore wing in area posterior to vein R₁ relatively abundant, forming two irregular rows across disc of wing, and with many located close to outer wing margin, fig. 222..............2
2. Costal margin of fore wing hyaline......
   ....**2. ferrugineus**
   Costal margin of fore wing partly or completely shaded with brown or tan....3
3. Fore wing shaded with brown at base only..............**3. brevicostatus**
   Fore wing shaded with light tan in costal area from base to apex of wing........
   ....**4. skokianus**

MATURE MALE NYMPHS

1. Length of body 12–13 mm..**4. skokianus**
   Length of body not over 10 mm.......2
2. Claw of fore leg relatively long and slender, the length more than five times greatest thickness, fig. 265; abdominal gills of seventh pair double..........
   ....**2. ferrugineus**
   Claw of fore leg relatively short and stout, the length less than four times greatest thickness, fig. 264; abdominal gills of seventh pair single.......**1. fluctuans**

1. *Callibaetis fluctuans* (Walsh)

Cloe fluctuans Walsh (1862:379).

This species was described from the female only, and the types are lost.

MALE.—Length of body 5.5–7.0 mm., of fore wing 6–8 mm., of caudal filaments 10–12 mm. Head, fig. 257, brown, light yellow around bases of antennae and on

lateral areas of frontal shelf; eyes brown when insect is alive; each antennal scape and pedicel brown, flagellum light yellow. Dorsum of thorax brown, venter almost entirely light yellow, with only a few brown, punctate dots on mesosternum; wings hyaline, fig. 221, without any coloration, all veins and crossveins hyaline; costal crossveins of fore wing wanting entirely or vestigial, stigmatic crossveins not anastomosed, slanting, 5–7 in number; crossveins in disc of wing relatively few, fig. 221, marginal intercalaries on outer margin usually single, sometimes double; legs light yellow, fore leg faintly stained with brown near apex of femur and at base and apex of tibia; middle and hind legs each with faint brown staining near apex of femur and with a minute, brown dot at apex of each tarsal segment. Abdomen light yellow, with brown shading: tergite 1 dark brown on meson; tergites 2–9 each completely shaded with light brown except for a narrow, longitudinal, pale streak on meson, a fairly large spot at anterior margin near each anterolateral angle, and a narrow line crossing tergite at posterior margin; a pair of dark brown, submedian dots at anterior margins of tergites 4–10, a pair of short, curved, dark brown dots near anterolateral angles of each abdominal tergite, and a pair of longitudinal, dark brown marks at lateral margins of tergites 1–7; abdominal sternum light yellow, with a pair of short, curved, submedian, dark brown marks near anterior margin of each sternite, and usually a minute, dark brown dot at anterolateral angles of each sternite. Genitalia light yellow; caudal filaments white, articulations not darkened.

FEMALE.—Length of body 6–8 mm., of fore wing 7–9 mm., of each caudal filament 9–10 mm. Coloration much as in male, but brown shading of dorsum of thorax more restricted. When insect is alive, pink staining visible on vertex, on dorsal area of pronotum, and on mesonotum anterior to wing bases; wings hyaline, brown stained in costal, subcostal, and first radial interspaces, this staining often extending on membrane slightly posterior to vein R₂ at apex and in basal area of wing, and on veins as far back as M₁; brown staining interrupted around crossveins; in living insect, pink staining present in wing on basal two-thirds of veins Sc and R₁ and on crossveins in this area, costal crossveins present but irregular and often broken, stigmatic crossveins slanting, irregular, sometimes partly anastomosed, 8 to 12 in number; femur of each leg usually with faint brown shading extending from base to apex, on outer side. Abdomen with dark brown spots as in male, but dorsal, light brown shading faint or wanting; dorsum usually with many minute, punctate, brown dots scattered over surface; sternum punctate, but brown dots few; caudal filaments as in male.

NYMPH.—Length of body 8–9 mm. Head brown, with a white spot just dorsal to each antennal socket and on meson between sockets; each antenna as long as fore leg. Thorax brown, with minute, white mottling on mesonotum; legs uniformly light brown. Abdominal dorsum mostly brown, with small area at base and apex of lateral projection of each segment white, a single median spot or two submedian, coalescing spots on each tergite, and a longitudinal, white stripe on either side of median spot on each of tergites 2–8; gills semihyaline, tracheae lavender-brown; each gill borne by segments 1 and 2 triple, gills of segments 3–6 double, seventh gill single; caudal filaments each with a subapical, dark brown crossband.

Known from Illinois, Iowa, New York, and Wisconsin.

Illinois Records.—Specimens, collected April 23 to November 3, are from Belleville, Brussels, Cairo, Chambersburg, Collinsville, Grand Tower, Greenville, Havana, Herod (pool near Gibbons Creek), Jonesboro, Morris, Mount Carmel, Muncie, Oakwood, Peoria, Pingree Grove, Quincy, Rantoul, Rock Island, Rosiclare, Springfield, St. Jacob, St. Joseph, Sterling, Urbana, Waukegan, and Western Springs.

2. Callibaetis ferrugineus (Walsh)

Cloe ferruginea Walsh (1862:379).

There is at present in the Museum of Comparative Zoology a single male specimen, determined as of this species by Walsh, which was collected at Rock Island, Illinois, a year after the original description was published. Unfortunately, it is badly broken but, insofar as the characters can be seen on this fragmentary specimen, it is in agreement with the current concept of the species. The female was unknown to Walsh, but the

association of the correct female to be placed with the male has been arrived at through the rearing of adults of both sexes from lots of nymphs that almost certainly represented pure cultures of the species.

MALE.—Length of body 7.0–8.5 mm., of fore wing 8.0–9.5 mm. Color extremely variable, ranging from almost completely light yellow, with a few minute, brown spots, to almost completely brown, with darker brown or black markings. Head brown, with face below antennae white; each antennal scape and pedicel white, with apexes brown, flagellum white at base, gray or tan distad; upper eyes tan, lower gray, each with a brown, longitudinal stripe. Thorax brown to yellow, with dark, punctate spots on pleura and venter; wings hyaline, three costal veins faintly yellow; marginal intercalaries of outer margin of each fore wing usually double; legs white to yellow, femora often with dark red, punctate spots, all femora vaguely darkened at apexes, each fore tibia red-brown at apex. Abdomen varying from yellow to almost black; dark, punctate spots present over most of surface, a dark, longitudinal, median, dorsal stripe often present; genitalia and caudal filaments white.

FEMALE.—Length of body 8–9 mm., of fore wing 9–10 mm. Body invariably darker than that of male, varying from tan to almost black. Head usually grayish tan or brown, face below antennal sockets white to tan, eyes gray, with brown band. Thorax usually dull gray-brown on dorsum; wings hyaline, each typically with dark brown shading occupying entire anterior three interspaces and usually extending posteriorly onto R, sometimes reaching almost to vein M, fig. 222; veins often alternately brown and white and outer margin of wing often spotted with brown; hind wing usually brown at base; legs yellow to tan or light brown; all femora always with broad, darkened areas at apexes. Abdomen usually dark brown to almost black; caudal filaments white, occasionally with some basal articulations darkened.

NYMPH.—Fig. 266. Length of body 8–9 mm. Head brown, with a white spot just dorsal to each antennal socket, on either gena just ventral to each compound eye, and on meson between antennal sockets; each antenna slightly longer than fore leg. Thorax brown, slightly mottled with darker brown

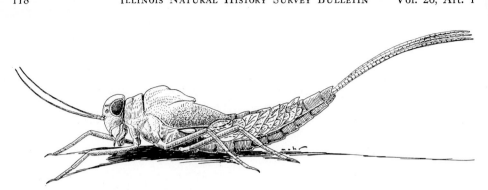

Fig. 266.—*Callibaetis ferrugineus,* mature nymph, lateral aspect.

and white on notum; legs tan, each femur with a subapical, dark brown ring. Abdominal dorsum usually mostly tan or white, with a median, dark brown spot at either lateral margin and a pair of large, sublateral, dark brown spots on each tergite 2–7; median area of tergites 2 and 3 dark brown; tergites 8–9 mostly light brown; gills white, tracheae purple-brown, those on segments 1 and 2 triple, and on segments 3–7 double; caudal filaments each with a dark brown, subapical crossband.

Known from Illinois, Iowa, Maine, Maryland, Michigan, New York, and Wisconsin.

Illinois Records.—Specimens, collected April 18 to September 23, are from Antioch, Channel Lake, East Dubuque, Elgin, Fox Lake, Freeport, Giant City State Park, Havana, Kickapoo State Park, Lake Bluff, McHenry, Richmond, Rockford, Rock Island, Rosecrans, Spring Grove, St. Charles, Waukegan, Wolf Lake, and Zion.

3. *Callibaetis brevicostatus* Daggy

Callibaetis brevicostatus Daggy (1945:388).

This species may prove to be a synonym of *semicostatus* Banks (1914:614), which was described from Manitoba.

MALE.—Length of body and of fore wing 8 mm. Fore wing stained with brown at bases of veins Rs to M, but otherwise hyaline; longitudinal veins brown, crossveins hyaline; costal crossveins weak but present, stagmatic crossveins slanted and partly anastomosed; fore wing with numerous crossveins and paired marginal intercalary veins. Abdomen brown, and densely covered with dark brown, punctate dots; genitalia and caudal filaments white.

Known from Minnesota and Saskatchewan.

4. *Callibaetis skokianus* Needham

Callibaetis skokianus Needham (1903:215).

This species might eventually prove to be a synonym of *ferrugineus,* as the two are separated principally on the color of the wings, a character that is known to vary in other species of the genus. Long series of specimens of both *skokianus* and *ferrugineus* have, however, been studied and no intergrades between the two have as yet been found.

Our recent collecting in Illinois has failed to produce the nymph of this species. The only specimens of the nymph of *skokianus* I have seen are those in the Cornell University collection, and they are in very poor condition.

MALE.—Length of body and of fore wing 9–10 mm. General color bright yellow-brown or tan, shaded with dark red-brown. Head yellow-brown, face tan below antennal sockets; each antennal scape yellow, brown at apex, pedicel usually entirely brown, flagellum yellow at base, shaded with brown in middle, white at apex; upper portion of each eye yellow, lower tan, with a brown band across middle. Thorax yellow or tan; a broad, median, longitudinal, dark brown stripe present on mesonotum; pleura and sternum usually with brown, punctate dots; legs yellow or white, with apexes of all femora lightly shaded with tan, and each fore tibia and tarsus entirely tan to brown; wings hyaline, three costal interspaces of each fore wing washed with tan or light yellow-brown, stigmatic crossveins anastomosed, marginal intercalaries of outer wing margin usually double. Abdomen chestnut brown, with numerous dark brown, punctate dots, a dark brown, longitudinal, median band usually extending the length of the abdominal tergum, this band often interrupted at each suture, a large, dark brown spot present near either anterolateral angle of each tergite; abdominal sternum lighter yellow-brown than dorsum, each sternite with a pair of submedian, parenthesis-shaped, brown marks; genitalia, fig. 260, yellow to white; caudal filaments white.

FEMALE.—Length of body 9–10 mm., of fore wing 10–11 mm. Head and body generally lighter in color than those of male. Head yellow to almost white, with tan shading, face below antennal sockets always white. Thorax yellow to tan, dorsal, longitudinal, median stripe light brown, legs white; all femora usually vaguely washed with tan, brown shading at apexes of all tarsal segments; wings hyaline, longitudinal veins of each fore wing anterior to M tan, first three interspaces of fore wing shaded with chestnut brown, this shading interrupted at crossveins, brown shading also extending over small part of basal area of posterior radial, median, and cubital interspaces; hind wing not shaded. Dorsum of abdomen light brown, thickly sprinkled with dark brown, punctate dots; venter tan to almost white; caudal filaments white.

NYMPH.—Length of body 12–13 mm. Head light yellow to white, with brown shading on frons, between eyes, and ventral to ocelli between antennal sockets. Thorax light yellow, with brown shading on mesonotum and on wingpads; legs white, tarsi tinged with brown. Abdomen mostly light yellow on dorsum, a median, interrupted, longitudinal, brown stripe usually present, this brown area spreading on tergites 2 and 3 to cover most of exposed dorsal area; gills hyaline, tracheae brown; gills on segments 1 and 2 triple, those on 3–7 double, ventral lobe on seventh gill greatly reduced; caudal filaments darkened near apexes.

Callibaetis skokianus is known from Illinois, Minnesota, Missouri, New York, North Dakota, and Ontario. It was originally described from a very long series of specimens collected over a 4-year period at the turn of the century from a pond on the campus of Lake Forest College located on the shore of Lake Michigan north of Chicago. This pond has now disappeared, and intensive, recent collecting and rearing in that area has failed to yield additional specimens, but a few scattered ones have been taken elsewhere in the state.

Illinois Records.—AURORA: July 17, 1927, Frison & Glasgow, 1♀. EAST DUBUQUE: at light, July 21, 1927, Frison & Glasgow, 1♀. FREEPORT: at light, June 10–11, 1948, Burks, Stannard, & Smith, 1♀; Aug. 4, 1948, Burks & Stannard, 1♀. HAVANA: April 14, 1894, Hart & Hempel, 1♂; April 17, 1894, Hart & Hempel, 1♂; April 18–19, 1894, C. A. Hart, 2♂. LAKE FOREST: in pond on college campus, April–May, 1899, J. G. Needham, 29♂; May, 1902, J. G. Needham, ♂♂, ♀♀, nymphs. PINGREE GROVE: May 9, 1939, Ross & Burks, 4♂.

30. *CENTROPTILUM* Eaton

Centroptilum Eaton (1869:132).

In this genus, the stalk of each turbinate eye in the males is quite low, so that the lower portion of the eye is almost in contact with the faceted upper portion. The width of the vertex separating the compound eyes in the females is only slightly greater than the length of one eye. Typically, the head and thorax in the males are dark brown, with the abdomen light, but often strikingly marked on the dorsum with red or red-brown; the females are generally light in color, the abdominal tergites heavily marked with black tracheal lines. Each fore wing has relatively few crossveins and the marginal intercalary veins are single. The hind wing is long and slender, with a hooked, costal, subbasal projection; with two longitudinal veins and sometimes with vestigial third vein; the crossveins are wanting or vestigial. The male genital forceps are four segmented, the first segment being short and broad, the second also short and broad with, typically, a prominent tubercle on the mesal margin, the third segment is slender and usually not strongly bowed, and the fourth segment is often three or four times as long as broad; occasionally, the fourth segment is only as long as broad. There is a large, variously shaped penis cover between the bases of the forceps arms.

The nymphs are streamlined, vigorously swimming forms, typically developing in the shallow, rapidly flowing water of brooks and creeks. The nymphal labial and maxillary palps have each three segments. The claws are long and slender, and lack ventral denticles. The abdominal gills are platelike and usually single on all segments; in some species each gill borne by the basal segments has a recurved, dorsal flap. There are three well-developed caudal filaments.

Centroptilum includes 22 Nearctic species, only 3 of which occur in Illinois.

Characteristics for the separation to species of the females and nymphs in this genus have not yet been found.

KEY TO SPECIES

ADULT MALES

1. Entire mid-dorsal area of the abdomen shaded with dark red....**3. quaesitum**
   Dorsal area of abdomen not entirely shaded with red; tergites 2–6 white or faintly yellow, with small spots or lines of red or black.................2
2. Posterior margin of each of middle abdominal tergites with a pair of transverse, sublateral, red lines; longitudinal, black spiracular lines present at lateral margins of tergites 2–6..............
   ....**2. rufostrigatum**
   Abdominal tergites without transverse, red lines; only black spiracular lines present...................**1. walshi**

1. *Centroptilum walshi* McDunnough

Centroptilum walshi McDunnough (1929:173).

MALE.—Length of body and of fore wing 6 mm. Head pale yellow, upper portion of each turbinate eye yellow in life, each antenna yellow, flagellum faintly tinted with tan in basal third. Thorax pale yellow, shaded with light brown on median dorsal area; legs light yellow to white, a minute, longitudinal, black line present on ventral side of each posterior femur; wings completely hyaline, hind wing five times as long as broad, with two longitudinal veins. Abdomen pale yellow, almost white, with irregular, longitudinal, black spiracular line at each lateral margin of tergites 1–7, and tergites 8–10 entirely but lightly shaded with tan; forceps pale yellow, second forceps segment with a prominent tubercle on mesal margin, and apical margin of forceps baseplate evenly rounded from side to side, fig. 267; caudal filaments white.

FEMALE.—Length of body and of fore wing each 5.5–6.5 mm. Entire body pale green when insect is alive, quickly fading to light yellow; thorax lacking brown shading, but abdominal tergites heavily blotched laterally with black and apical abdominal tergite shaded with brown.

Known from Illinois, Iowa, and Kansas. **Illinois Records.**—BEARDSTOWN: at light,

June 13, 1946, Mohr & Burks, 2 ♀. HOMER: June 30, 1925, T. H. Frison, 1 ♂, 3 ♀. OAKWOOD: June 8–9, 1926, Frison & Auden, 1 ♂, 3 ♀. STERLING: at light, June 22, 1948, L. J. Stannard, 1 ♂. URBANA: June 1, 1941, T. H. Frison, 2 ♀.

2. *Centroptilum rufostrigatum* McDunnough

Centroptilum rufostrigatum McDunnough (1924b:95).
Centroptilum bistrigatum Daggy (1945:389). New synonymy.

I have studied types of both *rufostrigatum* and *bistrigatum* and find them unquestionably of the same species.

MALE.—Length of body and of fore wing 4.5–5.0 mm. Head very dark brown, antennae light brown; upper portion of each compound eye bright yellow in life. Thorax dark brown, lighter brown on venter; wings hyaline, hind wing four and one-half times as long as broad, with two longitudinal veins, three or four faint crossveins, and with the costal projection long and hooked; legs yellowish white, coxae light brown. First abdominal segment tan, segments 2–6 white, segments 7–10 dark brown above, white below, a pair of transverse, sublateral, red lines present at posterior margin of each of tergites 2–8, usually these red marks on tergites 7 and 8 visible only on living specimens or specimens preserved in alcohol; longitudinal, black spiracular lines usually present at lateral margins of tergites 2–6; genitalia, fig. 268, white, apical forceps segment minute; caudal filaments white.

FEMALE.—Length of body 4.5–5.0 mm., of fore wing 5–6 mm. Head and thorax light brown, dorsum of abdomen uniformly light brown, with transverse, red lines at posterior margins of tergites as in male; longitudinal, black spiracular lines of male replaced by heavy, black spiracular blotches at lateral margins; legs and caudal filaments white.

Known from Illinois, Manitoba, Minnesota, New Brunswick, and Wisconsin.

Illinois Records.—KANKAKEE: July 10, 1925, T. H. Frison, 1 ♂. OAKWOOD: July 30, 1939, B. D. Burks, 1 ♂.

3. *Centroptilum quaesitum* McDunnough

Centroptilum quaesitum McDunnough (1931b:87).

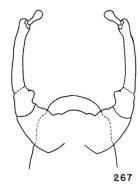

267

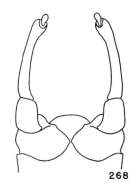

268

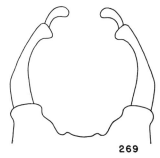

269

Fig. 267.—*Centroptilum walshi*, male genitalia. (After McDunnough.)
Fig. 268.—*Centroptilum rufostrigatum*, male genitalia.
Fig. 269.—*Heterocloeon curiosum*, male genitalia.

MALE.—Length of body and of fore wing 6–7 mm. Head light yellow; scape and pedicel of each antenna yellow, flagellum gray-tan. Thorax tan, pleura faintly stained with red, sternum mostly yellow; legs light yellow, almost white, with each fore femur faintly stained with tan, and each fore tibia and tarsus slightly darkened with gray; wings hyaline, hind wing three and one-half times as long as broad. Entire abdominal tergum, except for narrow area at lateral margins, uniformly shaded with dark red; abdominal venter faint yellow, almost white; genitalia and caudal filaments white.

Known from Alberta and Illinois.

Illinois Record.—CAIRO: at light, July 17, 1947, L. J. Stannard, 1 ♂.

31. *HETEROCLOEON* McDunnough

Heterocloeon McDunnough (1925*b*:175).

In this genus, each fore wing has the marginal intercalary veins arranged in pairs. Each hind wing is reduced to a narrow, almost threadlike vestige which is either entirely without venation or with faint traces of a single longitudinal vein, and is also without a costal projection. The fore tarsus in the males is from one-half to two-thirds as long as the fore tibia. Each arm of the male genital forceps has four segments: the first segment is short and broad, with a small protuberance near the mediobasal angle; the second segment is narrower and conical; the third is long, slender, and bowed, with the medioapical angle produced; and the fourth segment is small, about two and one-half times as long as broad. The penis cover is emarginate on the meson, with each lateral angle conically produced.

In the nymphs, the labial palp, fig. 261, has the suture between segments 2 and 3 partly or completely obliterated and segment 2 is not expanded at the mesoapical angle; the gills are single and platelike, with well-marked, pinnately branched tracheae; there are but two caudal filaments that are well developed.

Heterocloeon curiosum (McDunnough)

Centroptilum curiosum McDunnough (1923:43).
Heterocloeon curiosum (McDunnough). McDunnough (1925*b*:175).
Baetis (*Acentrella*) *curiosum* (McDunnough). Ide (1937*b*:235).

MALE.—Length of body 4.5–5.0 mm., of fore wing 5.0–5.5 mm. Head and thorax very dark brown to black; legs white, with the fore femur shaded with gray and all coxae dark brown; wings hyaline, stigmatic crossveins of fore wing partly anastomosed; hind wing reduced to a narrow vestige, usually entirely without venation, but sometimes with traces of one longitudinal vein. Abdominal segments 2–6 white, or stained taintly with yellow or brown, apical abdominal segments chestnut brown, genitalia and caudal filaments white; apical forceps segments, fig. 269, each three times as long as wide.

Known from Maryland, New York, Ontario, and Quebec.

32. *BAETIS* Leach

Baetis Leach (1815:137).
Brachyphlebia Westwood (1840:25).
Acentrella Bengtsson (1912:110).

In *Baetis,* the stalk of each turbinate eye of the male is relatively high, so that

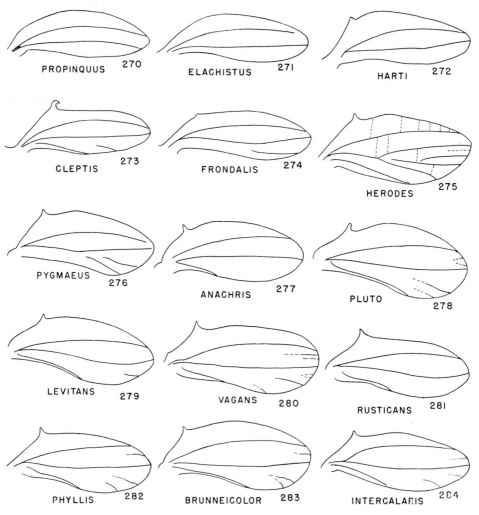

Fig. 270.—*Baetis propinquus,* hind wing of adult male.
Fig. 271.—*Baetis elachistus,* hind wing of adult male.
Fig. 272.—*Baetis harti,* hind wing of adult male.
Fig. 273.—*Baetis cleptis,* hind wing of adult male.
Fig. 274.—*Baetis frondalis,* hind wing of adult male.
Fig. 275.—*Baetis herodes,* hind wing of adult male.
Fig. 276.—*Baetis pygmaeus,* hind wing of adult male.

Fig. 277.—*Baetis anachris,* hind wing of adult male.
Fig. 278.—*Baetis pluto,* hind wing of adult male. (After McDunnough.)
Fig. 279.—*Baetis levitans,* hind wing of adult male.
Fig. 280.—*Baetis vagans,* hind wing of adult male.
Fig. 281.—*Baetis rusticans,* hind wing of adult male. (After McDunnough.)
Fig. 282.—*Baetis phyllis,* hind wing of adult male.
Fig. 283.—*Baetis brunneicolor,* hind wing of adult male.

Fig. 284.—*Baetis intercalaris,* hind wing of adult male.

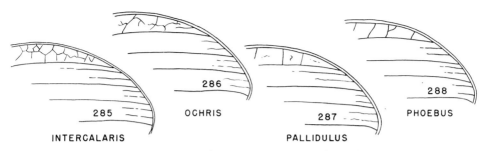

Fig. 285.—*Baetis intercalaris,* anteroapical area of fore wing of male.
Fig. 286.—*Baetis ochris,* anteroapical area of fore wing of male.
Fig. 287.—*Baetis pallidulus,* anteroapical area of fore wing of male.
Fig. 288.—*Baetis phoebus,* anteroapical area of fore wing of male.

the upper faceted portion of the eye is widely separated from the lower portion, figs. 255, 256. The width of the vertex separating the compound eyes in the females is usually three times as great as the width of one eye. Each fore tarsus in the males varies from slightly shorter to slightly longer than each fore tibia. Each fore wing has relatively few crossveins, and the marginal intercalaries are paired, figs. 31, 220. Each hind wing is relatively long and narrow, with or without an acute costal projection, sometimes with two but usually with three longitudinal veins, of which the third is always shortest, virtually or quite without crossveins, and with intercalary veins commonly present, figs. 270–284. The male genitalia, figs. 289–297, consist of a pair of four-segmented forceps and a penis cover. This cover varies from a well-developed, flaplike lobe to a small, extremely inconspicuous, membranous papilla. In some specimens, it is quite difficult to demonstrate the penis cover. The saclike and amorphous penis is membranous and internal, but can be seen extruded on an occasional specimen. The cerci are longer than the body.

The nymphs, fig. 298, are streamlined, and live in shallow running water. They are most commonly found under stones and among debris or emergent vegetation along the banks of brooks or creeks. The maxillary palp has two or three segments; the labial palp has three. The legs are relatively long and slender, with long, narrow claws, each of which bears minute denticles on the inner ventral surface. In all known species, the gills are single and platelike, fig. 228. The median caudal filament is either shorter than the cerci or vestigial.

In order to study the species in this genus, it is necessary to make dry mounts of the wings, especially the hind wings, and to clear, stain, and make slide mounts of the male genitalia. The latter operation must be done with extreme care, as these genitalia are quite fragile and, also, easily distorted while being manipulated.

The genus *Baetis* is one of the largest and most difficult genera of the mayflies. It includes about 50 Nearctic species, of which 17 are at present known to occur in Illinois; 25 species are treated in this report.

Ide (1937b:219) has published complete descriptions and figures of the nymphs of a large number of species of *Baetis.*

Reliable characteristics for the separation to species of adult females and female nymphs of this genus have not yet been found.

KEY TO SPECIES

ADULT MALES

1. A prominent, pointed, mesal projection at apex of second segment of genital forceps arm, fig. 289 . . . . . **1. spinosus**
   No apicomesal projection on second segment of genital forceps arm, figs. 290–297 2

2. Hind wing without a costal projection, figs. 270, 271 3
   Hind wing with a costal projection, figs. 272–284 5

3. Abdominal tergites 2–6 white, or very faintly stained with yellow . . . . **2. propinquus**
   Abdominal tergites 2–6 mostly or entirely dark brown 4

4. Fourth segment of genital forceps arm as long as wide, fig. 290 . . . . **3. elachistus**
   Fourth segment of genital forceps arm three times as long as wide, fig. 291 . . . .**4. frivolus**

5. Hind wing entirely without a third longitudinal vein, figs. 272, 274, 276 6
   Hind wing with a third longitudinal vein, figs. 273, 275, 277–284 9

6. Hind wing relatively narrow and lacking marginal intercalary veins posterior to

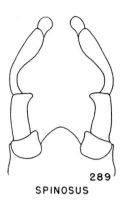

289
SPINOSUS

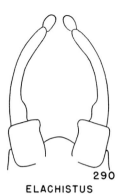

290
ELACHISTUS

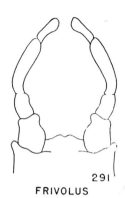

291
FRIVOLUS

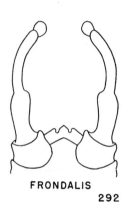

FRONDALIS
292

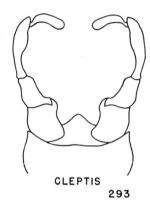

CLEPTIS
293

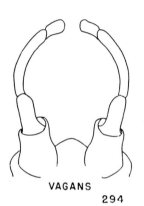

VAGANS
294

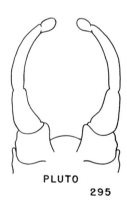

PLUTO
295

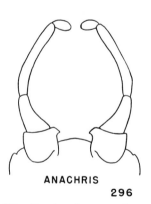

ANACHRIS
296

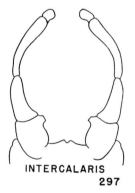

INTERCALARIS
297

Fig. 289.—*Baetis spinosus,* male genitalia.
Fig. 290.—*Baetis elachistus,* male genitalia.
Fig. 291.—*Baetis frivolus,* male genitalia.
Fig. 292.—*Baetis frondalis,* male genitalia.
Fig. 293.—*Baetis cleptis,* male genitalia.
Fig. 294.—*Baetis vagans,* male genitalia.
Fig. 295.—*Baetis pluto,* male genitalia. (After McDunnough.)
Fig. 296.—*Baetis anachris,* male genitalia.
Fig. 297.—*Baetis intercalaris,* male genitalia.

second longitudinal vein, fig. 272.....
......................................**5. harti**
Hind wing relatively broad and with one or two long, marginal intercalary veins posterior to second longitudinal vein, figs. 274, 276 7
7. Abdominal tergites 2–6 entirely white...
......................................**6. pygmaeus**
Abdominal tergites 2–6 almost entirely dark brown, or faintly stained with brown, and with a red crossband at posterior margin of each tergite... 8
8. Abdominal tergites 2–6 uniformly dark brown except for anterolateral angles of each tergite ....**7. frondalis**
Abdominal tergite 2 mostly covered by brown staining, tergites 3–6 faintly stained with brown at posterolateral angles and with a medially interrupted, red crossband at posterior margin of each tergite ....**8. baeticatus**
9. Second longitudinal vein of hind wing forked, fig. 275 ....**9. herodes**
Second longitudinal vein of hind wing not forked, figs. 273, 277–284.........10
10. Hind wing with an unusually long costal hook, fig. 273.............**10. cleptis**
Hind wing with a relatively short costal projection, figs. 277–284..........11
11. Male genitalia of the *vagans* type; that is, with the second forceps segment cylindrical, fig. 294 12
Male genitalia of the *intercalaris* type; that is, with the second forceps segment frustate, fig. 297.................13
12. Fore wing 6.5–7.5 mm. long; abdominal sternites 2–6 smoky gray, with a tan tinge...................**11. vagans**
Fore wing 4.5–5.0 mm. long; sternites 2–6 white or faintly yellow.**12. incertans**
13. Head and thorax bright yellow-tan; abdominal segments 2–6 lightly stained with tan or yellow, apical abdominal tergites light orange-brown or tan ...14
Head and thorax chestnut brown, dark brown, or black, with apical abdominal tergites same color or lighter red-brown; abdominal segments 2–6 white, pale yellow, or partly or completely brown 15
14. Marginal intercalaries in second interspace of fore wing shorter than those in first or third interspace, fig. 287; abdominal tergites 2–6 yellow........
......................................**13. pallidulus**
Marginal intercalaries in first three interspaces of fore wing equal in length, fig. 286; abdominal tergites 2–6 white
......................................**14. ochris**
15. First genital forceps segment without a mesoapical papilla, fig. 295 16
First genital forceps segment with a mesoapical papilla, figs. 296, 297.......17
16. Thorax very dark brown to black; abdominal tergites 2–6 dark red-brown, with narrow, yellow or tan areas at anterior and lateral margins of each...
......................................**15. pluto**
Thorax dark chestnut brown; abdominal tergites 2–6 light yellow, with trans-

verse, narrow, red-brown line at posterior margin of each...**16. levitans**
17. Abdominal tergites 2–6 partly or entirely brown or red-brown...............18
Abdominal tergites 2–6 entirely white or pale yellow......................23
18. Compound eyes small, fig. 256........
......................................**17. flavistriga**
Compound eyes larger, fig. 255.......19
19. Abdominal tergites 2–6 white, with a red-brown band at posterior margin of each
......................................**18. cingulatus**
Abdominal tergites 2–6 mostly or entirely dark brown....................20
20. Hind wing without marginal intercalary veins, figs. 277, 281...............21
Hind wing with marginal intercalary veins, figs. 282, 283...............22
21. Fore wing 5 mm. long, with numerous anastomosed crossveins in stigmatic area; abdominal sternites 2–6 white...
......................................**19. anachris**
Fore wing 4 mm. long, stigmatic area with only two to six nonanastomosed crossveins; abdominal sternites 2–6 a faint smoky gray.......**20. rusticans**
22. Longitudinal veins of fore wing brown..
......................................**21. phyllis**
Longitudinal veins of fore wing hyaline or only veins Sc and R₁ stained a faint yellow.............**22. brunneicolor**
23. Fore wing with marginal intercalaries of subcostal interspace two or three times as long as those in first and second R₁ interspaces, fig. 285...**23. intercalaris**
Fore wing either with marginal intercalaries absent in subcostal interspace, or, if present, almost as long as those in first and second R₁ interspaces, fig. 288...........................24
24. Compound eyes small, as in fig. 256; forewing 4 mm. long..........**24. nanus**
Compound eyes large, as in fig. 255; forewing 5 mm. long........**25. phoebus**

MATURE MALE NYMPHS

1. Caudal filaments either entirely dark or entirely light, or basal half to two-thirds of each filament uniformly dark and with apical half or third of filament light, but never with a dark brown crossband at middle and at apex in addition to this shading.............2
Caudal filaments relatively light in color, with dark brown crossband at or near the middle and at apex of each......5
2. Abdominal tergites 2–10 each uniformly brown on disc, with lateral margins white or light tan..................3
Abdominal tergites 5, 9, and 10 mostly or completely white, others brown, with lateral margins largely white........4
3. Each cercus having its basal two-thirds shaded with brown; median caudal filament five-sixths as long as cercus..
......................................**22. brunneicolor**
Each cercus uniformly shaded with brown from base to apex; median caudal filament two-thirds to three-fourths as long as cercus.................**21. phyllis**

4. Median caudal filament two-fifths as long
   as each cercus.........**20. rusticans**
   Median caudal filament three-fifths as
   long as each cercus......**11. vagans**
5. Gills with prominently darkened, pinnately branching tracheae.............6
   Gills either without visible tracheae or
   with at most a single, median trachea
   in each gill, these tracheae never conspicuous........................10
6. Each gill of seventh pair relatively slender,
   with apex pointed.................7
   Each gill of seventh pair oval, with apex
   rounded........................8
7. Abdominal tergites 2–10 each brown on
   disc, lateral margins of tergites 2–7
   white; a median, light tan or cream-
   colored, longitudinal stripe extending
   the length of the abdominal dorsum...
   ....**6. pygmaeus**
   Abdominal tergites 2–4 and 6–7 mostly
   dark brown, tergites 5 and 8–10 mostly
   white; no longitudinal, pale, median
   stripe present on abdominal dorsum...
   ....**9. herodes**
8. Abdominal tergites 2–10 each uniformly
   dark brown on disc, lateral margins
   white.................**7. frondalis**
   Abdominal tergites 2–10 each variegated with brown and white on disc;
   tergites 5 and 8–10 usually largely or
   entirely white.....................9
9. Each caudal filament with brown shading
   at base, in addition to dark brown crossbands at middle and at apex........
   ....**23. intercalaris**
   Each caudal filament white at base,
   darkened only at middle and at apex..
   ....**16. levitans**
10. Abdominal tergites 2–9 each uniformly
    brown on disc, with lateral margins
    white, but tergite 5 usually somewhat
    lighter in color than others; each caudal
    filament shaded with brown in basal
    half, in addition to dark brown cross-
    bands at middle and at apex.........
    ....**15. pluto**
    Abdominal tergites 2–9 more or less varie-
    gated with brown and white, tergites
    5 and 9 mostly or entirely white; each
    caudal filament white or pale cream
    colored in basal half..............11
11. Median caudal filament almost as long
    as cerci, at least five-sixths as long....
    ....**17. flavistriga**
    Median caudal filament not more than
    three-fourths as long as cerci........12
12. Tenth abdominal tergite entirely white..
    ....**18. cingulatus**
    Tenth abdominal tergite mostly brown,
    white only along anterior margin.....
    ....**25. phoebus**

1. Baetis spinosus McDunnough

Baetis spinosus McDunnough (1925b:174).

MALE.—Length of body 4.0–4.5 mm., of
fore wing 5.0–5.5 mm. Head very dark
brown to black, yellow-brown at lateral

angles of frontal shelf; each antenna light
gray-brown, shading to yellow at apex of
flagellum; eyes in life dark red-brown. Thoracic notum black, yellow-brown along anterolateral margins of mesoscutum, pleura
and sternum dark brown, light red-brown or
yellow-brown at wing bases and bordering
sutures; all coxae brown, each fore leg
faintly yellow, middle and hind leg white;
wings hyaline, veins Sc and R_1 of fore wing
faintly stained with yellow-brown, stigmatic
crossveins 7–9 in number, uniformly slanting, not anastomosed; hind wing with only
two longitudinal veins, costal projection
either absent or vestigial. Abdominal segments 2–6 white or stained with tan, black
spiracular markings present; apical tergites
chocolate brown, sternites white. Genitalia,
fig. 289, distinctive, second segment of each
forceps arm with a prominent, apicomesal
tubercle, fourth segment as long as wide;
genitalia and caudal filaments white or,
occasionally, faintly tinged with yellow.

Known from Illinois, Indiana, Manitoba,
New York, Ontario, and Quebec.

Illinois Records.—AROMA PARK: Kan-
kakee River, July 8, 1948, Ross & Burks,
1 ♂. MUNCIE, Stony Creek: May 24, 1914,
1 ♂; July 3, 1929, Frison & Park, 1 ♂.
OAKWOOD, Salt Fork River: May 29, 1948,
B. D. Burks, 1 ♂; June 5, 1948, Burks &
Sanderson, 3 ♂. PRINCETON: Big Bureau
Creek, May 23, 1941, Ross & Burks, 1 ♂.
PROPHETSTOWN: July 7, 1925, T. H. Frison,
1 ♂.

2. Baetis propinquus (Walsh)

Cloe vicina Walsh (1862:380), not Hagen.
  Misidentification.
Cloe propinqua Walsh (1863:207). New name.
Baetis propinquus (Walsh). Eaton (1871:121).
Acentrella propinqua (Walsh). Traver
  (1937:83).
Baetis dardanus McDunnough (1923:41).

I have studied the lectotype of propinquus and a paratype of dardanus and I can
find no specific differences between them;
McDunnough (1925b:172) long ago concluded that the two species were probably
synonymous.

MALE.—Length of body 4–5 mm., of fore
wing 4.5–5.5 mm. Head dark brown to
black; antennae yellow-brown to tan; eyes
dark brown. Thorax dark brown to black,
yellow-brown at anterolateral margins of
mesoscutum, at apex of scutellum, on pleural

sutures and at wing bases, and usually on mesosternum; wings hyaline, veins Sc and R_1 of fore wing faintly tinged with tan near wing base; hind wing, fig. 270, narrow, costal projection absent, only two longitudinal veins present; all coxae yellow-brown, legs otherwise usually white, femora sometimes faintly shaded with gray-brown. Abdominal segments 2–6 white or very faintly stained with tan, black spiracular marks present; apical tergites yellow-brown, sternites white; genitalia and caudal filaments white or faintly stained with yellow.

FEMALE.—Length of body 4.5–5.5 mm., of fore wing 5.5–6.5 mm. Head, thorax, and abdominal tergites chestnut brown, legs light yellow, wings hyaline, longitudinal veins yellow; hind wing as in male; abdominal sternites white, often with a faint pinkish tinge; caudal filaments light yellow.

Known from Illinois, Manitoba, and Ontario.

Illinois Records.—AROMA PARK: July 8, 1948, Ross & Burks, 1 ♂; Aug. 6, 1947, Burks & Sanderson, 2 ♂. EAST DUBUQUE: at light, July 21, 1927, Frison & Glasgow, 18 ♂, 6 ♀. ELIZABETHTOWN: at light, July 14, 1948, Mills & Ross, 1 ♂. ERIE: Rock River, June 26, 1947, B. D. Burks, 1 ♂. MILAN: Rock River, June 4, 1940, Mohr & Burks, 7 ♂. MOMENCE: Aug. 5, 1938, Burks & Boesel, 2 ♂; Aug. 16, 1938, Ross & Burks, 1 ♂. MUNCIE: June 8, 1927, Frison & Glasgow, 1 ♀. OAKWOOD: May 29, 1948, B. D. Burks, 1 ♂. ROCK ISLAND: 7 ♂, 16 ♀ (Walsh 1862:380). ROCKTON: Rock River, Aug. 4, 1948, Burks & Stannard, 1 ♂, 1 ♀. SHAWNEETOWN: 1 ♂ (Traver 1935a:699). URBANA: July 11, 1898, C. A. Hart, 1 ♀.

3. *Baetis elachistus* new species

This species agrees with *amplus* (Traver), described from North Carolina, in having the wing veins yellow-brown, the hind wing without a costal angulation, and abdominal tergites 2–6 brown. The two differ in that the fourth genital forceps segment in *amplus* is three times as long as wide, while this structure in *elachistus* is only as long as wide; the genitalia of *elachistus* likewise are distinct in that the basal forceps segment is more prominently produced at the mesoapical angle than it is in *amplus*.

MALE.—Length of body and of fore wing each 5.0–5.5 mm. Head dark brown, with a white spot at each lateral margin of frontal shelf; each antennal scape and pedicel white, shaded with tan at apexes, flagellum tan. Thorax very dark brown, almost black; legs white, coxae shaded with brown, fore femur stained with tan, which becomes slightly darker toward apex, middle and hind femora shaded with tan at apexes, all tibiae brown shaded at bases; wings hyaline, veins yellow-brown, fore wing with 10 to 12 highly anastomosed, stigmatic crossveins, no marginal intercalaries in subcostal interspace, usually none also in first R_1 interspace, although one short intercalary sometimes present here; hind wing, fig. 271, narrow, only two longitudinal veins present, costal angulation absent. First abdominal segment dark brown, segments 2–6 uniformly dark yellow-brown, tergites 7–10 brown, sternite 7 yellow-brown, 8 and 9 brown, lateral margins of 9 dark brown; first genital forceps segment brown, mesoapical angle slightly produced, fig. 290, second segment tan, third segment tan, slightly bowed, one and one-half times as long as second segment, fourth segment tan, globose, as long as wide; each caudal filament tan in basal fifth, gradually merging into white distad.

Holotype, male.—Duncans Mills, Illinois, Spoon River, October 20, 1941, B. D. Burks. Specimen in alcohol.

Paratypes.—Same data as for holotype, 5 ♂. Specimens in alcohol, wings and genitalia on microscope slides.

4. *Baetis frivolus* McDunnough

Baetis frivolus McDunnough (1925b:174).

MALE.—Length of body 4–5 mm., of fore wing 5–6 mm. Head very dark brown, lighter at lateral angles of frontal shelf; antennae brown; eyes brown, each stalk shorter than in most species of genus. Thorax very dark brown dorsally, slightly lighter laterally and ventrally; anterolateral margins of mesoscutum, area along outer parapsides, and margins of mesosternum yellow-brown; wings hyaline, anterior longitudinal veins of fore wing shaded with tan, base of fore wing brown, stigmatic crossveins anastomosed, hind wing long, narrow, costal projection and third longitudinal vein lacking; legs with all coxae brown, fore leg faintly shaded with gray-brown, middle and hind legs white or faintly yellow. Abdominal tergites uniformly brown, black tracheal markings at spiracles, sternites 1–6 deep yellow; apical sternites tan; genitalia and caudal filaments white; genitalia, fig. 271, with fourth forceps segment three times as long as wide.

Known from Illinois, Ontario, Quebec.

Illinois Record.—GOLCONDA: April 30, 1940, Mohr & Burks, 1 ♂.

5. *Baetis harti* McDunnough

Baetis harti McDunnough (1924a:7).

MALE.—Length of body 2.5–3.5 mm., of fore wing 3–4 mm. Head very dark brown; each antenna brown, shading to yellow at apex of flagellum; eyes brown. Thorax dark brown, yellow-brown near apex of mesoscutellum; all coxae brown, legs otherwise light yellow to white; wings hyaline, stained with brown at base of vein Sc of fore wing; stigmatic crossveins of fore wing only three or four in number, not anastomosed; hind wing, fig. 272, relatively broad at base, costal projection well developed, third longitudinal vein absent. Abdominal segments 2–6 white to yellow, with black spiracular marks; apical tergites dark brown, sternites tan; genitalia and caudal filaments white.

Known from Illinois.

Illinois Records.—KANKAKEE: Aug. 16, 1938, Ross & Burks, 1 ♂; May 17, 1938, H. H. Ross, 2 ♂; July 21, 1935, Ross & Mohr, 1 ♂. URBANA: West Branch Salt Fork River, July 11, 1898, C. A. Hart, 8 ♂.

6. *Baetis pygmaeus* (Hagen)

Cloe pygmaea Hagen (1861:54).
Baetis pygmaeus (Hagen). McDunnough (1925a:214; 1925b:172).

This species was described from a single very small female specimen collected in eastern Canada along the St. Lawrence River. When Eaton examined this type (1885:170), it was fragmentary. When McDunnough saw it (1925b:172), it was badly broken, consisting only of one fore wing and part of the mesothorax with the legs attached. At present, the type is but a bare pin, with the label.

McDunnough (1925b:172), however, secured male and female specimens of a common species of *Baetis* from Ontario and Quebec, along the St. Lawrence River, and

was able to match the female of this species with the few fragments of Hagen's type which were still preserved at that time. McDunnough based his conception, and redescription, of this species on these specimens and the associated males. I have seen some of this topotypic material of *pygmaeus,* as named by McDunnough, and I follow his identification of it.

MALE.—Length of body and of fore wing 3.0–3.5 mm. Head very dark brown to black; each antennal scape and pedicel dark brown, flagellum lighter smoky brown. Thorax very dark brown to black; coxae yellow-brown, fore femur faintly stained with smoky tan, all legs otherwise white; wings hyaline, veins C, Sc, and R_1 stained with tan near bases, stigmatic crossveins six to eight in number, uniformly slanting, not anastomosed and usually none reaching vein Sc; hind wing, fig. 276, relatively narrow, with well-developed costal angulation, one or two long, marginal intercalaries present posterior to second longitudinal vein, and third vein absent. Abdominal segments 2–6 white or faintly yellow, black stigmatic markings present; apical tergites chestnut brown, sternites pale tan to white; genitalia and caudal filaments white.

Known from the midwestern and northeastern states and southern Canada.

Illinois Records.—HEROD: July 16, 1947, sweeping, L. J. Stannard, 1 ♂. JONESBORO: branch of Clear Creek, May 15, 1946, Mohr & Burks, 1 ♂. QUINCY: at light, May 18, 1940, Mohr & Burks, 1 ♂.

7. *Baetis frondalis* McDunnough

Baetis frondalis McDunnough (1925b:173).

MALE.—Length of body 4–5 mm., of fore wing 5–6 mm. Head very dark brown, almost black; each antennal scape and pedicel yellow-brown, flagellum tan; eyes in life red-brown. Thorax almost completely very dark brown to black, mesonotum marked with small, vaguely defined, yellow-brown streaks anteriorly on prescutum and laterally on anterior notal wing processes; all coxae brown, front leg faintly shaded with tan, middle and hind legs white to faint yellow-brown, fore tarsus slightly shorter than fore tibia. Wings hyaline, veins C, Sc, and R_1 faintly stained with brown near bases, otherwise all veins hyaline; stigmatic crossveins of each fore wing six to eight in number, obliquely slanting, not anastomosed, and most not reaching vein Sc; hind wing, fig. 274, long and narrow, costal projection minute but always clearly present, third longitudinal vein absent, usually a single long intercalary vein present posterior to second longitudinal vein. Abdominal tergites 2–6 dark brown, with anterolateral angles of each tergite pale yellow, a faint black circle at each spiracle; sternites 2–6 white or faintly yellow, sometimes with a brown, transverse streak at posterior margin of each sternite; apical tergites chocolate brown; sternites opaque white, shaded with brown on median area of basal half of each and laterally on apical sternite; genitalia white, with second segment of forceps semiquadrate, fig. 292; caudal filaments white.

Known from Illinois, Ontario, Quebec.

Illinois Records.—DES PLAINES: Fox River, May 26, 1936, H. H. Ross, 1 ♂. OAKWOOD, Salt Fork River: May 29, 1948, Burks & Evers, 2 ♂; June 5, 1948, Burks & Sanderson, 5 ♂. WEST CHICAGO: July 9, 1948, Ross & Burks, 2 ♂.

8. *Baetis baeticatus* new species

This species resembles *frondalis* in the structure of the hind wing and the male genitalia; the two differ in that the body of *baeticatus* is strikingly slender and long, being longer than the fore wing, while the body of *frondalis* is shorter than the fore wing. They also differ in color, *baeticatus* having a medially interrupted, transverse, red stripe at the posterior margin of each of abdominal tergites 2–6, and these tergites are faintly stained with brown; in *frondalis,* abdominal tergites 2–6 are almost completely dark brown, with the red stripes wanting.

MALE.—Length of body 6.0 mm., of fore wing 5.5 mm. Head dark brown, white at lateral angles of frontal shelf; each antennal scape white, with faint brown shading at base and at apex, pedicel tan, flagellum white at base, shading to tan at apex; ocelli white; each compound eye with stalk relatively low, lateral margin of upper faceted portion almost touching upper margin of lower portion, upper portion golden brown, lower black. Thorax dark brown, mesonotum yellow at anterior end of each outer parapsidal furrow, on prescutum, and on scutellum; pleural sutures white, mesosternum light yellow-brown in center, dark brown at margins; all coxae partly shaded with dark brown, legs otherwise white, except that fore tibia is slightly darkened at apex; fore tibia one and one-half times as long as fore femur, fore tarsus four-fifths as long as fore tibia, second fore tarsal segment one and one-half times as long as third, fourth and fifth segments equal in length and each one-half as long as third segment; wings hyaline, base of each fore wing and basal halves of veins Sc and R_1 stained with brown, stigmatic crossveins six in number, all uniformly slanted, not anastomosed and most not quite reaching vein Sc; no marginal intercalaries in subcostal interspace, a single, short intercalary present in first R_1 interspace, two well-developed ones in each of the two following interspaces; hind wing long, narrow, with a minute costal projection, third longitudinal vein absent, a single marginal intercalary vein present posterior to second longitudinal vein. Abdominal segments 2–6 with white ground, tergite 2 suffused with brown stain over all but anterolateral triangles, tergites 3–6 faintly brown stained on posterior third of each, a medially interrupted, red band at posterior margin of each of tergites 2–7, and a pair of large, black tracheal marks in a cluster covering most of posterolateral area of each of tergites 2–6; a double, longitudinal, black spiracular line extending length of abdomen on either side, and a prominent, black spot at each spiracle on segments 1–6; apical tergites dark yellow-brown, sternite 7 white, 8 brown stained, 9 with lateral margins dark brown; genitalia white, first forceps segment quadrate, second cylindrical, third slender and slightly bowed, fourth as wide as long; a prominent spine present in a median depression between bases of forceps; caudal filaments white.

FEMALE.—Size as in male. Head and thorax dark brown, similar to those of male, legs and wings as in male; abdominal tergites uniformly dark yellow-brown, red crossbands of male absent, but black spiracular and tracheal markings present; sternite 1 faint brown, sternites 2–8 white, with a pair of submedian, brown dots at anterior margin of each; sternite 9 brown at lateral margins; caudal filaments tan at bases.

Holotype, male.—Wichert, Illinois, June 11, 1947, L. J. Stannard. Specimen in alcohol.

Allotype, female.—Same locality as for holotype, June 9, 1948, Burks & Stannard. Specimen in alcohol.

Paratypes.—Oakwood, Illinois, June 4, 1948, B. D. Burks, 1 ♂ adult, 3 ♂ subimagoes. Specimens in alcohol.

9. *Baetis herodes* new species

This species agrees with *parvus* Dodds, described from Colorado, in having the second vein of the hind wing forked, and the first segment of the male genital forceps nontuberculate. The two differ in that the abdominal tergites 2–6 are uniformly white in *parvus,* but are shaded with red-brown at the posterior margins in *herodes;* in *parvus,* the hind wing has a single marginal intercalary vein between the branches of the second vein, while, in *herodes,* the hind wing has one long and two short marginal intercalaries between the branches of the second vein. McDunnough (1925*a*:214-5; 1925*b*: 172) recorded, as the Colorado species *parvus* Dodds, a Quebec species having the venation of the hind wing similar to that of *herodes.* He had female specimens only. As many of the southeastern Canadian species of mayflies also occur in southern Illinois, these Quebec female specimens actually might be the females of *herodes.* Unfortunately, female specimens of *herodes* have not yet been secured here in Illinois.

MALE.—Length of body and of fore wing 5.0–5.5 mm. Head very dark brown, almost black; scape and pedicel of each antenna dark brown, flagellum smoky yellow; each eye in life with upper facets dark brown, lower ones black. Thorax black, becoming dark brown after death; wings hyaline, stigmatic crossveins partly anastomosed, 9 to 10 in number; hind wing, fig. 275, with prominent costal projection, second vein branched, two strong and one vestigial marginal intercalaries present between branches, third vein reaching posterior margin of wing at a point three-fifths the distance from base to apex of wing; numerous vestigial crossveins present; fore leg smoky, fore tibia shaded with brown at apex, middle and hind legs faintly yellow, almost white, apexes of femora and bases of tibiae stained with brown. First abdominal segment brown; tergites 2–6 white, with a transverse, brown-shaded area at posterior margin of each tergite, small, black blotch at each spiracle; tergites 7–10

dark red-brown; sternites 2–6 white, 7–10 lightly shaded with red-brown, and 10 also shaded with dark gray-brown at lateral margins; caudal filaments white. Genitalia: first forceps segment nontuberculate, smoky brown, following segments faintly yellow; second and third segments equal in length, second segment frustate, third segment slender, bowed, and slightly enlarged toward apex; fourth segment globose, one and one-half times as long as wide.

FEMALE.—Unknown.

NYMPH, MALE.—Length of body 6–7 mm. Head and thorax mottled brown and white, legs white, with gray-brown shading on coxae, at base, middle, and apex of each femur, and in middle of each tibia and tarsus. Abdominal tergites 1–4 and 6–7 mostly brown, tergites 5 and 8–10 mostly white; abdominal sternites white, with longitudinal, brown streak near each lateral margin of sternites; each gill having a black, median trachea with a few short, lateral branches; each gill of seventh pair slender and pointed at apex; median caudal filament almost as long as cerci, caudal filaments light tan, with a brown crossband near tip of each.

A female nymph associated with these male nymphs, and apparently of the same species, has abdominal tergites 1–9 uniformly brown, tergite 10 white, and each caudal filament uniformly tan, without a brown crossband.

Holotype, male.—Herod, Illinois, Gibbons Creek, April 9, 1947, B. D. Burks. Specimen dry, on a pin.

Paratypes.—ILLINOIS.—Same data as for holotype, 1 ♂ ; April 10, 1947, 1 ♂ subimago. Nymphs and exuviae also were collected at Herod, Illinois, April 4–10, 1947.

INDIANA. — SPENCER: McCormick's Creek, April 27, 1948, W. E. Ricker, 3 ♂ . Of these paratypes, one adult male is dry, on a pin; three adult males and subimago male are in alcohol.

10. *Baetis cleptis* new species

This species is similar to *erebus* Traver, described from Arizona, in that the hind wing is long and narrow, and has a prominent, hooklike costal projection, the median abdominal segments are dark brown, and the first male genital forceps segment has a fairly large, rounded anteromedian projection. *B. cleptis* differs from *erebus* in that *cleptis* is smaller, the hind wing is relatively wider, the second forceps segment is semi-tuberculate, and there is a single acute projection between the bases of the genital forceps in *cleptis*, in contrast to two small projections in this position in *erebus*.

MALE.—Length of body 3.5 mm., of fore wing 4 mm. Head dark brown, with white spot at either lateral margin of frontal shelf; each antenna brown, with flagellum becoming white toward the apex. Thorax dark brown, lighter along dorsal sutures and at apex of mesoscutellum; wings hyaline, a brown spot at base of each fore wing, veins light brown; stigmatic crossveins in each of fore wings six or seven in number, slanting, not anastomosed, no marginal intercalaries in Sc interspace; hind wing, fig. 273, narrow, only one-third as wide as long, costal projection hooked, veins 1 and 2 slightly converging at apexes, a single marginal intercalary present between veins 2 and 3, third vein slightly more than half as long as wing; all coxae brown, fore leg entirely tan, middle and hind legs white. Abdominal tergites 1–6 yellow-brown, tergites 7–10 slightly darker brown; sternites yellow-brown, with lateral margins of sternite 9 dark brown; black, longitudinal tracheal lines at spiracles; genital forceps, fig. 293, yellow-brown at bases, graduating to very light tan at apexes; first segment with rounded anteromesal projection, second segment cylindrical, with a suggestion of a mesal tubercle at apex, third segment constricted at base, enlarged toward apex, twice as long as second segment, fourth segment long, slender, four times as long as wide; caudal filaments white, slightly shaded with tan in basal area.

Holotype, male.—Detroit, Illinois, September 15, 1939, Ross & Mohr. Specimen in alcohol.

Paratype.—Same data as for holotype, 1 ♂. Specimen in alcohol, wings and genitalia on microscope slides.

11. *Baetis vagans* McDunnough

Baetis vagans McDunnough (1925a: 219).

MALE.—Length of body 5.5–6.5 mm., of fore wing 6.0–7.5 mm. Head dark brown, yellow-brown at lateral angles of frontal shelf; each antennal scape usually entirely yellow, sometimes partly shaded with brown, pedicel brown, flagellum gray-tan, shading

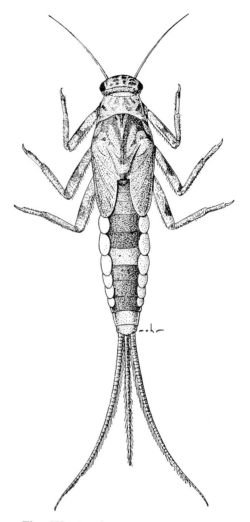

Fig. 298.—*Baetis vagans*, mature nymph.

to yellow at apex; eyes red-brown. Thorax dark brown, marked with yellow-brown at anterior and posterior ends of each outer parapsidal furrow, at wing bases, along pleural sutures, on entire prosternum, and on margins of mesosternum; wings hyaline, anterior longitudinal veins stained pale yellow, stigmatic crossveins of fore wing numerous, anastomosed to form a dense network; hind wing, fig. 280, with prominent costal projection, third longitudinal vein well developed, and marginal intercalary veins present between both veins 1 and 2, and 2 and 3; all coxae pale gray-tan, fore leg light gray-brown, fore tibia with a darker gray spot at apex, middle and hind legs pale yellow to white. Abdominal tergites 2–6 brown, with anterior fourth of each tergite yellow, sternites faint gray-brown or smoky tan; apical tergites bright chestnut or chocolate brown, sternites white, sometimes faintly stained with tan; genitalia, fig. 294, white; caudal filaments white.

Known from Illinois, New York, Ontario, Pennsylvania, and Quebec.

Illinois Records. — ELGIN: Botanical Garden, April 25, 1941, Ross & Burks, 1 ♂; April 19, 1939, Burks & Riegel, 8 N; May 9, 1939, Ross & Burks, 5 ♂, 5 ♀, 40 N; May 23, 1939, Burks & Riegel, 16 N.

12. *Baetis incertans* McDunnough

Baetis incertans McDunnough (1925a:220).

Ide (1937b:223) considers this species to be a synonym of *vagans*. It is quite closely related to *vagans*, differing only in that it is smaller, the length of the body being 4.5 mm. and the length of the fore wing 5 mm.; abdominal tergites 2–6 are uniformly brown, and abdominal sternites 2–6 are white or faintly tinged with yellow.

Known from Quebec.

13. *Baetis pallidulus* McDunnough

Baetis pallidula McDunnough (1924a:8).

Live specimens of *pallidulus* can be recognized at once by the brilliant lemon-yellow and tan head and thorax; the bright yellow-orange of the compound eyes also is distinctive. These colors, unfortunately, are quickly lost following death. Regardless of the method of preservation of specimens, all trace of the bright yellow coloration is soon lost. In alcohol, the eyes and thorax fade to a pale tan or deep cream color; dry specimens become light or dark tan.

MALE.—Length of body 4–5 mm., of fore wing 5–6 mm. Head tan and yellow; each antenna with scape and pedicel orange-tan, flagellum yellow; compound eyes orange-yellow. Thorax yellow and tan; legs yellow, with faint, brown shading at apex of each femur; wings hyaline, veins Sc and R_1 yellow at bases, seven to nine stigmatic crossveins, partly anastomosed; marginal intercalaries in second interspace short, fig. 287; hind wing broad, costal angulation well developed, third longitudinal vein only one-fourth as long as wing, usually one marginal intercalary vein posterior to second longitudinal vein. Abdominal segments 2–6 light tan or yellow, black spiracular marks usually present; apical tergites orange-brown or tan, sternites yellow, sometimes stained with tan or orange-brown; genitalia yellow, of the *intercalaris* type, a well-developed apicomesal tubercle on the first forceps segment, the second segment frustate; caudal filaments white.

Known from Illinois, Indiana, and Ontario.

Illinois Records. — ALDRIDGE: May 14, 1940, Mohr & Burks, 1 ♂. APPLE RIVER CANYON STATE PARK: June 6, 1940, Mohr & Burks, 1 ♂; July 12, 1938, Burks & Boesel, 1 ♂. MUNCIE, Stony Creek: May 24, 1914, 3 ♂; June 3, 1917, 1 ♀; June 8, 1927, Frison & Glasgow, 1 ♂. OAKWOOD: May 24, 1926, T. H. Frison, 1 ♂. ROCKFORD: June 13, 1931, Frison & Mohr, 1 ♂. SERENA: Indian Creek, May 12–16, 1938, Ross & Burks, 2 ♂.

14. *Baetis ochris* new species

This species is similar to *pallidulus* in being generally light in color, in having male genitalia of the *intercalaris* type, and in having three well-defined, longitudinal veins and a well-developed costal projection in the hind wing. The two differ in that *ochris* has the marginal intercalaries in the first three interspaces of the fore wing equal in length, while *pallidulus* has them unequal, with those in the second interspace shorter than those in the first and third interspaces; the apical segments of *ochris* are brown, of *pallidulus* yellow or tan; abdominal tergites 2–6 are white in *ochris*, yellow in *pallidulus*.

MALE.—Length of body and of fore wing each 4 mm. Head tan, shaded with light yellow-brown near compound eyes; antennae yellow-brown; eyes yellow-brown. Thorax yellow-brown; wings hyaline, stigmatic crossveins 7 to 10 in number, partly anastomosed, a strong crossvein usually present in second interspace below stigmatic area near margin of wing; marginal intercalaries in first three interspaces nearly equal in length, fig. 286; hind wing broad, with three longitudinal veins and usually no marginal intercalaries, sometimes a very short one present between first and second veins; legs white.

Abdominal segments 2–6 white, extensive black tracheal outlines at each spiracle; apical segments brown; genitalia and caudal filaments white; genitalia of the *intercalaris* type, with a papillate projection at mesoapical angle of first forceps segment, second segment conical, and fourth segment slightly longer than wide.

Holotype, male.—Richmond, Illinois, at light, June 24, 1938, B. D. Burks. Specimen in alcohol.

Paratypes.—Same data as for holotype, 2 ♂. Specimens in alcohol.

15. *Baetis pluto* McDunnough

Baetis pluto McDunnough (1925a: 218).

MALE.—Length of body and of fore wing 4.5 mm. Head and thorax dark brown; legs yellow, with femora somewhat darker yellow; stigmatic crossveins of each fore wing anastomosed; hind wing, fig. 278, with third longitudinal vein relatively long. Abdominal tergites 2–6 dark red-brown, with only anterior and lateral margins lighter in color; genitalia, fig. 295, tinged with brown and of the *intercalaris* type, with mesoapical papilla of first forceps segment wanting; caudal filaments white.

Known from Ontario and Quebec.

16. *Baetis levitans* McDunnough

Baetis laevitans McDunnough (1925a: 215).
Baetis levitans McDunnough (1925a: 216).

This species differs from *pluto* only in that the thorax is chestnut brown, the hind wing, fig. 279, usually has fewer marginal intercalary veins and a shorter third longitudinal vein, the abdominal tergites 2–6 are yellow, with a brown, transverse streak at the posterior margin of each, and the genitalia are white.

Known from New York, Pennsylvania, and Quebec.

17. *Baetis flavistriga* McDunnough

Baetis flavistriga McDunnough (1921: 120).

MALE.—Length of body 4.5 mm., of fore wing 5.0 mm. Head gray-brown, with face black and antennae dark smoky gray. Thorax dark gray-brown, with a greenish cast and yellow markings; legs white, femora shaded with yellow; each hind wing with third longitudinal vein well developed, one-half as long as wing, and a single marginal intercalary vein usually present between veins 2 and 3. Abdominal tergites 2–6 light yellow-brown, apical tergites dark yellow-brown, faint black spiracular markings present; genitalia and caudal filaments white.

Known from New York, Ontario, and Quebec.

18. *Baetis cingulatus* McDunnough

Baetis cingulatus McDunnough (1925a: 216).

MALE.—Length of body 5.0 mm., of fore wing 5.5 mm. Head very dark brown. Thorax dark gray-brown, with small, variable tan or yellow markings; legs light yellow, with all femora shaded with brown; each hind wing with third longitudinal vein well developed, one-half as long as wing, marginal intercalaries absent or extremely faint. Abdominal tergites 2–6 yellow, with a narrow, bright red-brown, transverse stripe at posterior margin of each, apical tergites darker red-brown; genitalia and caudal filaments white.

Known from New York, Pennsylvania, and Quebec.

19. *Baetis anachris* new species

This species is similar to *rusticans* in that the hind wing is relatively broad and has three longitudinal veins, abdominal tergites 2–6 are brown, and the genitalia are of the *intercalaris* type, having the mesoapical papilla of the first forceps segment present. The two differ in that the stigmatic crossveins of the fore wing in *anachris* are numerous and anastomosed, while there are only four to six nonanastomosed stigmatic crossveins in *rusticans;* abdominal sternites 2–6 in *anachris* are white, while they are a light smoky gray in *rusticans.*

MALE.—Length of body 4.5 mm., of fore wing 5.0 mm. Head chestnut brown, each antenna chestnut brown, blending into yellow at apex of flagellum; compound eyes light red-brown, extremely large. Thorax chestnut brown, with yellow areas along anterior lateral margins of mesoscutum, on prescuta, at wing bases, at apex of scutellum, and on anteromedian and dorsolateral areas of metanotum; all legs yellow, except coxae shaded with brown and each fore femur stained with tan; wings hyaline, veins colorless, stigmatic area of each fore wing with numerous partly incomplete and anastomosed crossveins; intercalary veins in first interspace long, obsolescent, those in following interspaces well developed but shorter; hind wing, fig. 277, wide, no marginal intercalaries present, third vein short. Abdominal tergites brown, with white area at anterolateral angle and anteromedian area of each; a darker brown, median spot at posterior margin of each tergite; basal sternites faint yellow, almost white; apical tergites opaque pinkish tan, sternites yellow, apical sternite with longitudinal, brown streak near each lateral margin; genitalia, fig. 296, and bases of caudal filaments tan, filaments white distad of bases.

NYMPH, MALE.—Length of body 6.0–6.5 mm. Head and thorax light brown, with median, longitudinal, dorsal stripe and apex of scutellum white or a faint tan; thoracic venter white; legs white, shaded with brown on coxae, basal three-fourths of femora, apexes of tibiae, and apical one-fourth of each tarsus. Abdominal tergites light brown, with traces of a median, longitudinal, white streak; tergites 5, 9, and 10 mostly white, lateral margins of each tergite white; gills white, with a single, well-marked, brown trachea on each gill of pairs borne by segments 2–6, anterior and posterior margins of these gills also brown; sternite 1 white, all following sternites tan, shading to brown at lateral margins, apical three sternites almost entirely brown; caudal filaments tan, median one three-fourths as long as each cercus.

Holotype, male. — Havana, Illinois, White Oak Creek near Matanzas Lake, June 13, 1946, Mohr & Burks. Specimen dry, on pin; hind wing and genitalia on microscope slides. Specimens of nymphs associated with this adult collected on same date.

20. *Baetis rusticans* McDunnough

Baetis rusticans McDunnough (1925a:217).

MALE.—Length of body 3.5 mm., of fore wing 4.0 mm. Head and thorax dark graybrown, with yellow-brown or red-brown markings on thoracic notum; legs very light yellow-brown, femora shaded with darker

brown; stigmatic crossveins of each fore wing only four to six in number and not anastomosed; hind wing, fig. 281, with third longitudinal vein short and marginal intercalary veins absent. Abdominal tergites 2–6 gray-brown, apical tergites dark walnut brown; genitalia and caudal filaments white.

Known from New York and Quebec.

21. *Baetis phyllis* new species

This species is similar to *brunneicolor* in having abdominal tergites 1–10 uniformly brown in color, the hind wing with three longitudinal veins and marginal intercalaries, and the first segment of the male genital forceps tuberculate, the second segment frustate. The two differ in that the wing veins are brown in *phyllis* but hyaline in *brunneicolor;* the third vein of the hind wing is shorter in *phyllis* than in *brunneicolor,* and the tubercle of the first forceps segment is somewhat better developed in *brunneicolor* than it is in *phyllis.*

MALE.—Length of body and of fore wing each 6 mm. Head dark brown; each antennal scape tan, pedicel brown, flagellum smoky tan. Thorax brown, with yellow along anterolateral margins of mesoscutum, along outer parapsides, and on meson at posterior margin of mesoscutum; legs white, except coxae shaded with brown, each fore femur tan, fore tibia with brown spot at apex; wings hyaline, veins of fore wing tan, crossveins hyaline, stigmatic crossveins numerous, anastomosed; hind wing, fig. 282, broad, two and one-third times as long as broad, costal angulation prominent, veins brown at bases, tan distad, first and second veins converging slightly at wing margin, a single marginal intercalary between veins 1 and 2, and two marginal intercalaries between veins 2 and 3, a single, irregular crossvein extending from vein 2 to 3, vein 3 reaching wing margin slightly basad of the middle. All abdominal tergites brown, tenth tergite darker brown; narrow, transverse, yellow stripe at anterior and posterior margins of each tergite; dark brown, almost black, spiracular lines present; entire abdominal sternum white, apical sternite shaded with brown at lateral margins; first forceps segment shaded with brown, bearing a minute tubercle at inner apical angle, second segment frustate, third segment two and one-half times as long as second, fourth segment twice as long as broad; caudal filaments white, basal five or six articulations stained with red-brown.

NYMPH, MALE.—Length of body 6–8 mm. Head and dorsum of thorax dark brown; thoracic sternum white; legs white, all femora almost completely shaded with brown, fore tibia and .tarsus brown, middle and hind tibiae shaded with brown at bases, tarsi darkened at apexes; tergum of abdomen uniformly dark brown, sternum also usually uniform brown, but basal sternites sometimes lighter brown toward meson; each gill with median, black trachea only; median caudal filament two-thirds to three-fourths as long as cerci.

Holotype, male. — Vandalia, Illinois, April 16, 1946, Mohr & Burks. Specimen dry, on pin.

Paratypes.—ILLINOIS.—Same data as for holotype, 1 ♂. PORT BYRON: May 15, 1942, Ross & Burks, 8 ♂. Specimens in alcohol. Numerous specimens of nymphs collected at both above localities, along with adults.

22. Baetis brunneicolor McDunnough

Baetis brunneicolor McDunnough (1925b:173).

MALE.—Length of body 6.0–6.5 mm., of fore wing 6.5–7.5 mm. Head brown, lighter at lateral angles of frontal shelf; each antennal scape brown, pedicel tan, flagellum tan at base, shading to yellow at apex. Thorax dark brown, marked with yellow or tan at anterior apex of mesoprescutum, at posterior ends of outer parapsidal sutures, on mesoscutellum, on pleural sutures, and on sternal sutures; legs with all coxae yellow-brown, fore leg light yellow-brown, middle and hind legs white, often with femora faintly stained with tan; wings hyaline, veins Sc and R_1 of fore wing usually stained with pale yellow, stigmatic crossveins numerous, anastomosed; hind wing, fig. 283, with third longitudinal vein long, marginal intercalaries present between veins 1 and 2, and 2 and 3. Abdominal tergites 2–6 brown, sternites tan; apical tergites opaque brown, sternites yellow-brown; genitalia pale tan, of the *intercalaris* type, with a well-developed anteromesal tubercle on each first forceps segment; caudal filaments pale tan at bases, white distad.

Known from Illinois and Ontario.

Illinois Records. — DUNDEE: May 23, 1939, Burks & Riegel, 1 ♂. HAVANA: Matanzas Lake, Nov. 5, 1939, Ross & Burks, 2 ♂.

23. Baetis intercalaris McDunnough

Baetis intercalaris McDunnough (1921:118).
Baetis lasallei Banks (1924:425).

McDunnough (1938:25) stated that *intercalaris* and *lasallei* were so close as scarcely to warrant a separation. I have studied types of both species at the Museum of Comparative Zoology, and find no characters that serve to separate the two.

MALE.—Length of body 4.0–4.5 mm., of fore wing 4.5–5.0 mm. Head very dark brown, antennae brown, each becoming yellow toward tip of flagellum; eyes in life red-tan. Thorax very dark brown to black, with yellow-tan on lateral margins of mesonotum, at wing bases, and lateral margins of mesosternum; all coxae brown, legs otherwise white; wings hyaline, veins Sc and R_1 of fore wing stained with faint yellow near bases, stigmatic crossveins numerous, anastomosed marginal intercalaries in subcostal interspace extremely long, fig. 285; hind wing, fig. 284, with marginal intercalaries present, and with third longitudinal vein short. Abdominal segments 2–6 snow-white, black spiracular dots present; apical tergites dark chestnut brown, sternites creamy white; genitalia, fig. 297, white, mesoapical papilla of first forceps segment well developed; caudal filaments white.

Known from Connecticut, Illinois, Indiana, Maryland, New York, Ontario, and Pennsylvania. This is by far the commonest Illinois *Baetis*.

Illinois Records.—ALTO PASS: May 14, 1940, Mohr & Burks, 1 ♂. AMBOY: Green River, July 7, 1939, Mohr & Riegel, 1 ♂. CEDARVILLE: May 26, 1938, Ross & Burks, 1 ♂. DES PLAINES: Fox River, May 26, 1936, H. H. Ross, 1 ♂. FREEPORT: at light, June 10–11, 1948, Burks, Stannard, & Smith, 2 ♀. KANKAKEE: Kankakee River, June 6, 1935, Ross & Mohr, 1 ♂; June 29, 1939, Burks & Ayars, 1 ♂; July 21, 1935, Ross & Mohr, 1 ♂; Aug. 1, 1933, Ross & Mohr, 1 ♂; Aug. 16, 1938, Ross & Burks, 1 ♂; at light, July 9, 1948, Ross & Burks, 3 ♂. MILAN: Rock River, June 4, 1940, Mohr & Burks, 1 ♂. MOMENCE: June 15,

1938, Ross & Burks, 1♂; Aug. 16, 1938, Ross & Burks, 4♂. MOUNT CARROLL: July 14, 1944, Frison & Sanderson, 1♂. MUNCIE: June 8, 1927, Frison & Glasgow, 1♂. OAKWOOD: June 2, 1927, T. H. Frison, 1♂; June 14, 1935, C. O. Mohr, 4♂, 1♀; July 4, 1946, Mohr & Burks, 1♂; Salt Fork River, June 23, 1948, B. D. Burks, 5♂. OREGON: July 4, 1946, Mohr & Burks, 3♂; July 9, 1925, T. H. Frison, 1♂, 1♀. ROCK CITY: May 24, 1938, Ross & Burks, 1♂; May 30, 1938, Mohr & Burks, 2♂, 1♀. ST. CHARLES: Fox River, July 8, 1948, Ross & Burks, 50♂. STERLING: at light, May 22, 1941, Ross & Burks, 40♂. WILMINGTON: at light, Aug. 6, 1947, Burks & Sanderson, 5♂.

24. Baetis nanus McDunnough

Baetis nanus McDunnough (1923:42).

MALE.—Length of body 3 mm., of fore wing 4 mm. Head and thorax light gray-tan, with dorsal thoracic sutures shaded with brown; legs white, each fore femur yellow; hind wing with third longitudinal vein long, extending slightly beyond the middle of wing, and with a single marginal intercalary vein usually present posterior to vein 2; abdominal tergites 2–6 light yellow, apical tergites dark grayish yellow-brown; genitalia and caudal filaments white.

Known from Indiana and Ontario.

25. Baetis phoebus McDunnough

Baetis phoebus McDunnough (1923:41).

MALE.—Length of body 5.0 mm., of fore wing 5.5 mm. Head and thorax brown, with yellow or yellow-brown shading on thoracic notum; legs faintly yellow, almost white; each fore wing with intercalary vein of first three interspaces of approximately equal length, fig. 288; hind wing with third longitudinal vein slightly more than one-half as long as wing, a single marginal intercalary vein present posterior to vein 2; abdominal tergites 2–6 white, apical tergites bright chestnut brown; genitalia and caudal filaments white.

Known from Ontario.

33. PSEUDOCLOEON Klapálek

Pseudocloeon Klapálek (1905:105).

In this genus, the upper portion of each compound eye in the adult males is set on a high stalk. In Illinois species, the height of this stalk is at least as great as the diameter of the lower portion of the eye. In some South American species of *Pseudocloeon,* the height of the stalk supporting the upper portion of the eye is several times as great as the diameter of the lower portion. Each compound eye in the females is small, and the two eyes are separated by a space at least three times as great as the width of one eye. In each fore wing, there are relatively few crossveins, the stigmatic area is usually milky, the stigmatic crossveins are slanting and partly anastomosed, and the marginal intercalary veins occur in pairs. The hind wings are wanting. The male in all species known to me has a small, brown, or black, median dorsal dot on the base of the second abdominal segment; the basal 6 or 7 abdominal sternites in both sexes almost always show the blackened outlines of vestigial tracheae decurrent from the spiracles. In the male genitalia, the first forceps segment is about as broad as long, with the inner apical angle usually produced toward the meson, the second segment is narrower than the first, the third is long, slender, and usually bowed, and the fourth is small; there is a small penis cover between the bases of the forceps.

The nymphs are streamlined and typically live in the shallow, fairly rapidly flowing water along the banks of brooks or creeks. The maxillary palps have two segments, and the labial palps three. There is but a single pair of wingpads, the gills are single and platelike, and there usually are but two well-developed caudal filaments. The nymph of one species, *minutum* Daggy (1945:395), is described as having three well-developed caudal filaments.

Pseudocloeon has 19 described Nearctic species, 5 of which occur in Illinois. Reliable characteristics for the separation of the females of these species have not yet been found.

KEY TO SPECIES

ADULT MALES

1. Abdominal tergites 2–6 entirely white, or white faintly tinged with tan, and without any markings except black spiracular dots or lines...................2
   Abdominal tergites 2–6 light tan or white, with red, tan, or brown spots or dots..3
2. Abdominal sternites with midventral, brown or black dots..**1. punctiventris**

Abdominal sternites without midventral dots....................**2. dubium**
3. Abdominal tergites 2–6 or –7 with large, bright red spot near either lateral margin of each tergite.......**3. parvulum**
Abdominal tergites 2–6 lacking large, sublateral, bright red spots; sometimes with a brown spot near each spiracle. . 4
4. Abdominal tergites 2–6 each with a pair of submedian, red-brown dots and a short, median, longitudinal, red line . . .
.......................**4. myrsum**
Abdominal tergites 2–6 each with a pair of submedian, red or red-brown dots, but longitudinal, median, red line wanting.....................**5. veteris**

MATURE MALE NYMPHS

1. Caudal filaments each with alternating brown and white crossbands from base to apex; each abdominal gill with a subapical, brown spot. . .**3. parvulum**
Caudal filaments white, each with a brown crossband in middle; apex of each filament may or may not be brown.....2
2. Caudal filaments each with a brown crossband in middle and another at apex. . .
.......................**4. myrsum**
Caudal filaments each with a brown crossband in middle only...............3
3. Abdomen with median, ventral, dark spots on posterior margins of sternites 3– or 4–6 or –7..........**1. punctiventris**
Each abdominal sternite may have a pair of submedian, dark dots or streaks, but middle sternites lack median dots at posterior margins..........**2. dubium**

1. *Pseudocloeon punctiventris* (McDunnough)

Cloeon punctiventris McDunnough (1923: 45).

MALE.—Length of body 3.5–4.5 mm., of fore wing 4–5 mm. Head very dark brown, almost black; antennae dark brown, becoming light toward apex of flagellum of each. Thorax dark brown, almost black; legs very light yellow, with coxae and trochanters brown, and each fore femur shaded with smoky brown; wings hyaline. First abdominal segment light brown; tergites 2–6 white, without markings, except that there may be faint, black spiracular dots; sternites 2–6 white, with a black or dark brown, median dot on posterior margin of each; apical abdominal segments red-brown, median black or dark brown dot may be present on posterior margin of each of segments 7 and 8; genital forceps faintly tinted with tan, apical forceps segment twice as long as broad; cerci white.

FEMALE.—Length of body 4.0–4.5 mm., of fore wing 5.0–5.5 mm. Head and thorax light brown. Abdomen tan, heavily shaded with black tracheal markings; median ventral dots large, black, present on posterior margins of sternites 1–6 or –7.

NYMPH.—Length of body 4.5–5.5 mm. Head and thorax brown and tan; legs white or light yellow, with femora vaguely shaded with brown in the middle and at apex of each. Abdomen tan, with brown markings: first and second tergites almost entirely brown, lighter on posterior margins near posterolateral angles; tergites 3 and 4 tan, each with a pair of brown, submedian dots and lateral margins shaded with brown; tergite 5 mostly brown, lighter on meson; tergite 6 entirely brown except for a pair of submedian, tan dots at posterior margin; tergite 7 mostly brown, tan on meson and at posterolateral angles; tergites 8 and 9 tan, with a pair of submedian, brown dots on each, tergite 9 often also vaguely shaded with brown at lateral margins; abdominal sternum tan, a pair of submedian, brown dots present on each sternite, a median, dark brown dot on posterior margin of each of sternites 3–7, apical sternites vaguely shaded with brown; gills with pinnately branched, black or dark brown tracheae; caudal filaments light yellow, with a dark brown crossband in middle of each.

Known from Illinois, Ohio, and Ontario.

Illinois Records.—MOUNT CARMEL: at light, June 18, 1947, Burks & Sanderson, 1 ♂. OAKWOOD, Salt Fork River: May 2, 1943, H. H. Ross, 2 ♂; May 6, 1936, Ross & Mohr, 1 ♂; June 4, 1948, B. D. Burks, 2 ♂. SERENA, Indian Creek: May 12, 1938, Ross & Burks, 1 ♂, 1 ♀; May 16, 1938, B. D. Burks, 3 N. WOLF LAKE: Hutchins Creek, May 12, 1939, Burks & Riegel, 3 N.

2. *Pseudocloeon dubium* (Walsh)

Cloe dubia Walsh (1862: 380).

The male lectotype of this species is in the Museum of Comparative Zoology. This specimen is somewhat broken, but agrees quite well with the current concept of the species.

MALE.—Length of body 3–5 mm., of fore wing 4.0–5.5 mm. Head brown; each antennal scape and pedicel brown, flagellum white; each compound eye, in life, with faceted area of upper portion tan, columnar area yellow, and lower portion brown. Thorax dark, rich brown, almost black;

legs white, with coxae shaded with brown, a minute red-orange or red-brown mark at apex of each trochanter, each femur stained with orange-red or tan in middle and at apex, and each tibia stained with same color at apex; wings hyaline, with brown spot at base of each extending from vein Sc to 1A. First abdominal tergite brown, tergites 2–6 white, often very faintly stained with tan, apical tergites brown; black spiracular dots and, usually, longitudinal lines present on tergites 2–6 or –7, each spiracular dot almost always white in center; abdominal venter with first sternite variably stained with tan, sternites 2–6 white, apical ones stained with tan of varying intensity; genitalia with first forceps segment tan, other segments white, apical forceps segment two to two and one-half times as long as broad; the two cerci white.

NYMPH.—Length of body 4–5 mm. Head and thorax mottled brown and white; legs white, with brown shading in middle and at apex of each femur. Dorsum of abdomen with strongly contrasting areas of brown and white: first and second tergites mostly brown, with white spot at each posterolateral angle; tergites 3 and 4 white, usually with a pair of submedian dots on each tergite, lateral margins brown; tergite 5 with broad middle area white, the lateral areas brown; tergites 6 and 7 almost entirely brown, usually with a pair of submedian, white dots at the posterior margin of each; tergite 8 marked like tergite 4; tergite 9 with a pair of large, brown, lateral spots; basal abdominal sternites white, apical ones shaded with brown, in some specimens each sternite bearing a pair of submedian dots; gills with tracheae obscure, almost invisible; caudal filaments white, with a broad, brown crossband at middle of each.

Known from Illinois, Ontario, and New York.

Illinois Records.—AROMA PARK: Kankakee River, June 4, 1947, B. D. Burks, 1 ♂. BAKER: Indian Creek, May 12, 1938, Ross & Burks, 2 N. EDDYVILLE: Lusk Creek, May 28, 1946, Mohr & Burks, 1 ♂. MAZON: Mazon Creek, May 16, 1938, Ross & Burks, 10 N. MUNCIE: May 13, 1941, H. H. Ross, 1 ♂. OAKWOOD, Salt Fork River: May 2, 1943, H. H. Ross, 5 ♂; May 6, 1936, Ross & Mohr, 60 N. ROCK ISLAND: 18 ♂, 24 ♀ (Walsh 1862:380). SERENA, Indian Creek: May 12, 1938, Ross & Burks, 2 N; May 16,

1938, B. D. Burks, 2 N. WATSON: April 23, 1932, Ross & Mohr, 1 N. WOLF LAKE, Hutchins Creek: May 12, 1939, Burks & Riegel, 2 N; May 15, 1940, Mohr & Burks, 3 N.

3. *Pseudocloeon parvulum* McDunnough

Pseudocloeon parvulum McDunnough (1932b:210).

MALE.—Length of body 3.0–3.5 mm., of fore wing 3.5–4.0 mm. Head dark brown; antennae brown, apex of each flagellum slightly lighter. Thorax dark brown; legs light yellow, with coxae brown, each fore femur uniformly shaded with smoky tan, and middle and hind femora with red shading near apexes and on ventral margins; wings hyaline, longitudinal veins faintly stained with tan, vein Sc with dark tan spot at base. First abdominal tergite tan, tergites 2–6 white, very faintly suffused with tan, each with a pair of large, red spots and black spiracular lines at lateral margin; apical tergites dark tan; first abdominal sternite suffused with tan, sternites 2–6 lighter, almost white, no ventral markings present, apical sternites same color as basal one; genitalia white, with first segment of forceps suffused with tan, apical forceps segment only slightly longer than broad; cerci white or light yellow.

FEMALE.—Length of body 3–4 mm., of fore wing 3.5–4.5 mm. Head and entire body tan or light brown, with abdominal sternites slightly lighter in color; red shading of abdominal tergites sometimes visible; wings tan at bases, wing veins faintly yellow; cerci white or faintly yellow.

NYMPH.—Length of body 4.0–4.5 mm. Head and body brown, without large, contrastingly colored areas on dorsum of abdomen; legs each with femur brown on basal half to two-thirds, tibia brown at apex, tarsus brown in apical half; abdominal tergites typically each with a median anterior, two submedian, and a pair of posterolateral, light colored spots; gills platelike, each with a subapical, brown spot and black tracheae, the latter usually with a single stem and only one lateral branch; caudal filaments with narrow, alternating brown and white crossbands from base to apex of each.

Known from Alberta, Illinois, Quebec, and Ontario.

Illinois Records.—MAZON: Mazon Creek, May 16, 1938, Ross & Burks, 2 N. OAKWOOD: May 24, 1926, T. H. Frison, 2 ♂, 3 ♀. SERENA, Indian Creek: May 12, 1938, Ross & Burks, 17 N; May 16, 1938, B. D. Burks, 2 N.

4. *Pseudocloeon myrsum* new species

This species resembles *rubrolaterale* Mc-Dunnough (1931*b*:86) in possessing large, lateral, red-brown spots on the abdominal tergites and median, brown dots on the sternites, but differs in that each tergite possesses a pair of submedian, red-brown dots with a median, longitudinal, red mark between each pair of dots. This species likewise resembles *anoka* Daggy (1945:391) in possessing a large, brown mark on abdominal tergites 2 and 6, but differs in possessing the submedian dots and median marks mentioned above, as well as the lateral, red-brown spots.

MALE.—Length of body 3.5–4.0 mm., of fore wing 5.0–5.5 mm. Head dark yellow-brown, bright yellow around bases of antennae and at lateral angles of frontal shelf; ocelli yellow; faceted portion of upper section of each compound eye (in life) tan, columnar portion yellow, lower section of eye brown; scape and pedicel of antenna brown, flagellum yellow. Thorax dark, rich brown dorsally, with bright yellow markings along sutures; pleura and sternum mostly yellow, with dark brown shading only in central areas of scleromes; wings hyaline, each stained with red at base of costa and on anterior wing sclerites, base of subcosta dark brown, this color extending across wing base to vein 1A; veins and cross-veins hyaline, stigmatic crossveins slanting, partially anastomosed; legs very light yellow, almost hyaline, with coxae brown and each femur stained with red-brown at base, middle, and apex; fore femur shaded with dull brown in basal three-fourths; each tibia shaded with rose red in middle, fore tibia shaded with dull brown at apex. First abdominal segment dull brown: tergites 2–6 hyaline, faintly stained with tan; tergites 2 and 6 each with a large, median, brown spot, each of tergites 2–6 with a pair of submedian, dark red or red-brown dots and a vague, longitudinal, median, red mark between each pair of dots; a large, vague, red-brown spot near lateral margin of each of

these tergites and a pair of longitudinal, black spiracular lines on each tergite; apical four tergites yellow-brown, tergite 7 shaded on all but lateral and apical margins with dark red-brown; abdominal sternum hyaline, with a median, dark red-brown dot at posterior margin of sternites 2–, 3–, or 4–8; sternites 7 and 8 faintly shaded with red or red-brown; genitalia and caudal filaments white.

FEMALE.—Length of body 5 mm., of fore wing 6 mm. Body bright golden-brown; yellow at apex of mesoscutellum, on apical abdominal tergite, and on venter of entire body; lateral, brown shading of tergites 2–6 faintly indicated; legs as in male, except that each fore femur is almost completely brown; costal vein of fore wing red at base, but basal, brown area of male wanting; caudal filaments white, each faintly stained with yellow at base.

NYMPH.—Length of body 5.0–5.5 mm., caudal filaments 3.0–3.5 mm. Head yellow, faintly shaded with brown on face. Thorax white or tan, mottled with brown on dorsum and pleura, venter white; wingpads with pattern of future adult, longitudinal veins shown by brown lines; legs white, each femur with a brown cloud at base and middle and, sometimes, at apex. Abdomen white or tan, with brown shading: tergites 1 and 2 almost completely shaded with brown, a pair of submedian, circular, light spots at posterior margin of each, tergite 2 with a large, median, brown spot; tergites 2–5 mostly light, with a pair of submedian, brown dots on each; tergites 6 and 7 shaded like tergite 1; tergites 8 and 9 light, each with two pairs of submedian, brown dots; tergite 10 light, with a pair of submedian, brown streaks and a pair of dark spots at lateral margins; in dark individuals, these dark areas and spots tend to spread and almost coalesce; gills hyaline, tracheae pinnate, brown; abdominal sternum white or tan, with a pair of submedian, brown dots on each of sternites 2–7; apical sternites usually uniformly shaded with brown; cerci white, except for a brown apex and a brown crossband at middle of each.

Holotype, male. — Eddyville, Illinois, Lusk Creek, May 16–17, 1947, B. D. Burks. Specimen dry, on pin.

Allotype, female.—Wolf Lake, Illinois, Hutchins Creek, April 2–3, 1946, Burks & Sanderson. Specimen dry, on pin.

Paratypes.—Same data as for allotype, 2 ♂. Specimens dry, on pins.

NYMPHS.—ILLINOIS.—EDDYVILLE: Lusk Creek, May 16–17, 1947, B. D. Burks, 2 N. WOLF LAKE, Hutchins Creek: April 3, 1946, Burks & Sanderson, 10 N; May 12, 1939, Burks & Riegel, 1 N.

5. *Pseudocloeon veteris* McDunnough

Pseudocloeon veteris McDunnough (1924a:8).

As McDunnough remarked when describing this species, he based his description on old, faded specimens. Recently collected specimens, which could be studied while still alive, necessitate considerable change in the description of the color of this species.

MALE.—Length of body 4.5–5.0 mm., of fore wing 5.0–6.0 mm. Head dull brown, frontal shelf hyaline; antennae dull tan, each scape white at apex; eyes orange-yellow. Thorax bright orange-brown; yellow on prescutum, along outer parapsidal furrows, on scutellum, on pleural sutures, and covering the entire thoracic sternum; wings hyaline, faintly stained yellow at bases, stigmatic crossveins numerous, anastomosed; legs pale yellow, apex of each femur shaded with orange. Abdominal segments 2–6 yellow, sometimes faintly suffused with tan, each tergite with a pair of submedian, red or red-brown dots, and usually a short, median, red or red-brown, transverse stripe at posterior margin; segments 1–6 with black spiracular dots and longitudinal spiracular line usually present on either side; sternites 1–6 yellow, with a median, dark brown spot at posterior margin of each and often black tracheal outlines extending ventrad from spiracles on anterior sternites; apical tergites dark orange-brown, sternites suffused with pale pink; genitalia white, apical forceps segment two and one-half times as long as broad; caudal filaments white.

FEMALE.—Length of body 5 mm., of fore wing 6 mm. Head and thorax light brown; legs light yellow, femora shaded toward apexes with pink or tan; abdominal dorsum light brown, with a pair of submedian, red-brown dots faintly visible on each tergite; abdominal sternum light tan, with a fairly large, black dot on meson of posterior margin of thoracic metasternum and abdominal sternites 1–7.

Known from Illinois.

Illinois Records.—MUNCIE: May 13, 1931, H. H. Ross, 1 ♂. OAKWOOD: April 24, 1925, T. H. Frison, 10 ♂, 4 ♀; May 18, 1926, T. H. Frison, 2 ♂; June 5, 1948, Burks & Sanderson, 2 ♂. URBANA: Salt Fork River, May 13, 1898, 4 ♂, 3 ♀.

34. *NEOCLOEON* Traver

Neocloeon Traver (1932b:365).

The genus *Neocloeon* is similar to *Cloeon* in that the fore wings have relatively few

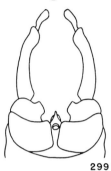

299

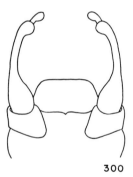

300

Fig. 299.—*Neocloeon alamance*, male genitalia.
Fig. 300.—*Cloeon mendax*, male genitalia. (After Traver.)

crossveins and single marginal intercalary veins, and the hind wings are wanting.

In the males of *Neocloeon,* the fore tarsus is approximately as long as the fore tibia, and either one is one and one-third to one and one-half times as long as the fore femur; the partly differentiated first tarsal segment of the hind leg is as long as the three apical, clearly differentiated segments combined. The stalk supporting the upper faceted portion of each compound eye is quite low; its height is only about one-half as great as the diameter of the lower portion. Each arm of the genital forceps, fig. 299, has the basal segment broad and short, the second segment semiquadrate, with a large projection on the median margin, the third segment long, relatively stout, and very little bowed, and the fourth segment two and one-half to three times as long as broad. The penis cover is produced on the meson.

The nymphs have long, three-segmented maxillary palps, the labial palps are broad, truncate, and three-segmented; the tarsal claws are long, slender, and nondenticulate; the abdominal gills are single and platelike, and they bear blackened tracheae which branch on the inner sides only; the three caudal filaments are well developed.

Neocloeon alamance Traver (1932:365), originally described from North Carolina and later recorded from Tennessee, is the only known species in this genus.

35. *CLOEON* Leach

Cloeon Leach (1815:137).
Cloea Billberg (1820:97). Emendation, unnecessarily proposed.
Cloe Burmeister (1839:797). Emendation, unnecessarily proposed.
Chloeon Lubbock (1863:61). Emendation, unnecessarily proposed.
Cloeopsis Eaton (1866:146).
Procloeon Bengtsson (1915:34). New synonymy.

In species of the genus *Cloeon,* the fore wing has relatively few crossveins and only a single marginal intercalary vein in each interspace of the outer wing margin; the hind wings are wanting.

In the males, the fore tarsus is approximately equal in length to the fore tibia, and either one is slightly longer than the fore femur. The hind tarsus has the partly differentiated first segment as long as, or slightly longer than, the two apical tarsal segments combined. Each arm of the genital forceps, fig. 300, has the first segment broad, short, and narrowed at the apex; the second segment is somewhat conic and not clearly separated from the third segment; the third segment is long, slender to fairly stout, and usually slightly bowed; the fourth segment is about as broad as long. The penis cover is usually truncate at its apex.

Both maxillary and labial palps in the nymphs have three segments; each tarsal claw is relatively short, broad at the base, and slender at the tip, and bears a single row of minute ventral denticles; the abdominal gills are sheetlike, undulated, and double, with the tracheae usually branching palmately; the three caudal filaments are well developed. Mayflies of this genus, so far as is known, develop in small ponds or in the still eddies along the banks of streams.

Cloeon includes 10 described Nearctic species, 8 of which are treated here. The species of *Cloeon* all seem to be quite rare in eastern North America.

Reliable characteristics for the specific determination of the females and nymphs in this genus have not yet been found.

KEY TO SPECIES

ADULT MALES

1. Abdominal tergites 2–6 almost entirely dark red-brown or gray-brown......2
   Abdominal tergites 2–6 hyaline, white or faintly yellowish, often with red or dark brown markings..............3
2. Caudal filaments white, stained with red-brown at bases, but articulations not darkened; abdominal tergites 2–6 light gray-brown...............**1. ingens**
   Caudal filaments white, with articulations dark brown or black throughout; abdominal tergites 2–6 almost entirely dark red-brown.........**2. dipterum**
3. Abdominal tergites 2–6 uniformly stained with light red, without darker markings ....**3. mendax**
   Abdominal tergites 2–6 hyaline, white, or faintly yellow, often with red or dark brown markings..................4
4. Abdominal tergites 2–6 hyaline, with a pair of large, lateral, dark brown blotches on each, these blotches on segments 3 and 6 so large as to coalesce on meson...................**4. minor**
   Abdominal tergites 2–6 white or light yellowish, without large, lateral, dark brown blotches, but sometimes with small, red marks on some or all of these tergites...........................5
5. At least some of abdominal tergites 2–6 with small, red markings............6
   Abdominal tergites 2–6 white, without red markings......................7
6. Abdominal tergites 2 and 3 each with a faint, median, red streak...........
   ....**5. insignificans**
   Posterior margin of each of abdominal tergites 2–6 with a pair of submedian, red dashes.........**6. rubropictum**
7. Mesonotum light yellow-brown.........
   ....**7. vicinum**
   Mesonotum dark brown, shading to creamy white at lateral margins and at apex of scutellum.........**8. simplex**

1. *Cloeon ingens* McDunnough

Cloen ingens McDunnough (1923:44).

MALE.—Length of body 8 mm., of fore wing 9 mm. Thoracic notum black, pleura light brown, legs gray-brown, and wings hyaline. Basal abdominal tergites uniformly shaded with gray-brown, with a narrow, transverse, black-shaded area at posterior margin of each tergite; apical abdominal tergites chocolate brown; venter dirty white, apical sternites faintly stained with brown; and genital forceps and caudal filaments white, the latter stained with red-brown at bases.

Known from Alberta, Maine, and Quebec.

2. *Cloeon dipterum* (Linnaeus)

Ephemera diptera Linnaeus (1761:377).

A single female specimen of this Palearctic species has been taken in Illinois. Comparison of this specimen with the description given in Eaton (1885:182) and with reliably determined European material of *dipterum* leaves no doubt of the correctness of the identification. *C. dipterum* is a common and widespread European species which has not heretofore been proved to be present in North America. The description given below of the male was written from specimens collected in Switzerland, which were sent to the Illinois Natural History Survey collection by Dr. F. Schmid of Lausanne.

MALE.—Length of body 6–7 mm., of fore wing 7–8 mm. Thorax very dark brown to black, legs tan or brown, wings hyaline, with longitudinal veins and most crossveins light brown. Basal abdominal tergites heavily shaded with dark red-brown, basal sternites tan, with dark red-brown shading at lateral margins; apical abdominal segments dark chocolate brown, almost black; genital forceps light yellow, almost white; caudal filaments white, articulations throughout dark brown to black.

FEMALE.—Length of body 8 mm., of fore wing 9 mm. Head light yellow, almost white, without darker markings; bases of ocelli and entire compound eyes very dark gray, almost black; antennae light yellow, each with a narrow band of dark brown shading at apex of scape and of pedicel. Thorax uniformly light tan, shaded with brown along posterior margin of mesotergum; small, brown spot on meso- and metasternum near base of each coxa; legs light yellow, with a small, brown spot on ventral side of each middle and hind coxa, at apex on dorsal side of each fore trochanter, and near apex of each femur; minute, brown dot on dorsal side at apex of each of three apical tarsal segments. Wings hyaline, with light brown shading in costal and subcostal interspaces of each fore wing, this shading extending from base to apex of wing, but interrupted at each crossvein in these interspaces; 5 crossveins in costal interspace basad of bulla, 10 crossveins present in this interspace distad of bulla; crossveins partly anastomosed in stigmatic area; all longitudinal veins brown; crossveins in costal, subcostal, first radial, cubital, and anal interspaces hyaline, all other crossveins brown; marginal intercalary veins brown except in anal region. Abdomen light tan, tergites faintly shaded with brown; tergites 2–8 each with small, longitudinal spot on meson at anterior margin, a pair of rather broad, sublateral, curved spots on basal two-thirds, a pair of very small spiracular dots and a pair of narrow, lateral, longitudinal lines; tenth tergite with a pair of large, lateral, triangular marks; abdominal sternum pale yellow, with dark red-brown markings; sternites 2–8 each with a pair of sublateral, longitudinal bars, a minute, transverse line at posterior margin at point where each longitudinal, red-brown bar ends, a transverse bar at anterior margin extending laterally from each longitudinal bar to anterolateral angle of sternite, and a short, narrow, longitudinal mark extending posteriorly from a point near each anterolateral angle of sternite; sternites 2 and 3 also each with a pair of short, longitudinal lines at basolateral angles; sternite 9 with a pair of vague, fairly large, red-brown spots near anterolateral angles; each paraproct with a submedian, longitudinal, brown line. Caudal filaments white, articulations dark red-brown; in basal area of each filament, alternating articulations with broader color band.

In Europe, *Cloeon dipterum* has long been recorded as ovoviviparous and the length of life of an adult female may be as much as 3 weeks. The single female specimen from Illinois is probably an adventive.

Illinois Record.—CHAMPAIGN: at light, Aug. 26, 1939, C. O. Mohr, 1♀.

3. *Cloeon mendax* (Walsh)

Cloe mendax Walsh (1862:381).

The lectotype of this species, now in the Museum of Comparative Zoology, is a female and, as it is a true *Cloeon,* it must be a different specimen from the one McDunnough found labeled as *mendax* in the M.C.Z. about 1928 (McDunnough 1929: 173). The specimen he saw was a *Centroptilum,* with two pairs of wings, while the specimen I saw there in 1942 clearly had only one pair of wings. Walsh's male type, unfortunately, is lost, and additional male specimens from Illinois are yet to be taken. I have seen males of this species from New York.

MALE.—Length of body 4 mm., of fore wing 6 mm. Thoracic notum and pleura stained with red; thoracic venter white, with a gray-green cast; legs white, often with a green cast; wings hyaline. Abdominal tergites stained with red; sternites greenish white; apical segments opaque, basal ones translucent; genital forceps, fig. 300, and the three well-developed caudal filaments white.

FEMALE.—Length of body 5 mm., of fore wing 7 mm. General color very light yellowish, lacking red staining of male. Thorax, wings, and legs in fresh specimens variably stained with bright green.

This species is known from Illinois, Massachusetts, Michigan, New York, and Ontario.

Illinois Records.—BARRY: sweeping willows near pond, Aug. 12, 1948, Sanderson & Stannard, 2♀. ROCK ISLAND: 2♂, 4♀ (Walsh 1862:381).

4. *Cloeon minor* McDunnough

Cloeon minor McDunnough (1926:190).

MALE.—Length of body and of fore wing 3 mm. Thoracic notum black; each pleuron very dark brown, with pink staining near wing base; sternum dark brown; legs white, with a faint, red spot near middle of each femur; wings hyaline. Abdominal tergites 2–6 hyaline, with a pair of large, brown, sublateral blotches on each segment, these blotches coalescing on meson of tergites 3 and 6; sternites 2–6 hyaline, each with a pair of lateral, brown triangles and a pair of submedian, black dashes at posterior margin; apical tergites dark brown and sternites

light brown; genital forceps and caudal filaments white.

Known from Ontario.

5. *Cloeon insignificans* McDunnough

Cloeon insignificans McDunnough (1925b:186).

MALE.—Length of body 3 mm., of fore wing 4 mm. Thoracic notum and pleura dark brown, sternum lighter brown, legs white, and wings hyaline, with longitudinal veins faintly yellowish. Basal abdominal tergites white, with a faint red, median streak on tergites 2 and 3, sternites 2 and 3 white; apical tergites light brown, sternites tan; genital forceps and caudal filaments white.

Known from Ontario.

6. *Cloeon rubropictum* McDunnough

Cloeon dubium Clemens (1913:341),
       not Walsh. Misidentification.
Cloeon rubropicta McDunnough (1923:43).

MALE.—Length of body 3–4 mm., of fore wing 4–5 mm. Head dark brown, shading to yellow below ocelli, antennae yellow, flagella slightly dusky. Thorax dark brown on dorsum and pleura, venter yellow; legs light yellow; wings hyaline. Abdominal segments 2–6 white, translucent, each tergite with a pair of submedian, red, transverse marks at posterior margin and a pair of sublateral, red dots; tergites 2, 3, and 6 with vague, median, longitudinal, red marks; a longitudinal, black spiracular hairline present at lateral margins of these segments; sternites 2–6 unmarked, apical tergites red-brown, apical sternites vaguely shaded with brown; genital forceps and caudal filaments white.

FEMALE.—Length of body 4–5 mm., of fore wing 5–6 mm. Head and thorax yellow to dull tan, thoracic venter dirty white to yellow. Abdominal tergites translucent tan, faint, submedian, red marks often present on posterior margins of all tergites, extensive, black tracheal and spiracular marks present; venter pale yellow to white; legs and caudal filaments white or faint yellow.

Known from Illinois, New York, Ontario, Ohio, and Quebec.

Illinois Records. — EDDYVILLE: Lusk Creek, June 19, 1940, Mohr & Riegel, 1♂. OAKWOOD: June 6, 1925, T. H. Frison, 1♂, 1♀; June 9, 1926, Frison & Auden, 1♂, 3♀.

7. *Cloeon vicinum* (Hagen)

Cloe vicina Hagen (1861:56).

MALE.—Length of body 4 mm., of fore wing 4–5 mm. Thorax almost completely light yellow-brown, only slightly darker on mesoscutum; wings hyaline and in stigmatic area of each wing five or six crossveins which do not quite reach subcostal vein; legs white, with each fore coxa and base of fore femur stained with tan. Abdominal segments 2–6 white, without red or brown markings; apical tergites dull brown; apical sternites very light tan; genital forceps white, basal segment of each bearing a very small, mesoapical projection; caudal filaments white.

Known from the District of Columbia, New York, and West Virginia.

8. *Cloeon simplex* McDunnough

Cloeon simplex McDunnough (1925b:185).

MALE.—Length of body 4.5–5.5 mm., of fore wing 5–6 mm. Head brown, scape and pedicel of each antenna yellow, flagellum brown. Thoracic notum brown, with posterior margins of pronotum and mesoscutum and entire mesoscutellum white; pleura white, with vague, tan shading; prosternum white; meso- and metasternum brown, with white markings; wings hyaline; legs white. Abdominal segments 1–6 translucent white, with longitudinal, black spiracular lines; apical tergites yellow-brown, sternites white, opaque; genital forceps and caudal filaments white.

FEMALE.—Length of body 5–6 mm., of fore wing 6–7 mm. Head and thorax yellow or light tan, sometimes tinged with green; abdomen yellow, with extensive, black tracheal and spiracular markings; legs white, wings hyaline, both sometimes with green stain.

Known from Illinois, Indiana, Ontario, Quebec, and Wisconsin.

Illinois Records.—RICHMOND: June 13, 1938, B. D. Burks, 1 ♂. ROSECRANS: Des Plaines River, June 14–21, 1938, B. D. Burks, 1 ♂, 3 ♀, 5 N. SPRING GROVE: June 9, 1938, Mohr & Burks, 1 ♂.

AMETROPIDAE

This family as treated here includes some of the genera that were placed in the fami-lies Ametropodidae and Ecdyonuridae in Ulmer's classification (1933), and repre-sents a combination of the subfamilies Ame-tropinae and Metretopinae of Traver's clas-sification (1935a:429, 433). Lestage (1938: 180) has divided these genera among three families, Ametropodidae, Metretopodidae, and Siphloplectonidae.

The Ametropidae contain forms which are interstitial between the typical heptageniids and the typical baetids. The wing venation is quite similar to, or identical with, that of the Heptageniidae, but the hind tarsus in all ametropid genera has only four clearly differentiated segments. The nymphs re-semble either the heptageniid or baetid form, but the tarsal claws of the middle and hind legs are slender (at least six times as long as greatest width) and usually longer than their respective tibiae, figs. 25, 301, 302, 312.

In the male ametropid adults, the fore tarsus is from two and one-half to nearly five times as long as the fore tibia, the first fore tarsal segment varies from three-fourths to one and one-half times as long as the second segment; the compound eyes are almost or quite contiguous on the meson, each eye is obscurely divided into an upper area of larger facets and a lower area of slightly smaller facets, and the living insect often has a faint color band crossing the eye at the boundary line between the two areas of facets in the eye. In both sexes, the cubital intercalaries, fig. 308, consist of two or, more often, four straight veins which are detached at the bases; the hind wing always has a broadly angulate costal projection; and vein M of the hind wing is forked near the wing base, or, farther dis-tad, near the middle of the wing. In both sexes of all members of this family, the abdomen is markedly long and slender, with the apical segments more slender and elon-gate than are the basal segments. There may be either two or three well-developed caudal filaments.

In the nymphs, which are quite hetero-geneous, the head is flattened laterally or dorsoventrally and the eyes are directed an-teriorly, dorsally, or laterally. The fore tarsal claw is single, slender, and long in *Pseudiron, Ametropus,* and *Metreturus,* but bifid in *Siphloplecton,* fig. 303, and *Metre-topus;* the claws of the middle and hind legs are longer than their respective tibiae, figs. 301, 302. In the nymphs of most genera, the body is elongate and fishlike, but in *Pseudiron,* it is flattened dorsoventrally.

Gills are borne by abdominal segments 1–7, but the structure of these gills varies greatly

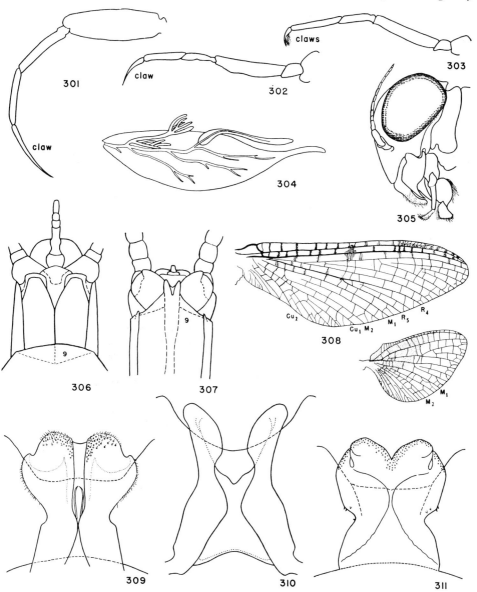

Fig. 301.—*Pseudiron centralis,* hind leg of mature nymph.
Fig. 302.—*Siphloplecton interlineatum,* middle leg of mature nymph.
Fig. 303.—*Siphloplecton interlineatum,* fore leg of mature nymph.
Fig. 304.—*Pseudiron centralis,* gill of third abdominal segment.
Fig. 305.—*Siphloplecton interlineatum,* head of adult male, lateral aspect.
Fig. 306.—*Siphloplecton interlineatum,* terminal abdominal sternites of female.
Fig. 307.—*Pseudiron centralis,* terminal abdominal sternites of female.
Fig. 308.—*Siphloplecton basale,* wings.
Fig. 309.—*Siphloplecton interlineatum,* male genitalia.
Fig. 310.—*Pseudiron centralis,* male genitalia.
Fig. 311.—*Siphloplecton basale,* male genitalia.

among the various genera in this family. All known nymphs have each three relatively short caudal filaments, with each outer filament bearing a dense fringe of setae on the mesal margin only.

KEY TO GENERA

Adults

1. Median caudal filament well developed, almost as long as cerci. . . . .**36. Ametropus**
   Median caudal filament vestigial or represented by only a one- to four-segmented stub. .2
2. One pair of cubital intercalary veins present in fore wing. . . . .**38. Metretopus**
   Two pairs of cubital intercalary veins present in each fore wing. 3
3. First segment of fore tarsus of male three-fourths as long as second segment; ninth abdominal sternite of female extended caudad and with a pronounced median notch on posterior margin, fig. 307. . . . .**39. Pseudiron**
   First segment of fore tarsus of male slightly longer than second segment; ninth abdominal sternite of female not produced caudad and without a median notch on posterior margin, fig. 306. . . . .**40. Siphloplecton**

Mature Nymphs

1. Eyes directed anteriorly; each fore coxa with a large, lobelike, median appendage. . . . .**36. Ametropus**
   Eyes directed laterally or dorsally. 2
2. A median, ventral, hooklike spur present on each thoracic segment and a median, dorsal, hooklike spur present on each abdominal segment, fig. 312. . . . .**37. Metreturus**
   Thoracic and abdominal segments without median, dorsal, or ventral, hooklike spurs. 3
3. Head flattened dorsally, prognathous; eyes dorsal; claw of fore tarsus single. . . . .**39. Pseudiron**
   Head not flattened dorsally, hypognathous; eyes lateral; claw of fore tarsus double, fig. 303. 4
4. Maxillary palp two-segmented. . . . .**38. Metretopus**
   Maxillary palp three-segmented. . . . .**40. Siphloplecton**

36. *AMETROPUS* Albarda

Ametropus Albarda (1878:129).

In the genus *Ametropus,* the wing venation differs only slightly from that of the typical heptageniid form, the fore wing having two pairs of cubital intercalary veins. The hind wing has an acute costal angulation, vein M is forked at a point halfway

from its base to the apex, and veins R_4 and R_5 are fused throughout their length. The male penis lobes are fused to form a conical structure with a narrow, V-shaped apical cleft, somewhat as in *Baetisca* and some species of *Ephemerella.* The apical abdominal sternite of the female has a median caudal cleft. The median caudal filament is long, nearly as well developed as are the cerci.

The nymphs are fishlike in general body form, with the eyes directed anteriorly. The head is small, with the frontal margin cut away almost to the antennal sockets, nearly completely exposing the mouth-parts. The pronotum is wider than the head, and a projecting, membranous flap is borne by the prosternum. Each fore coxa has a fleshy, lobelike mesal projection. All the tarsal claws are single, slender, and pointed, and are much longer than the tibiae. The abdominal gills are single and platelike.

Only two Nearctic species are known in the genus *Ametropus.* One of these, *neavei* McDunnough (1928a:9), was described from Alberta. The other, *albrighti* Traver (1935a:431), described from the nymph only, was first collected in New Mexico and has subsequently been found in Utah.

37. *METRETURUS* new genus

The genus *Metreturus* is here erected for a new species, known only in the nymphal stage, fig. 312, which is radically different from that of all other North American mayflies. It is referred to the family Ametropidae because of the very short tibiae, the long, slender tarsal claws, and the three short caudal filaments, each of the outer ones of which bears a dense fringe of setae on the mesal margin only. The wing venation, visible in the nymphal wingpads, can be seen with sufficient clarity to show that the adult of this form could be referred to the Ametropidae, as there are two pairs of parallel and basally detached cubital intercalary veins.

The nymph of *Metreturus* has the head small and hypognathous, with the eyes lateral; there is a pair of short, submedian horns near the antennal bases, on the face above the clypeus; there is a median, acute projection on the margin of the clypeus, and there is an oblique, laterally projecting lobe arising at the lower margin of each com-

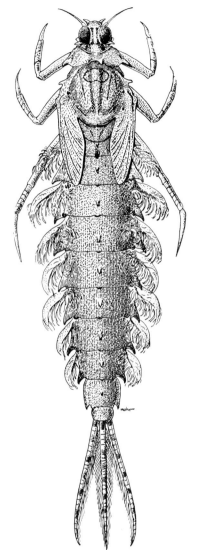

Fig. 312.—*Metreturus pecatonica*, mature nymph, dorsal aspect.

pound eye. The thorax has a pair of finger-like, projecting lobes on the lateral margins of both the pronotum and the mesonotum; a hooklike, median projection is borne on the sternum of each thoracic segment; each fore tarsal claw is single, long, and slender, and each of the middle and hind tarsal claws is single, slender, and much longer than the respective tibia. Abdominal segments 1 to 9 have each a pair of acute projections at the posterolateral angles; a single, hooklike, median projection is present in the center of each tergite 1 to 9, and also a small, acute,

median papilla on the posterior margin of each of these tergites; each of segments 1 to 7 bears a pair of deeply fissured gills, each gill consisting of a dorsal member with a recurved, ventral flap and a smaller, ventral member; the three caudal filaments are relatively short, the outer ones bearing a fringe of setae on the mesal margins only. The available material of this genus shows that the adult would have three well-developed caudal filaments.

Genotype: *Metreturus pecatonica* new species.

Metreturus pecatonica new species

FEMALE NYMPH.—Fig. 312. Length of body 20.0 mm., of caudal filaments 5.5 mm. Entire head and body creamy white, virtually without dark markings except for the brown lines on wingpads indicating paths of adult wingveins, vague, brown crossband at base of each front wingpad, and light brown crossbands on caudal filaments. Head with eyes separated on vertex by space three-fourths as great as the width of one eye; a pair of large and hornlike projections arising on face between the antennal sockets; an oblique and finger-like, projecting lobe arising from each gena just ventral to compound eye; antennal scape about three-fifths as long as pedicel, flagellum five times as long as pedicel; clypeus with margin acutely projecting on meson, laterally cut away so as to expose bases of mandibles; labrum, fig. 313, with broad, mesal emargination on outer margin; mandibles, fig. 314, each with apex broad, bearing seven teeth in three groups, molar area wanting; maxilla, fig. 316, with apex broad, mandible-like, bearing five or six slender, acute teeth; palp reduced to a single, small, lobelike segment; hypopharynx bilobed; labium, fig. 315, with glossa large, rounded, each paraglossa with mesal margin straight, lateral margin convex, palp three-segmented, with apical segment relatively minute, apposed to an apicomesal projection of second segment. Pronotum with a finger-like projection at each posterolateral angle and a pair of small, submedian projections at posterior margin; hooklike projection in center of prosternum; fore femur as long as tibia and tarsus (without claw), fore tarsal claw single, slender at apex, enlarged near base, one and one-half times as long as tibia; mesonotum with finger-like, lateral projection at base of each front wingpad, mesosternum with hook-like projection in center of basisternum, middle femur as long as tibia and tarsus without claw, middle tarsal claw long, slender, more than twice as long as middle tibia;

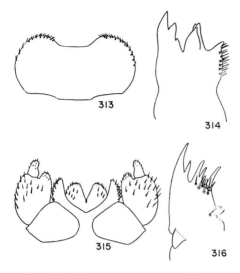

Fig. 313.—*Metreturus pecatonica*, labrum of nymph.

Fig. 314.—*Metreturus pecatonica*, mandible of nymph.

Fig. 315.—*Metreturus pecatonica*, labium of nymph.

Fig. 316.—*Metreturus pecatonica*, maxilla of nymph.

metasternum with hooklike projection in the center, hind femur one and one-half times as long as tibia and tarsus without claw, hind tarsal claw five times as long as tibia. Abdomen three times as long as thorax, tapering toward posterior end, so that segment 9 is two-fifths as wide as segment 1; prominent, flat flange and projecting, posterolateral angle present at each lateral margin of tergites 1–9; a hooklike projection present in center of each tergite 1–9, with a small, median papilla present at posterior margin of each of these tergites; segments 1–7 each bearing a pair of gills, each gill composed of a dorsal member, with a recurved ventral flap, and a ventral member, each gill member having a stout, median tracheal trunk from which arise numerous, pinnately branching tracheae; lamina of gill partly eroded away between these tracheae, so that ventral member of each gill almost assumes appearance of a cluster of branching filaments; nearly all or smaller part of

upper member of each gill having a similar appearance, this atrophy of gill lamina of upper member almost complete in first gill, but becoming progressively less so in more posterior gills, the upper member of seventh gill with lamina eroded only at apex and on mesal side; caudal filaments alternately banded with faint yellow and light brown.

Holotype, female nymph. — Taken by seining in Sugar River, one-fourth mile above mouth, near Harrison, Illinois, July 6, 1926, R. E. Richardson. Specimen mature, taken just as molt to subimaginal stadium was beginning. Specimen in alcohol.

Paratype. — ILLINOIS: Taken with dip-net near water's edge at mouth of Pecatonica River, near Rockton, May 8, 1927, R. E. Richardson, 1 ♀ nymph, about half grown. Specimen in alcohol.

The streams in which these specimens were found have subsequently been dredged and straightened, but the available information indicates that, at the time the collecting was done, they were fairly rapid, shallow, and moderate-sized streams with sand and rock bottoms. Since being dredged, they have been sluggish, heavily silted streams with mud bottoms. Intensive collecting in these rivers in recent years has failed to produce additional specimens; the species probably has disappeared completely from them.

38. *METRETOPUS* Eaton

Metretopus Eaton (1901:253).

In *Metretopus*, the wing venation is similar to that of *Siphlonurus*, fig. 219, but differs principally in that the cubital intercalary veins of the fore wing consist of two long, straight veins instead of a series of short, sinuate veins decurrent from vein Cu_1. The hind wing in *Metretopus* also has a relatively acute costal projection. The median caudal filament is vestigial in the adult. Characters of the nymph, used in the above key to genera, are from Bengtsson (1909:16).

One Holarctic species, *norvegicus* Eaton (1901:254), has been reported from Alberta by McDunnough (1925b:187).

39. *PSEUDIRON* McDunnough

Pseudiron McDunnough (1931b:91).

Pseudiron is clearly an interstitial genus showing similarity to the members of both the Heptageniidae and Baetidae. The adult wing venation is typical for the Heptageniidae, resembling most closely that of *Rhithrogena,* fig. 320, but the hind tarsus has only four clearly differentiated segments. The abdomen is quite long and slender, and the ninth sternite of the female has a median indentation on the posterior margin, fig. 307. The male genitalia are quite similar in structure to the genitalia of the *jejuna* group of species in the genus *Rhithrogena.* The median caudal filament is vestigial in the adults of *Pseudiron.*

Each of the nymphs has a somewhat flattened, heptageniid-like head, with the eyes dorsal in position; in addition, the head is slightly elevated on the meson between the antennal bases, faintly suggesting the baetine head form. The mouth-parts evidently are fitted for predatism (Spieth 1938a:3). The tarsal claws, fig. 301, are longer than the tibiae. The gills, fig. 304, are elongate and slender, with a small fibrillar tuft near the base on the ventral side and a narrow, flagellum-like appendage near the center of the posterior margin, also on the ventral side of the gill. This type of gill occurs in no other known member of the order Ephemeroptera.

Spieth (1938a:3) described a nymph which he had collected in southwestern Indiana and thought to be the nymph of *Pseudiron.* A nymph, apparently of the same species, has been collected in Illinois. In my opinion, this nymph can be accepted as that of *Pseudiron* without further question despite the fact that an actual rearing from nymph to adult has not yet been accomplished. The tarsal characters place it in the Ametropidae and the wing venation visible in the nymphal wingpads, as well as the form of the abdomen, are in agreement with the adult *Pseudiron* characteristics. Furthermore, this nymph and the adults of *Pseudiron* have been collected at the same location in Illinois.

Pseudiron centralis McDunnough

Pseudiron centralis McDunnough (1931b:91).

MALE.—Length of body and of fore wing 12 mm. Head light yellow, with red-brown shading on face and on vertex between compound eyes; each antennal scape and pedicel red-tan, flagellum light yellow; compound eyes of living insect dark gray. Dorsum of thorax reddish brown, with narrow, yellow stripes on lateral sutures of mesonotum; pleura and sternum yellow. Legs markedly long and slender, yellow-brown, with a dark, red-brown crossband present at middle and at apex of each femur, each tibia and tarsal segment darkened at apex; wings hyaline, each brown stained in stigmatic area, veins and crossveins brown. Abdomen with broad, dorsomedian, longitudinal, brown stripe, edges of this stripe extended to lateral margin at posterior margin of each tergite; abdominal sternum yellow, with ganglionic areas faintly brown-stained; genitalia, fig. 310, yellow-brown; caudal filaments yellow, with basal two or three segments shaded with brown, articulations light brown, becoming colorless toward the apexes of the filaments.

FEMALE.—Length of body 12–13 mm., of fore wing 13–14 mm. Coloration almost identical with that of male except that dorsum of thorax is mostly yellow-brown; stigmatic area of fore wing not brown stained, veins and crossveins of hind wing almost or quite hyaline; ninth abdominal sternite incised on meson of posterior margin, fig. 307; caudal filaments white, basal articulations light brown.

NYMPH.—Length of body 12–13 mm. Head broad, flattened dorsally, but slightly elevated on median area between antennal bases, thus somewhat intermediate in form between typical heptageniid and baetid nymphal heads. Pronotum with lateral margins expanded laterally as thin, platelike projections, reminiscent of the pronotum in some Palaeodictyoptera. Legs long, slender, fig. 301, with each tarsal claw longer than respective tibia; wingpads showing venational pattern typical for this genus. Abdomen long, slender, with lateral margins flaring; platelike, posterolateral angles acute on segments 8 and 9, rounded on more anterior segments; gills, fig. 304, with a small fibrillar tuft near base and a flagellum-like projection near middle of posterior margin; apex of gill lanceolate; cerci each with a fringe of long setae on inner side only, median caudal filament bearing long setae on both sides.

Known from Illinois, Indiana, Iowa, Kansas, Manitoba, and Missouri. Evidently develops in rivers that are fairly rapid and of moderate size.

Illinois Records.—CENTRALIA: at light,

June 17, 1947, L. J. Stannard, 2 ♀. DIXON: at light, June 26, 1947, B. D. Burks, 1 ♀. KEITHSBURG: at light on Mississippi River, June, 1932, 1 ♂. MOUNT CARMEL: at light, Burks & Sanderson, 1 ♀. PROPHETSTOWN: dredging sandy bottom of Rock River 15 yards from bank, May 21, 1925, R. E. Richardson, 1 N; sweeping vegetation on bank of Rock River, June 26, 1947, B. D. Burks, 1 ♂, 2 ♀. QUINCY: at light, June 8, 1939, Burks & Riegel, 1 ♀ ; July 6, 1939, Mohr & Riegel, 1 ♀. ROCK FALLS: at light, June 26, 1947, B. D. Burks, 1 ♀. ROCKFORD: June 2, 1944, H. S. Dybas, 1 ♀.

40. *SIPHLOPLECTON* Clemens

Siphloplecton Clemens (1915a:258).

The adult wing venation in *Siphloplecton* is very similar to that in the heptageniid type, differing principally in that the cubital intercalary veins of the fore wing are partly or completely joined by crossveins to the branches of Cu, fig. 308. In the true heptageniid wing, these intercalary veins are free at the bases, as in fig. 317. In the hind wing of *Siphloplecton,* vein M is forked near the base, fig. 308. The hind tarsus has four clearly differentiated segments. The abdomen is relatively long and slender, and the ninth sternite of the female is entire and not greatly produced posteriorly, fig. 306. There are two long caudal filaments, with the median one represented by a three- to six-segmented stub.

In the nymphs, which are streamlined and fishlike, the eyes are placed laterally on the head, which is typically baetid in form, as in fig. 305. The mouth-parts are evidently not fitted for predatism. Each fore tarsal claw is bifid, fig. 303; all other claws are slender and longer than the tibiae, fig. 302. Gills are present on abdominal segments 1–7; these gills are single and plate-like on segments 4–7 and double on segments 1–3, except in *interlineatum,* where the gills of segments 1–3 have merely a small, recurved ventral flap. There are three well-developed caudal filaments; each of the cerci has long, dense setae on the mesal side only.

This genus includes four species, only one of which has been taken in Illinois. Another species occurs in Indiana and it will probably eventually be found to occur in Illinois.

KEY TO SPECIES

ADULTS

Crossveins in entire fore wing brown, hind wing with entire basal third shaded with brown..**1. basale**
Crossveins in fore wing brown only near costal margin and along stem of R_{4+5}; hind wing with brown shading confined to area around bases of C, Sc, and R...**2. interlineatum**

MATURE NYMPHS

Median ventral stripe of abdomen continuous, uninterrupted; claw of middle leg 10 times as long as wide at base.........**1. basale**
Median ventral stripe of abdomen interrupted at each intersegmental suture; claw of middle leg 7 to 8 times as long as wide at base.................**2. interlineatum**

1. *Siphloplecton basale* (Walker)

Baetis basalis Walker (1853:565).
Siphlurus flexus Clemens (1913:338).

MALE.—Thorax dark brown, with white areas on pleura. Each fore wing with a prominent, brown color pattern in anterior basal area; hind wing with an intense, brown cloud at base, fig. 308. Abdominal dorsum dark brown; sternum white, with sternites 1 and 9 dark brown and intermediate sternites each with three brown dots, one median and two lateral; genitalia as shown in fig. 311.

FEMALE.—Similar to the male but color lighter in tone, with brown shading at wing bases faint or absent.

N Y M P H.—Three longitudinal, brown bands present on abdominal venter; gills on segments 1–3 double; each caudal filament with a dark, broad, vaguely defined cross-band located just distad of middle.

Known from Indiana, Manitoba, Michigan, New York, North Carolina, Ontario, and Quebec.

2. *Siphloplecton interlineatum* (Walsh)

Baetis femorata Walsh (1862:386), not Say.
    Misidentification.
Baetis interlineata Walsh (1863:190).
    New name.

MALE.—Length of body and of fore wing 12–14 mm. Head and thorax dark brown to black, with light yellow markings on pleura and sternum. Wings hyaline, small area of brown staining at each wing base, costal crossveins margined with brown; short, longitudinal, brown dash in second interspace below bulla; fore leg gray-tan, femur with broad, brown band at base and near apex, dark brown mark at apex of tibia and each tarsal segment; middle and hind legs yellow, with brown markings as in fore leg. Abdominal tergum white, with faint, brown shading on tergites 2–6, tergites 1 and 7–9 almost completely dark brown; venter white, ganglia and two small, lateral spots on each sternite may be brown stained; sternites 1 and 9 completely dark brown; male genitalia, fig. 309; the two well-developed caudal filaments white, the articulations brown.

FEMALE.—Length of body 12–14 mm., of fore wing 14–15 mm. Color almost identical with that of male, except that wings completely lack brown staining at bases, and brown shading of abdominal tergites is usually more intense; abdominal sternum almost completely white, with sternite 1 largely brown and sternites 7–9 heavily shaded with brown; caudal filaments similar to those of the male.

NYMPH.—Length of body 15–16 mm., of caudal filaments 5–6 mm. Thorax yellow and brown on dorsum, yellow on venter. Abdominal tergum yellow, with varying brown markings and with tergites 1, 6, and 9 almost completely brown; venter yellow, with three longitudinal, brown stripes; gills single on segments 4–7, those on segments 1–3 each with a minute, recurved ventral flap; caudal filaments each with a brown crossband near tip.

Known from Illinois, Indiana, Manitoba, and Minnesota. Develops in fairly rapid, moderate-sized rivers.

Illinois Records.—ILLINOIS: 1 ♂. HAVANA: Chautauqua Park, April 29, 1914, 1 ♂. MOMENCE, Kankakee River: May 5, 1938, Ross & Burks, 3 ♀; May 8, 1940, Mohr & Burks, 2 ♀, 1 N. ROCK ISLAND: 5 ♂, 2 ♀ (Walsh 1862:369).

HEPTAGENIIDAE

This family, as limited here, corresponds to the family Heptageniidae in Traver's classification (1935a:293) and is similar to, but not identical with, Ulmer's family Ecdyonuridae (1933:212).

In the family Heptageniidae, the male adults have large compound eyes, but these eyes show no obvious division into upper and lower portions. The upper portion of each eye is, however, composed of facets which are slightly smaller and usually less heavily pigmented than are those in the lower. The line that divides these two portions is quite obscure. The fore tarsus in the males is longer than the fore tibia in all genera except the rare Anepeorus. In all genera, the hind tarsus has five clearly defined segments, figs. 17, 19. Each fore wing invariably has four cubital intercalary veins, all of which are free at the bases, figs. 317–321. The hind wing has veins R_4 and R_5 diverging at or near the center of the wing, figs. 317–320, except in Arthroplea, in which R_4 and R_5 are fused for their entire length, fig. 321. The male genitalia consist of a pair of four- or five-segmented forceps and a pair of variously modified penis lobes, figs. 331–356, 363–380, 387–389, 391–393. The median caudal filament is invariably vestigial in the adults.

In the nymphs, figs. 360, 383–386, 390, 394, 395, the body is flattened, the head is broad, almost or quite flat, with the eyes located dorsally and the mouth-parts ventrally. The labrum is always much wider than long, each mandible has two canines, and each maxillary and labial palp has two segments. The mouth-parts in all but one genus are fitted for a diet of vegetable matter, such as diatoms, and animal and vegetable detritus. The exception is a nymph, tentatively placed as that of Anepeorus, fig. 394, which has mouth-parts clearly fitted for predacity. Almost all heptagenine nymphs have seven pairs of dorsal abdominal gills, each of which is composed of a dorsal, platelike member and a ventral, filamentous tuft. In Rhithrogena, however, the filamentous tuft is dorsal and the platelike element ventral, in Arthroplea the filamentous portion of each gill is wanting entirely, and in the supposed nymph of Anepeorus the gills are ventral and each gill consists of a narrow, elongate, anterior member and a posterior, fimbriate member. All heptagenine nymphs have three well-developed caudal filaments.

The differentiation and separation of the genera in this group can hardly be said to be on a firm and rational basis. The nymphs and the male adults can be segregated generically, but the females usually cannot. Pending the discovery of generic characters which will serve for the segregation of the females, also, I am using the generic characters employed by Traver (1935) and Ulmer (1933).

KEY TO GENERA

ADULT MALES

1. Fore tarsus not more than three-fourths as long as fore tibia....**47. Anepeorus**
   Fore tarsus longer than fore tibia.......2
2. Vein R_{4+5} of hind wing unbranched, fig. 321..................**48. Arthroplea**
   Vein R_{4+5} of hind wing forked at or near center of wing, figs. 317–320.......3
3. First segment of fore tarsus as long as or longer than second.......**43. Epeorus**
   First segment of fore tarsus shorter than second..........................4
4. Stigmatic crossveins of fore wing anastomosed, figs. 320–323; penes a pair of relatively undifferentiated, elongate or stubby lobes, fig. 391....................
   ....**46. Rhithrogena**
   Stigmatic crossveins of fore wing not anastomosed, or only partly so, as in fig. 319, or anastomosed and forming two rows of cellules, as in fig. 322; penis lobes variously modified, with lateral or apical expansions, figs. 331–356, 363–380, 389..........................5
5. Penis lobes divided to base or fused on meson at base only, fig. 388..........
   ....**44. Cinygmula**
   Penis lobes fused on meson for at least their basal halves, figs. 331–356, 363–380, 389..........................6

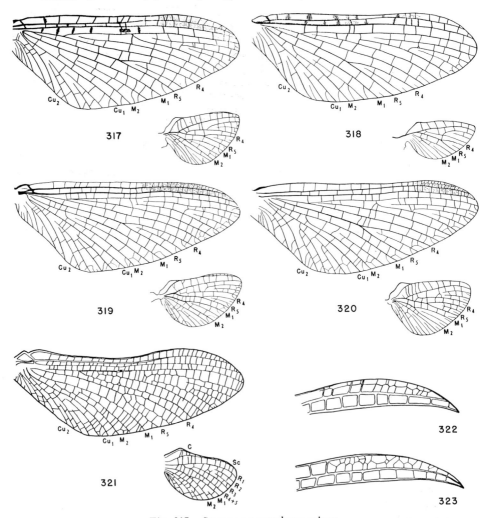

Fig. 317.—*Stenonema canadense,* wings.
Fig. 318.—*Heptagenia maculipennis,* wings.
Fig. 319.—*Epeorus pleuralis,* wings.
Fig. 320.—*Rhithrogena morrisoni,* wings.
Fig. 321.—*Arthroplea* sp., wings. (After Blair.)
Fig. 322.—*Cinygma integrum,* stigmatic area of fore wing.
Fig. 323.—*Rhithrogena morrisoni,* stigmatic area of fore wing.

6. Stigmatic area of fore wing with an irregular, longitudinal line dividing the stigmatic crossveins into two rows of cellules, fig. 322 . . . . **45. Cinygma**
Stigmatic area of fore wing not divided into two rows of cellules, fig. 318, although the crossveins are often partly or greatly anastomosed 7

width of each less than one-third the length . . . . **41. Stenonema**
Small subapical spine absent; large mesal spines robust, greatest width of each more than one-third the length, fig. 363 . . . . **42. Heptagenia**

12. Small, mesal subapical spine present, figs. 348, 350, 352 . . . . **41. Stenonema**

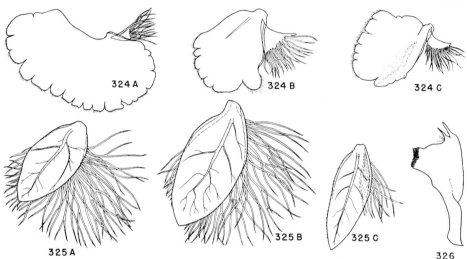

Fig. 324*A*.—*Rhithrogena* sp., gill of first abdominal segment.
Fig. 324*B*.—*Rhithrogena* sp., gill of fifth abdominal segment.
Fig. 324*C*.—*Rhithrogena* sp., gill of seventh abdominal segment.
Fig. 325*A*.—*Heptagenia diabasia,* gill of first abdominal segment.
Fig. 325*B*.—*Heptagenia diabasia,* gill of fifth abdominal segment.
Fig. 325*C*.—*Heptagenia diabasia,* gill of seventh abdominal segment.
Fig. 326.—*Heptagenia maculipennis,* mandible of mature nymph.

7. Outer lateral margin of each penis lobe with a cluster of spines which may be either large, figs. 333–339, or small, fig. 332 . . . . **41. Stenonema**
Outer lateral margin of each penis lobe without a cluster of spines 8
8. Apex of each penis lobe transversely or obliquely truncate, with a rounded or angulate, apicomesal corner, figs. 331, 340–347, 349–351, 353–356, 363–366 . . 9
Apex of each penis lobe not transversely or obliquely truncate 12
9. Apex of each lobe with the mesal angle lower than the apicolateral angle, figs. 331, 364–366 10
Apex of each lobe with mesal angle nearly on a level with, figs. 340–346, 349, 351, 353–356, 363, or higher than, figs. 347, 350, apicolateral angle 11
10. Apical angle of each penis lobe with a cluster of rather large spines, fig. 331 . . . . **41. Stenonema**
Apical angle of each penis lobe without a cluster of spines, figs. 364–366 . . . . **42. Heptagenia**
11. Small subapical spine, fig. 10*b*, present at apicomesal angle of each penis lobe, figs. 340–347, 349–351, 353–356; large mesal spines, fig. 10*g*, slender, greatest

Small, mesal subapical spine absent, figs. 366–370, 375, or, if present, greatly enlarged as in figs. 374, 376–380 . . . . **42. Heptagenia**

MATURE NYMPHS

1. Second segment of maxillary palp extremely long, recurved over dorsum of thorax, fig. 395 . . . . . . **48. Arthroplea**
Second segment of maxillary palp not recurved over dorsum of thorax 2
2. Mouth-parts fitted for predatism, with maxillae and mandibles fanglike . . . . **47. ?Anepeorus**
Mouth-parts fitted for a diet of vegetable matter or plant and animal detritus; each mandible with a broad molar area, figs. 326, 330 3
3. Median caudal filament vestigial, fig. 386 . . . . **43. Epeorus**
Median caudal filament well developed, figs. 360, 383–385 4
4. Gills on seventh abdominal segment slender and semifilamentous, either entirely without tracheae or with one, two, or three simple tracheae without lateral branches, fig. 360 . . . . **41. Stenonema**

Gills on seventh abdominal segment platelike and bearing tracheae which possess lateral branches, figs. 325C, 328C, 329C.................5
5. Front of head incised on meson, so as to expose a portion of labrum when viewed from dorsal aspect....**44. Cinygmula**
Front of head not incised on meson, labrum not exposed dorsally........6
6. Gills of first and seventh pairs enlarged, converging beneath abdomen to form, with intermediate gills, an adhesive disc, fig. 390; fibrillar portion of each gill dorsal, fig. 324...**46. Rhithrogena**
Gills of firᵉt and seventh pairs not converging beneath abdomen, gills not forming an adhesive disc; fibrillar portion of each gill ventral, figs. 325, 329..
...............................7

7. Each mandible with a heavily chitinized lobe on mesal margin, basad of molar area, fig. 330; platelike component of first abdominal gill only two-thirds as long as that of seventh gill, fig. 329...
.......................**45. Cinygma**
Mandibles without a chitinized lobe basad of molar area, fig. 326; platelike components of first and seventh gills subeqial in length, fig. 325..........
...................**42. Heptagenia**

41. *STENONEMA* Traver

Stenonema Traver (1933a:173).

Most of the American species in the genus *Stenonema* were formerly placed under the

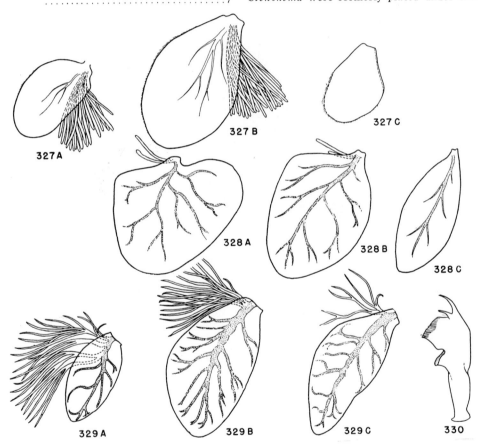

Fig. 327A.—*Epeorus* sp., gill of first abdominal segment.
Fig. 327B.—*Epeorus* sp., gill of fifth abdominal segment.
Fig. 327C.—*Epeorus* sp., gill of seventh abdominal segment.
Fig. 328A.—*Cinygmula* sp., gill of first abdominal segment.
Fig. 328B.—*Cinygmula* sp., gill of second abdominal segment.
Fig. 328C.—*Cinygmula* sp., gill of seventh abdominal segment.
Fig. 329A.—*Cinygma integrum*, gill of first abdominal segment. (After McDunnough.)
Fig. 329B.—*Cinygma integrum*, gill of second abdominal segment. (After McDunnough.)
Fig. 329C.—*Cinygma integrum*, gill of seventh abdominal segment. (After McDunnough.)
Fig. 330—*Cinygma integrum*, mandible of mature nymph. (After McDunnough.)

generic names *Heptagenia* and *Ecdyonurus* (or the emended form *Ecdyurus*). As Traver (1933a: 173) pointed out, however, these American species actually represent a discrete generic unit, differing from either *Heptagenia* or *Ecdyonurus* both in type of male genitalia and in nymphal characters.

Stenonema is the most difficult genus in the order Ephemeroptera. This is due to the fact that the male genitalia are quite similar throughout the genus, and the genitalic differences between closely related species are, therefore, obscure. The principal bases for the separation of species are the colors of the males and the relative proportions of the male first and second fore tarsal segments. Unfortunately, both vary considerably. Specimens collected early in the season usually are much darker and larger than specimens of the same species collected in midsummer. In the large male specimens, the first fore tarsal segment tends to be relatively longer than it is in smaller specimens. However, a decision as to the proper limitation of a given species can be made when all available characters of the adult males and the mature nymphs are considered together. Fortunately, most of our species of *Stenonema* have been reared, and good series of nymphs and adult males are available for study. The females are usually separable at best to species groups only.

In the adult males, the large eyes, never contiguous on the meson, are usually separated on the vertex by a space at least one-half as wide as one eye. Each fore leg is, in all species except *integrum,* as long as, or slightly longer than, the body. In *integrum,* the fore leg is slightly shorter than the body. In all species, the fore tarsus is longer than the fore tibia and the first fore tarsal segment varies from one-third to four-fifths as long as the second segment. The wing venation in both sexes, fig. 317, is typical for the family; the species of *Stenonema* and *Heptagenia* cannot, unfortunately, be distinguished generically by the characteristics of the wings. The male genitalia, figs. 331–356, consist of a pair of four-segmented forceps, the second segment being longer than the other three combined, and a pair of Γ-shaped penis lobes.

Use of these generic characters will require the transfer of the species *Epeorus modestus* Banks (1910: 202) to the genus *Stenonema.*

The nymphs, fig. 360, are greatly flattened, the legs sprawling laterally; the anterior margin of the head is entire or only very slightly emarginate on the meson, and the eyes are dorsal and just touching the posterior margin of the head. Each tarsal claw is single, short, slightly hooked at the apex, and has a fairly large ventral tooth near the base; in many species, each claw also has two or three minute ventral denticles near the tip. The abdominal gills are of three types: those on segments 1–6 in the members of the *interpunctatum* group are pointed at the apexes, fig. 358, and each gill of the seventh pair has one longitudinal trachea; in the members of the *tripunctatum* group, the first six pairs of gills are rounded at the apexes, fig. 357, and each gill of the seventh pair has one or two longitudinal tracheae; in the *pulchellum* and *bipunctatum* groups, the gills on segments 1–6 are truncate at the apexes, fig. 359, and the gills of the seventh pair are without tracheae. The three caudal filaments are equally long, uniformly clothed with short setae, and, often, with alternating pairs of segments dark and light in color.

All Illinois species of *Stenonema* develop under stones in the shallower parts of creeks and rivers. Some, such as *tripunctatum,* can tolerate a great deal of silt; in many of the small, sluggish, heavily silted streams in central Illinois, *tripunctatum* is the only mayfly now found. The later instar nymphs of most species of *Stenonema* can be reared through to maturity in stagnant water.

Adult specimens of *Stenonema* should be pinned for preservation and study. Most specimens, if collected and preserved in alcohol, cannot reliably be determined to species, as the necessary color characters, at best impermanent, are quickly lost in alcohol; dry specimens retain their colors for several years if stored out of the light. Correct specific determinations can seldom be made from the male genitalia and fore tarsal characters alone. The color of the eyes in adult males should be recorded at the time of collection, as these eye colors are helpful in placing specimens in the proper species or species group.

My views as to specific limits in this genus most closely coincide with those of Traver (1935a: 297).

The adult females can be distinguished only to groups.

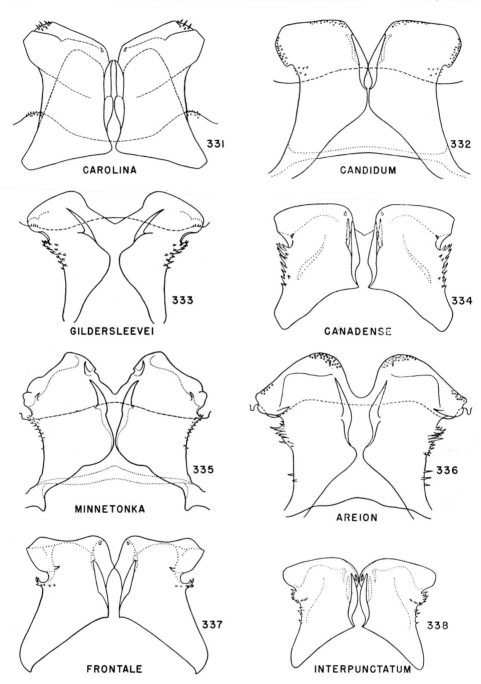

Fig. 331.—*Stenonema carolina*, male genitalia.
Fig. 332.—*Stenonema candidum*, male genitalia.
Fig. 333.—*Stenonema gildersleevei*, male genitalia.
Fig. 334.—*Stenonema canadense*, male genitalia.
Fig. 335.—*Stenonema minnetonka*, male genitalia.
Fig. 336.—*Stenonema areion*, male genitalia.
Fig. 337.—*Stenonema frontale*, male genitalia.
Fig. 338.—*Stenonema interpunctatum*, male genitalia.

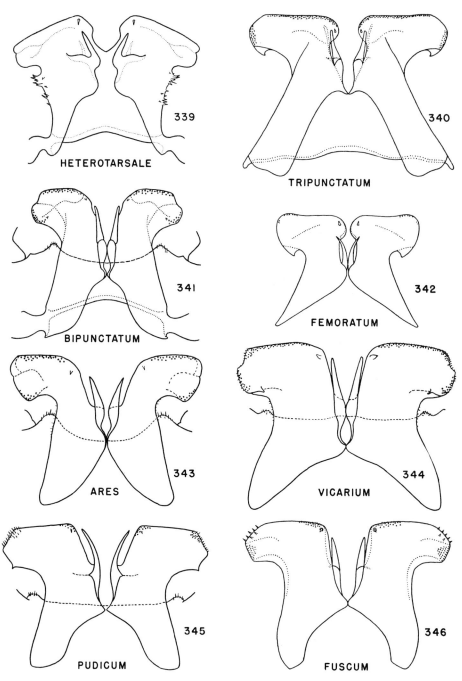

Fig. 339.—*Stenonema heterotarsale*, male genitalia.
Fig. 340.—*Stenonema tripunctatum*, male genitalia.
Fig. 341.—*Stenonema bipunctatum*, male genitalia.
Fig. 342.—*Stenonema femoratum*, male genitalia.
Fig. 343.—*Stenonema ares*, male genitalia.
Fig. 344.—*Stenonema vicarium*, male genitalia.
Fig. 345.—*Stenonema pudicum*, male genitalia.
Fig. 346.—*Stenonema fuscum*, male genitalia.

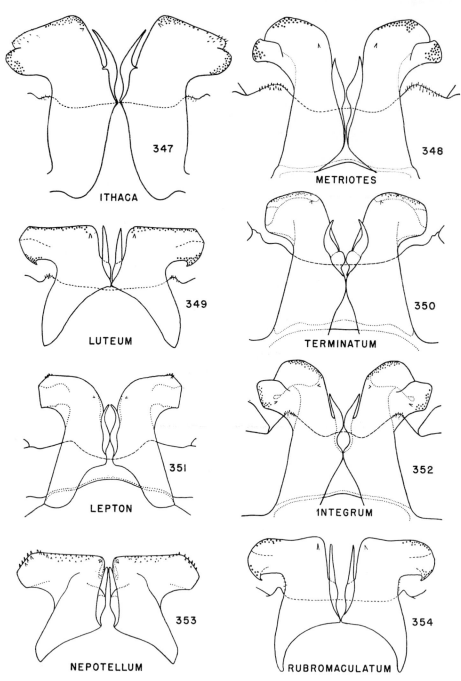

347

ITHACA

348

METRIOTES

349

LUTEUM

350

TERMINATUM

351

LEPTON

352

INTEGRUM

353

NEPOTELLUM

354

RUBROMACULATUM

Fig. 347.—*Stenonema ithaca,* male genitalia.
   Fig. 348.—*Stenonema metriotes,* male genitalia.
Fig. 349.—*Stenonema luteum,* male genitalia.
   Fig. 350.—*Stenonema terminatum,* male genitalia.

Fig. 351.—*Stenonema lepton,* male genitalia.
Fig. 352.—*Stenonema integrum,* male genitalia.
   Fig. 353.—*Stenonema nepotellum,* male genitalia.
   Fig. 354.—*Stenonema rubromaculatum,* male genitalia.

KEY TO SPECIES

ADULT MALES

1. Basal crossveins in first radial interspace of fore wing thickened and darkened in the middle, fig. 317; crossveins below bulla often connected by a longitudinal, black mark, always at least a black spot in the middle of one, two, or three crossveins below bulla...............2

   Basal crossveins in first radial interspace of fore wing not thickened and darkened, or thickened uniformly from end to end; longitudinal, black mark never connecting crossveins below bulla, and these crossveins never widened and darkened in the middle only........10

2. Caudal filaments and genital forceps uniformly gray-tan; genitalia as in fig. 331, forceps base with a pair of large, sublateral, setose projections..**1. carolina**

   Caudal filaments not uniformly gray-tan; filaments entirely white, without dark-colored rings at articulations, or these filaments white, yellow, or tan, with darker articulations; genital forceps white or yellow; eyes in life always light green.........................3

3. Outer lateral margin of each penis lobe with extremely minute spines, fig. 332 ....**2. candidum**

   Outer lateral margin of each penis lobe with a cluster of large spines, figs. 333–339............................4

4. Venter of abdomen with a median, longitudinal, dark brown or black line extending from anterior to posterior ends, this line slightly widened and with a narrow interruption at posterior margin of each sternite.......**3. gildersleevei**

   Venter of abdomen usually entirely immaculate, occasionally with a faint mark on meson of posterior margin of some sternites......................5

5. Abdomen white, with a bright Mars orange crossband at posterior margin of each tergite 1–7........**4. areion**

   Abdomen yellow, with black or very dark brown crossbands at posterior margins of tergites.........................6

6. Spiracular dots present on abdominal segments, in very dark specimens these dots tending to fuse with black crossbands at posterior margins of tergites; each penis lobe with an apicomesal and a mesially directed lateral spine, figs. 334, 335, 337......................7

   Spiracular dots absent from abdominal segments; each penis lobe with only the apicomesal spine, figs. 338, 339......9

7. Thoracic pleuron uniformly yellow, lacking an oblique, black streak ventral to base of fore wing.....**5. minnetonka**

   Thoracic pleuron with an oblique, black streak ventral to base of fore wing, also a similar streak sometimes present ventral to base of hind wing........8

8. Abdominal tergites with a relatively broad, black crossband at posterior margin of each; a longitudinal, black or dark gray line usually extending almost or quite the entire length of abdominal dorsum; in very dark specimens abdomen almost completely black on dorsum; meso- and metascutellum dark brown to black....**6. canadense**

   Abdominal tergites with a narrow, black line at posterior margin of each; median, dark, longitudinal line absent from abdominal dorsum, or, at most, with this line evident on basal one to three tergites only; meso- and metascutellum yellow....................**7. frontale**

9. Black marks always present on face ventral to antennal sockets; usually a dark brown, oblique streak present on pleuron ventral to base of fore wing; first segment of fore tarsus one-fourth to one-third as long as second segment ....**8. interpunctatum**

   Black marks usually absent from face ventral to antennal sockets; dark streak never present on pleuron; first fore tarsal segment one-half to two-thirds as long as second segment..... ....**9. heterotarsale**

10. Abdominal tergites 1– or 2–8 with three black marks at posterior margin of each: a median dot and a pair of submedian, transverse dashes.........11

    Abdominal tergites with a pair of submesal, transverse, black lines at posterior margin of each; or with a continuous, black or dark brown crossline at each posterior margin; or with dark brown shading covering posterior one-quarter to one-half of each tergite; or with tergites almost or entirely unmarked.........................12

11. Outer margin of hind wing hyaline; compound eyes separated on meson by a space as wide as a lateral ocellus...... ....**10. tripunctatum**

    Outer margin of hind wing shaded with dark brown; compound eyes separated on meson by a space two-thirds as wide as one lateral ocellus..**11. femoratum**

12. Abdominal tergites 3–8 each with a pair of submedian, transverse, short lines at posterior margin..................13

    Abdominal tergites with a continuous, dark brown or black crossline or color band at posterior margin of each or virtually without markings at posterior margin..........................14

13. Mesonotum dull gray-brown, first fore tarsal segment two-thirds as long as second............**12. bipunctatum**

    Mesonotum bright Mars orange, first fore tarsal segment one-half as long as second.....................**13. ares**

14. Ground color of abdomen dark tan or yellow-brown; each abdominal tergite 2–7 with a broad, dark brown, transverse color band occupying posterior one-fourth to one-half of tergite, apical three tergites almost completely shaded with dark brown.................15

Ground color of abdomen white or yellow; abdominal tergites 1–7 each with a narrow, black, transverse stripe at posterior margin or this margin virtually without markings..............18

15. Eyes of living insect brown; outer margin of hind wing shaded with brown ....**14. pudicum** Eyes of living insect gray; outer margin of hind wing not shaded..........16

16. Stigmatic area of fore wing shaded with dark red, this pigmentation concentrated in basal part of stigmatic area ....**15. vicarium** Entire stigmatic area of fore wing uniformly stained with yellow-brown..17

17. Abdominal tergites 8 and 9 dark brown, with lateral margins white........... ....**16. fuscum** Abdominal tergites 8 and 9 uniformly dark brown..............**17. ithaca**

18. Posterior margin of each tergite 2–7 with only minute, black dash on meson...19 Posterior margin of each tergite 2–7 with a black line extending completely across dorsum; occasional specimens entirely lacking marks at these posterior margins..........................20

19. Mesonotum dark brown; abdomen usually with spiracular dots.............. ....**18. mediopunctatum** Mesonotum chalky white; abdomen always lacking spiracular marks of any kind................**19. metriotes**

20. Abdomen lacking spiracular marks of any kind............................21 Abdomen with spiracular dots or oblique streaks on at least middle segments.. 23

21. Articulations of caudal filaments dark red-brown throughout; middle and hind femora each with a prominent, red-brown crossband in middle and at apex...................**20. luteum** Caudal filaments white or a faint yellow throughout, articulations not darkened, or sometimes basal 2 or 3 articulations only of each filament darkened; middle and hind femora each with red-brown shading at apex only, occasionally middle femur with faint shading in the middle..........................22

22. First fore tarsal segment two-fifths as long as second segment..**21. terminatum** First fore tarsal segment two-thirds to five-sixths as long as second segment.. ....**22. lepton**

23. Abdominal spiracular marks a series of short, oblique streaks....**23. integrum** Abdominal spiracular marks a series of dots..............................24

24. Mesonotum light clay-colored or yellow-brown..........................25 Mesonotum dark red-brown or black-brown..........................26

25. Pink-brown shading almost completely covering abdominal tergite 8; mesonotum yellow-brown....**24. nepotellum** Pink-brown shading of tergite 8 restricted to a median, longitudinal bar; mesono-

tum light clay-colored.............. ....**25. rubromaculatum**

26. Mesonotum red-brown; stigmatic area of fore wing stained a faint pink; only apex of mesoscutellum white........ ....**26. rubrum** Mesonotum dark, blackish-brown; stigmatic area of fore wing stained light brown; entire mesoscutellum white... ....**27. pulchellum**

KEY TO SPECIES GROUPS

ADULT FEMALES

1. Fore wing with basal crossveins in first radial interspace thickened and darkened in the middle, as in fig. 317; two or three crossveins below bulla thickened and blackened in the middle, these black spots often fused to form a short, longitudinal dash............. ....**interpunctatum group** Fore wing with basal crossveins in first radial interspace not thickened and darkened in the middle; two or three crossveins below bulla never thickened and darkened in the middle and never with a short, longitudinal black dash connecting these crossveins 2

2. Abdominal tergites 2–8 with three small, black marks at posterior margin of each: a median, black dot and a pair of submedian, transverse dashes........ ....**tripunctatum group** Abdominal tergites not with three small, black marks at posterior margin of each........................3

3. Abdominal tergites 3–8 each with a pair of short, submedian, transverse lines at posterior margin.................... ....**bipunctatum group** Abdominal tergites 2– or 3–7 or –8 each with a continuous, dark crossline or color band at posterior margin, or each of these tergites virtually or quite without markings at posterior margin....4

4. Ground color of abdomen tan or yellow-brown; tergites 2–7 each with a broad, dark brown, transverse band occupying posterior fourth to half of tergite.. ....**vicarium group** Ground color of abdomen light yellow or white; tergites entirely unmarked, with spiracular dots only, or with a narrow, black, transverse line at posterior margin of each of some or all tergites................................5

5. Abdomen lacking spiracular marks of any kind............**terminatum group** Abdomen with spiracular dots or oblique streaks....**mediopunctatum group, pulchellum group**

KEY TO SPECIES

MATURE NYMPHS

1. Gills borne by abdominal segments 1–6 pointed at apexes, fig. 358..........2 Gills borne by abdominal segments 1–6

rounded or truncate at apexes, figs. 357, 359.........................6
2. Dorsum of abdomen almost entirely brown, each tergite with only a pair of short, submedian streaks at anterior margin...................**1. carolina**
Dorsum of abdomen marked otherwise..3
3. Abdominal dorsal color pattern made up of a series of large, elongate and submedian, pale spots, with a transverse, black, median line at posterior margin of each of tergites 1–9...............
.....................**3. gildersleevei**
Abdominal dorsum with a pair of submedian, pale, longitudinal stripes, in most specimens these markings almost or quite continuous, but in others confined to tergites 4– or 5–10 and, in that case, stripes suddenly widened on tergites 8 and 9, constricted again on 10..............................4
4. Submedian, pale stripes present on tergites 3– or 4–10 only, these stripes suddenly widened on tergites 8 and 9, constricted again on tergite 10..........
.....................**9. heterotarsale**
Submedian, pale stripes extending length of abdomen and not suddenly widened on tergites 8 and 9.................5
5. Apex of ninth sternite entirely pale or very faintly stained with brown; anterior, dorsal margin of head usually with a pale, median spot..**7. frontale**
Apex of ninth sternite with a broad, dark brown crossband; anterior, dorsal margin of head without a pale, median spot or with such a spot only very faintly indicated......**2. candidum; 6. canadense; 8. interpunctatum**
6. Gills borne by abdominal segments 1–6 rounded at apexes, fig. 357..........7
Gills borne by abdominal segments 1–6 truncate at apexes, fig. 359.........8
7. Median, pale spot usually present on anterior margin of head; sublateral, brown spots present on sternites 2–8..
.....................**10. tripunctatum**
Median, pale spot absent from anterior margin of head; sublateral, brown spots present on sternites 5– or 6–8, sometimes these spots entirely absent......
.....................**11. femoratum**
8. Each tarsal claw with two minute ventral denticles near tip.................9
Tarsal claws without ventral denticles.13
9. Posterolateral angles of abdominal segments 3– or 4–9 spinelike.........10
Posterolateral angles of abdominal segments 6– or 7–9 spinelike..........11
10. Pale spot on meson of anterior margin of head; abdominal dorsum light brown, with large, median, pale spots on tergites 5, 8, and 9....**12. bipunctatum**
Head without median, pale spot on anterior margin; abdominal dorsum usually dark brown, apical tergites sometimes with small, faint, pale spots at anterior margins.......**25. rubromaculatum**
11. Each abdominal sternite 2–8 with somewhat irregular, brown shading extend-

ing across posterior margin and at each lateral margin..........**20. luteum**
Each abdominal sternite 1–7 entirely white, sternite 8 either entirely white, or with only a small, median, brown spot at anterior margin..........12
12. Normally exposed portions of abdominal tergites 1–10 uniformly brown, sometimes with faint, median, lighter spots on apical tergites........**26. rubrum**
Normally exposed portions of abdominal tergites 1–5, 7, and usually 9 mostly white, but with some small, brown markings; tergites 6, 8, and 10 almost entirely brown, tergite 9 sometimes also mostly brown........**27. pulchellum**
13. Posterolateral angles of abdominal segments 6– or 7–9 spinelike...........14
Posterolateral angles of abdominal segments 3– or 4–9 spinelike 15
14. Abdominal sternites 4– or 5–8 each with a medially sinuate, transverse, brown cross-stripe located near middle of sternite..................**17. ithaca**
Abdominal sternites 1–8 white, without markings except for slight, vague darkening along lateral margins of more posterior sternites......**13. ares**
15. Abdominal sternites 3– or 4–8 each with a brown crossbar located at posterior margin.........................16
Abdominal sternites 3– or 4–8 each with transverse, brown shading at anterior margin or in middle of sternite......17
16. Sternite 9 with entire posterior half dark brown.................**15. vicarium**
Sternite 9 with a pair of large, brown, sublateral spots at posterior margin...
.....................**16. fuscum**
17. Each sternite 3– or 4–8 with a transverse, brown bar occupying median two-thirds of anterior margin.**14. pudicum**
Each sternite 3– or 4–8 with a medially sinuate, brown crossband extending from side to side near anterior margin.
.....................**24. nepotellum**

INTERPUNCTATUM Group

1. *Stenonema carolina* (Banks)

Heptagenia carolina Banks (1914:616).

MALE.—Length of body 9–10 mm., of fore wing 11–12 mm. Compound eyes relatively small, separated on vertex by a space as wide as one eye; head and thorax yellow to tan-yellow; first fore tarsal segment one-half to three-fifths as long as second segment; wings hyaline, with veins and crossveins red-brown; basal subcostal and first radial crossveins blackened and widened at their anterior ends, and two crossveins in first radial interspace, below bulla, thickened and blackened in the middle, these marks sometimes fusing to form a longitudinal,

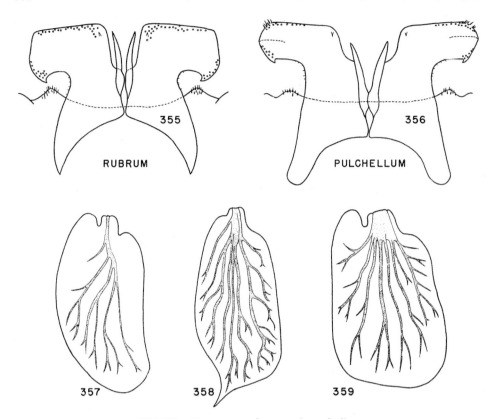

Fig. 355.—*Stenonema rubrum*, male genitalia.
Fig. 356.—*Stenonema pulchellum*, male genitalia.
Fig. 357.—*Stenonema tripunctatum*, fifth gill of nymph.
Fig. 358.—*Stenonema interpunctatum*, fifth gill of nymph.
Fig. 359.—*Stenonema nepotellum*, fifth gill of nymph.

black dash; stigmatic area stained yellow-brown; outer margin of hind wing dark brown. Abdomen white, with a faint gray or green tinge; posterior margin of each tergite bordered with brown or black; spiracular dots absent; penis lobes lacking lateral spines, fig. 331; genital forceps and caudal filaments uniformly gray-tan.

Nymph.—Length of body 10–12 mm. No pale, mesal spot on anterior margin of head. Abdominal dorsum almost uniformly brown, with only a pair of short, narrow, submedian, light-colored marks at anterior margin of each tergite; sternum uniformly light gray-tan, without darker markings; gills on segments 1–6 pointed at apexes, on 7 slender, with a single, longitudinal trachea in each; caudal filaments dark.

Known from New York, North Carolina, Quebec, South Carolina, Tennessee, and West Virginia.

2. *Stenonema candidum* Traver

Stenonema candidum Traver (1935a:308).

Spieth (1947:109) considers this to be a synonym of the form he designates *interpunctatum frontale* (Banks). I have seen the types of *candidum* and, as can be seen from fig. 332, the genitalia are quite distinct from those of all other species in the genus.

Male.—Length of body 7–8 mm., of fore wing 9–10 mm. Head deep yellow, a black mark on face below each antennal socket; antennal scape and pedicel dark yellow, flagellum gray-tan at base, fading to white at apex; eyes in life light gray-green. Thorax bright yellow, with mesonotum brown except on lateral margins and at apex of scutellum; legs yellow, with fore leg a little darker yellow, each femur with a median and an apical black band, median band of hind femur sometimes reduced or wanting; the first fore tarsal segment three-fifths to two-thirds as long as second; wings hyaline, veins light yellow-brown, crossveins in anterior two-thirds of fore wing dark brown, those in basal half of first radial interspace thickened and blackened in the middle, a longitudinal, black dash below the bulla usually not present, but occasionally faintly indicated, the stigmatic area faintly stained with brown, outer margin of hind wing slightly darkened. Abdomen bright yellow, with a black, transverse line at posterior margin of each tergite and with a slightly wider, black mark at meson and near either lateral margin, somewhat suggesting the color pattern of *femoratum*; large, black mark at each spiracle; tergites 8 and 9 suffused with red-brown; genitalia, fig. 332, light yellow, penis lobes with minute, lateral spines; caudal filaments entirely white.

NYMPH.—Length of body 8–9 mm. Head light brown, anterior, dorsal margin of head without a median, white spot, or with such a spot only faintly indicated, triangular, white mark in front of anterior ocellus. Pronotum with a pair of large, sublateral, white spots at anterior margin, thoracic notum otherwise light brown except on median line and on sutures anterior to wing bases; legs brown, each femur with a basal, median, and apical, white crossband, each tibia white near base and at apex. Abdomen dorsally light brown, with a pair of narrow, discontinuous, submedian, white lines; gills 1–6 pointed at apexes, seventh gill with a single trachea, venter of body white, lateral margins of sternite 8 and lateral and apical margins of 9 brown; caudal filaments white, alternating articulations faint brown.

Known from Illinois and Ohio.

Illinois Records. — EDDYVILLE: Lusk Creek, June 6, 1946, Mohr & Burks, 3 ♂ ; Belle Smith Spring, June 7, 1946, Mohr & Burks, 1 ♂. GRAYVILLE: Wabash River, April 10, 1946, Mohr & Burks, 1 ♂, 4 N.

3. *Stenonema gildersleevei* Traver

Stenonema gildersleevei Traver (1935a:315).

MALE.—Length of body 9–11 mm., of fore wing 10–12 mm. Head orange-tan, usually a continuous, black line crossing face below antennal bases; each antennal scape and pedicel white, flagellum gray-brown;

black markings on vertex between eyes. Thorax largely red-brown, pronotum mostly shaded with dark gray or black, dark shading along dorsal sutures of mesonotum and on mesoscutellum, metanotum black in median dorsal area; thoracic pleura with black shading below wing bases; sternum tan. All coxae light brown, the femora tan, with a median and an apical, dark brown crossband on each, tibiae yellow, tarsi tan, shaded with faint gray, first fore tarsal segment one-half as long as second; wings hyaline, veins and crossveins red-brown, crossveins below bulla usually connected by a black dash, stigmal area faintly stained with yellow, outer margin of hind wing shaded with light brown. Ground color of abdomen tan, overlaid with lavender-black shading: mid-dorsal line, posterior margin of each tergite, large sublateral area near posterior margin, and spiracular spot all dark shaded; on venter, mid-ventral line, anterior margin of each sternite, and, usually, a semitriangular, sublateral spot at anterior margin of each sternite dark shaded; genitalia, fig. 333, with penis lobes and forceps tan or yellow-brown; caudal filaments light yellow, articulations near apexes slightly darker.

NYMPH.—Length of body 11–13 mm. No pale spot on meson of anterior margin of head; gills borne by abdominal segments 1–6 pointed at apexes, seventh pair slender, with a single, longitudinal trachea in each; abdominal tergites each with a pair of submedian, longitudinal, light-colored streaks and a transverse, black crossband at posterior margin; abdominal venter white, with dark, longitudinal markings faintly indicated at lateral margins of sternites 7–9; caudal filaments yellow.

Known from Illinois, New York, and Ohio.

Illinois Record.—KANKAKEE: at light, June 6, 1935, Ross & Mohr, 2 ♂ .

4. *Stenonema areion* new species

This species resembles *interpunctatum* in lacking spiracular dots on the abdomen and *heterotarsale* in lacking the oblique, black mark below the base of the fore wing on either pleuron; *areion* differs from both those species in having a bright Mars orange crossband at the posterior margin of each abdominal tergite, and in having the ground color of the abdomen white rather than yellow, as in *interpunctatum* and *heterotarsale*.

MALE.—Length of body 7 mm., of fore wing 8 mm. Face below antennal sockets light yellow, a small, black mark on margin of frontal shelf ventral to each antennal socket; each antenna yellow, flagellum slightly grayed near base; area of face between antennal sockets and ocelli deep yellow; vertex Mars orange; eyes in life pale green. Pronotum yellow, black streak on either side; mesonotum amber-brown, with red-brown shading at posterior ends of outer parapsides and on lateral margins anterior to fore wing bases; mesoscutellum yellow in the center, Mars orange at margins, Mars orange shading also present on lateral margins of mesonotum posterior to fore wing bases; pleura bright yellow, minute, a dark brown point on each middle and hind coxal suture; thoracic sternum bright yellow. Wings hyaline, stigmatic areas stained with brown, veins light yellow-brown, crossveins black, two crossveins below bulla connected by black dash; veins and crossveins of each hind wing pale yellow, but crossveins and intercalaries at outer margin black; fore femur deep yellow, median and apical, dark brown crossbands present, tibia pale yellow, apex black, tarsus white, apexes of segments slightly darkened, first segment three-fifths as long as second segment; middle and hind legs white, each femur with a faint, dark brown shading in middle and a well-marked, brown band at apex. Abdomen white, posterior margin of each tergite 1–7 with Mars orange crossband; spiracular dots absent; apical three tergites yellow-orange, with overlying Mars orange shading; genitalia, fig. 336, white; caudal filaments white, articulations not darkened.

Holotype, male. — Oakwood, Illinois, June 25, 1948, B. D. Burks. Specimen dry, on pin.

Paratypes.—Same data as for holotype, 2 ♂. Specimens dry, on pins; genitalia on microscope slide.

5. *Stenonema minnetonka* Daggy

Stenonema minnetonka Daggy (1945:376).

MALE.—Length of body 9–10 mm., of fore wing 10–11 mm. Head yellow, a black mark on face ventral to each antennal base; vertex shaded with orange-brown on meson and at posterior margin; antennal scape and pedicel yellow, flagellum gray-tan at base, hyaline at tip. Mesonotum brown, scutellum yellow; semimembranous area anterior to base of fore wing shaded with light orange-brown, pleuron yellow, sternum yellow; fore leg deep yellow, apex of tibia and apexes of all tarsal segments black, first tarsal segment one-half as long as second; middle and hind legs yellow, tarsi shaded with gray-brown; middle and apex of each femur usually shaded with dark brown, middle band on hind femur sometimes obsolescent; wings hyaline, stigmatic area of fore wing stained with brown, veins yellow-brown, crossveins very dark brown or black; hind wing with all veins and crossveins yellow, outer margin dark brown. Abdomen yellow, black crossband at posterior margin of each tergite 1–8 with a black crossline; spiracular marks present; posterior half of tergite 8 and all of tergites 9 and 10 shaded with orange-brown; genitalia, fig. 335, yellow; caudal filaments gray-yellow, articulations brown.

Known from Illinois and Minnesota.

Illinois Records. — BENTON: at light, June 10, 1946, H. H. Ross, 2 ♂. FREEPORT: June 10–11, 1948, Burks, Stannard, & Smith, 1 ♂. QUINCY: July 6, 1939, Mohr & Riegel, 1 ♂. ROCKFORD: May 22, 1941, Ross & Burks, 1 ♂. SHAWNEETOWN: July 14, 1948, Mills & Ross, 1 ♂.

6. *Stenonema canadense* (Walker)

Baetis canadensis Walker (1853:569).
Stenonema interpunctatum canadense (Walker). Spieth (1947:107).
Stenonema conjunctum Traver (1935a:309).
Stenonema ohioense Traver (1935a:322).
Stenonema proximum Traver (1935a:325).

Spieth (1947:109) considers that *proximum* and *conjunctum* are synonyms of the form that he designates as *interpunctatum frontale* (Banks).

In the Museum of Comparative Zoology, the specimens determined as *interpunctata* (Say) by Walsh are clearly of the species we now are calling *canadense* (Walker). Walsh's redescription (1862:374) of *interpunctata* (Say) also obviously fits the present-day concept of *canadense,* and, as Spieth has shown (1940:333), the current concept of *canadense* is in agreement with Walker's type. The type of *interpunctata* is lost, but the original description of this species more nearly matches the species at present called by that name than it does *canadense*. Walsh's use of the name *interpunctata* thus may safely be considered to have been based on a misidentification.

Stenonema canadense is an extremely variable species, being almost entirely black in northern Ontario and grading to an almost entirely yellow form in southern Illinois.

MALE.—Length of body 8–10 mm., of fore wing 9–11 mm. Face white, with black streak ventral to each antennal base, in darkest specimens entire frontal shelf black; eyes in life light green; vertex red-brown, often with median, black shading. Mesonotum dark red-brown, often with a longitudinal, median, black stripe, scutellum sometimes yellow-brown, usually dark; semimembranous area anterior to base of fore wing yellow to bright red-brown; pleuron yellow, with variable amounts of dark brown and black shading, and an oblique, dark streak always present ventral to base of fore wing; sternum yellow, sometimes shaded with dark brown; fore leg deep yellow to light brown, apex of tibia black, tarsus largely shaded with gray, first tarsal segment from two-fifths to three-fifths as long as second; middle and hind legs yellow to tan; all femora with median and apical, dark brown or black shaded areas; wings hyaline, stigmatic area of fore wing stained with brown, veins yellow-brown, crossveins dark brown or black; veins and crossveins in hind wing yellow to yellow-brown, outer margin of hind wing shaded with brown. Abdomen with ground color yellow; tergites varying widely, in some specimens almost completely black, in others with only a black crossline at posterior margin of each tergite and a median, longitudinal, gray line extending partly or completely the length of dorsum; spiracular marks present in lighter specimens, these marks fusing with dark shading of tergites in darker specimens; apical three tergites shaded with red-brown; genitalia, fig. 334, yellow; caudal filaments gray-yellow to light yellow, articulations dark brown.

NYMPH.—Length of body 8–11 mm. Head anterior to eyes uniform brown, no median, pale spot on meson of anterior margin, but usually pale spot on this margin anterior to each antennal socket; large, pale spots lateral to compound eyes. Thoracic dorsum mostly brown, a pair of large, sub-

lateral, pale spots usually present on pronotum; tarsal claws without ventral denticles. Dorsum of abdomen brown, with a pair of somewhat variable, longitudinal, submedian, pale stripes and a row of pale spots near either lateral margin; gills borne by segments 1–6 pointed at apexes, gills of seventh pair each with one trachea; apical two or three abdominal sternites with longitudinal, brown band at lateral margins, more anterior sternites either unmarked or with vague, brown spots near lateral margins, entire apical fourth of ninth sternite brown; posterolateral angles of segments 7–9 spinelike; caudal filaments light brown, articulations perceptibly darker near apexes of filaments.

Known from Arkansas, Connecticut, Illinois, Indiana, Manitoba, Michigan, Minnesota, New Jersey, New York, North Carolina, Ohio, Oklahoma, Ontario, Tennessee, Quebec, West Virginia, and Wisconsin.

Illinois Records.—Adult specimens, collected April 10 to September 17, are from Alton, Apple River Canyon State Park, Aroma Park, Aurora, Carlinville, Cedarville, Charleston, Chicago, Crescent City, Erie, Fieldon, Freeport, Galena, Golconda, Havana, Kankakee, La Grange, Mahomet, Milan, Momence, Mount Vernon, Oakwood, Oregon, Palisades State Park, Prophetstown, Quincy, Richmond, Rock City, Rockford, Rock Island, Spring Grove, St. Charles, Starved Rock State Park, Sterling, Urbana, Waukegan, White Pines Forest State Park, and Wilmington.

7. *Stenonema frontale* (Banks)

Heptagenia frontalis Banks (1910:199).
Stenonema interpunctatum frontale (Banks). Spieth (1947:109).
Stenonema majus Traver (1935a:320).

The type of *frontale* is, unfortunately, in quite poor condition, but the species can be placed with reasonable certainty despite this.

MALE.—Length of body 8–10 mm., of fore wing 9–11 mm. Face below antennae yellow, black marks below each antennal socket; vertex yellow, shaded with brown near posterior margin; compound eyes in life light green; each antennal scape and pedicel yellow, flagellum gray-tan at base, becoming hyaline at apex. Mesonotum dark yellow-brown, shading to yellow at lateral margins and, usually, at ends of outer parapsides; scutellum yellow; semimembranous area anterior to base of fore wing shaded with rose-pink; pleuron yellow, an oblique, black streak ventral to base of fore and hind wings; sternum yellow; fore leg tan or deep yellow, apex of tibia black, first tarsal segment two-fifths to three-fifths as long as second segment; middle and hind legs yellow; each femur of all legs with a median and an apical, dark brown crossband; wings hyaline, stigmatic area of fore wing very slightly stained with brown, veins yellow-brown, crossveins very dark brown or black, veins and crossveins of hind wing yellow, outer wing margin shaded. Abdomen yellow, a black crossline at posterior margin of each tergite 1–8; spiracular marks present; apical three tergites shaded with orange-brown, sometimes with a rosy flush added; genitalia, fig. 337, light yellow; caudal filaments gray-yellow, articulations brown.

NYMPH.—Length of body 9–10 mm. Head anterior to compound eyes uniform brown, with a pale spot on anterior margin at meson and a lateral, pale spot on this margin anterior to each antennal socket; large, pale spots lateral to eyes. Thoracic notum mostly uniform brown, pronotum usually pale at lateral and posterior margins, and mesonotum pale on meson in anterior half; tarsal claws without ventral denticles. Abdomen dorsally mostly dark brown, a row of elongate, pale spots present on either side of the dorsal meson on tergites 4–10, sometimes these pale spots almost continuous; each sternite 1–9 with a pair of sublateral, brown spots, these spots more elongate and extending nearly the length of the segment on posterior two or three sternites, posterior margin of ninth sternite either entirely pale or faintly shaded with brown; caudal filaments light brown.

This species is known from Illinois, Kentucky, Massachusetts, Minnesota, and Ontario.

Illinois Records. — BENTON: at light, June 10, 1946, H. H. Ross, 1♂. KANKAKEE: June 6, 1935, Ross & Mohr, 1♂. MOMENCE: Aug. 16, 1938, Ross & Burks, 2♂. PALISADES STATE PARK: at light, June 16, 1948, Stannard & Smith, 1♂. WAUKEGAN: Aug. 15, 1938, Ross & Burks, 2♂. WHITE HEATH: Aug. 2, 1940, Ross & Riegel, 1♂.

8. *Stenonema interpunctatum* (Say)

Baetis interpunctata Say (1839:41).
Stenonema interpunctatum interpunctatum (Say). Spieth (1947:106).

MALE.—Length of body 7–9 mm., of fore wing 8–10 mm. Head yellow, face with prominent, black marks ventral to antennal sockets, vertex shaded with orange-brown; compound eyes in life light green; each antennal scape and pedicel yellow, flagellum gray-brown at base, becoming hyaline toward apex. Mesonotum dark yellow-brown; yellow at lateral margins, at posterior ends of outer parapsides, and on scutellum; semimembranous area anterior to base of fore wing yellow or flesh colored; each pleuron yellow, a gray-brown oblique streak present ventral to fore wing base, also, sometimes, ventral to hind wing base; sternum yellow; fore leg deep yellow, tibia black at apex, first fore tarsal segment from one-fourth to one-third as long as second segment; middle and hind legs light yellow; front and middle femora each with a median and an apical, dark brown crossband, hind femur with only apical band; wings hyaline, stigmatic area of fore wing faintly stained with brown, veins yellow-brown, crossveins dark brown or black; veins and crossveins of hind wing yellow, outer margin shaded with black. Abdomen yellow, posterior margin of each tergite 1–8 with a narrow, black cross-stripe; spiracular markings absent; apical three tergites shaded with pink- or orange-tan; genitalia, fig. 338, yellow; caudal filaments gray-yellow to almost hyaline, articulations not, or only obscurely, darkened.

NYMPH.—Length of body 8–10 mm. Head anterior to compound eyes uniform brown, anterior margin without light spots; relatively small, pale spots on margins lateral to eyes. Dorsum of thorax mostly uniform brown; pronotum with a pair of sublateral and a pair of anterolateral, pale spots, the latter sometimes extending along part or all of lateral pronotal margin; longitudinal, pale stripe on meson of anterior half of mesonotum; tarsal claws without ventral denticles. Gills borne by first six abdominal segments pointed at apexes, each gill of seventh pair with one trachea; dorsum of abdomen brown, a pair of submedian, longitudinal, pale streaks on each of tergites 1–9, these forming an almost continuous pair of stripes, usually of uniform width from end to end; sternites 7–9 each with a pair of sublateral, longitudinal, brown stripes, usually all anterior segments each with a pair of sublateral, brown spots; apex of ninth sternite with a broad, transverse, brown band, this fusing with sublateral, longitudinal marks; caudal filaments light brown, each filament with alternating articulations dark and light in the area near the tip.

Known from Georgia, Illinois, Indiana, Iowa, Kansas, Kentucky, Minnesota, Ohio, Ontario, Tennessee, and Wisconsin.

Illinois Records.—Adult specimens, collected May 5 to August 22, are from Aroma Park, Aurora, Batavia, Benton, Carbondale, Carlinville, Cowling, Dixon, East Dubuque, Effingham, Evanston, Freeport, Galesburg, Havana, Henry, Herod, Highland, Homer, Kankakee, Keithsburg, Mahomet, McHenry, Momence, Mount Carmel, Mount Carroll, Muncie, Murphysboro, Oakwood, Oregon, Ottawa, Pontiac, Rockford, Rock Island, Russellville, St. Charles, Savanna, Springfield, Spring Grove, Starved Rock State Park, Urbana, Waukegan, West Chicago, White Heath, and Wilmington.

9. *Stenonema heterotarsale*
(McDunnough)

Ecdyonurus heterotarsalis McDunnough (1933a:42).
Stenonema affine Traver (1933a:184).
Stenonema interpunctatum heterotarsale (McDunnough). Spieth (1947:110).

MALE.—Length of body 7–9 mm., of fore wing 9–11 mm. Face light yellow, usually unmarked; each antenna with scape and pedicel yellow, flagellum gray-tan at base, hyaline at apex; eyes in life light green; vertex orange-brown, usually a pair of black dots present, one near margin of each compound eye. Mesonotum light brown, lateral margins and scutellum yellow; semimembranous area anterior to base of fore wing yellow, occasionally edged with pink; pleuron yellow, unmarked; fore leg deep yellow, apex of tibia and apexes of tarsal segments dark brown or black, first tarsal segment two-fifths to three-fifths as long as second segment; middle and hind legs light yellow; fore and middle femora with median and apical, dark brown crossbands, hind femur with only apical dark band; wings hyaline, stigmatic area of fore wing

very faintly stained with brown, veins yellow-brown, crossveins black; hind wing with veins and crossveins yellow, outer wing margin darkened with black shading. Abdomen yellow, a narrow, black crossline at posterior margin of each of tergites 1–8; spiracular dots absent; apical three tergites shaded with pinkish brown; genitalia, fig. 339, light yellow; the caudal filaments pale yellow, the articulations not at all or only very faintly darkened.

NYMPH.—Length of body 9–10 mm. Head anterior to eyes a uniform brown; anterior margin usually with small, pale spot on meson, fairly large, pale spot on this margin just anterior to each antennal socket, two relatively small, pale spots on head margin lateral to each compound eye. Dorsum of thorax mostly uniform brown; pronotum with a pair of sublateral, round spots and lateral margin pale; anterior half of mesonotum pale on meson; tarsal claws without ventral denticles. Gills borne by abdominal segments 1–6 pointed at apexes, gills of seventh pair each with one trachea; dorsum of abdomen uniform brown, most specimens with a pair of submedian, longitudinal, pale streaks on each of tergites 4– or 5–10, those on 8 and 9 much the wider and those on 10 much reduced; sternites 5– or 6–9 each with a pair of sublateral, brown spots, those on sternite 9 extending almost the length of the segment and fused with a broad, transverse, brown band occupying apical fifth of sternite; caudal filaments light brown, in apical area these filaments with alternating articulations dark and light.

Known from Illinois, North Carolina, Ontario, and Quebec.

Illinois Records. — CHICAGO: July 3, 1940, J. J. Janacek, 1 ♂; July 8, 1937, Frison & Ross, 1 ♂; at light, July 13, 1931, T. H. Frison, 7 ♂; Sept. 2, 1902, Titus, 1 ♂. HOMER: at light, June 26, 1925, R. D. Glasgow, 1 ♂; Aug. 10, 1925, T. H. Frison, 1 ♂. KANKAKEE: June 6, 1935, Ross & Mohr, 3 ♂; June 15, 1938, Ross & Burks, 4 ♂; June 29, 1939, Burks & Ayars, 1 ♂; July 10, 1925, T. H. Frison, 10 ♂. MOUNT VERNON: Big Muddy River, April 10, 1946, Mohr & Burks, 1 ♂, 9 N. MUNCIE: Stony Creek, May 24, 1914, 1 ♂. OAKWOOD: May 24, 1926, T. H. Frison, 3 ♂; June 6, 1925, T. H. Frison, 1 ♂; June 16, 1925, 1 ♂; June 24, 1948, Mills & Ross, 2 ♂; June 25, 1948, B. D. Burks, 6 ♂. STERLING: at light, May

21, 1925, D. H. Thompson, 1 ♂. Waukegan: July 16, 1935, Ross & DeLong, 4 ♂; Aug. 4, 1926, 1 ♂. Wilmington: May 27, 1935, Ross & Mohr, 1 ♂.

TRIPUNCTATUM Group

10. *Stenonema tripunctatum* (Banks)

Heptagenia tripunctata Banks (1910:199).
Stenonema femoratum tripunctatum (Banks). Spieth (1947:99).

Male.—Length of body 8–11 mm., of fore wing 10–13 mm. Head tan to pale yellow, a brown line crossing face just ventral to antennal bases, vertex shaded with light brown along inner margins of compound eyes; each antennal scape brown, pedicel completely light brown or light brown only at apex and with basal part yellow, flagellum tan at base, becoming hyaline at apex; each compound eye in life pearl-gray, usually with a brown crossband, the two compound eyes separated on meson by a space as wide as lateral ocellus. Thoracic notum varying from uniform brown to pale yellow, almost white; each pleuron tan to pale yellow, always with darker shading ventral to base of fore wing; sternum tan to yellow; all coxae darkened with gray-tan, fore femur tan to yellow-brown, fore tibia and tarsus yellow to tan, apex of tibia and apexes of tarsal segments dark brown, first tarsal segment from two-fifths to one-half as long as second segment; middle and hind legs yellow to tan; each femur of all legs with median and apical, red-brown crossbands, each of middle and hind tibiae with a red spot near base; wings hyaline, stigmatic area shaded with brown, proximal part also with red suffusion, all veins and crossveins dark brown, outer margin of hind wing hyaline; in fore wing, veins crowded in region of bulla. Ground color of abdomen varying from light brown to light yellow, almost white; darker specimens with two mesal, dark brown stripes extending the length of abdomen and with a broad, dark brown crossband at posterior margin of each tergite; all specimens with a median, black dot and a pair of submedian, black dashes at posterior margin of each tergite 2–9; apical tergites uniformly shaded with dull tan or brown; abdominal sternites tan, with vague, brown shading near either lateral margin of each sternite, or uniform, light yellow;

genitalia, fig. 340, tan to yellow; caudal filaments tan or yellow, articulations brown.

Nymph.—Fig. 360. Length of body 9–12 mm. Head brown, freckled with pale dots, with a median and two sublateral, pale spots on anterior margin, a pair of large, pale

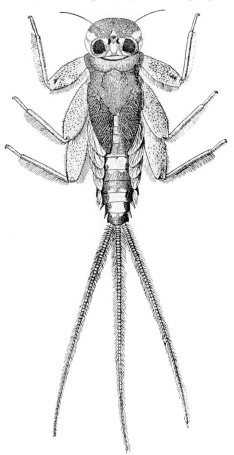

Fig. 360.—*Stenonema tripunctatum*, mature nymph, dorsal aspect.

spots lateral to each compound eye, and posterior margin of vertex mostly pale; pronotum brown, with relatively few pale dots and usually four pale spots near each lateral margin; each tarsal claw with two ventral denticles. Gills borne by abdominal segments 1–6 rounded at apexes, gills of seventh pair each with one or two longitudinal tracheae; abdominal tergites mostly brown, large, white, median spots on tergite 5, a pair of small, submedian, white spots at base and apex of tergites 6 and 7, and large, median, white spots on tergites 8 and 9; posterolat eral angles of segments 2–9 produced as large spines, apical angles slightly sinuate; sternites 2–8 each with a pair of sublateral, brown spots, sternite 9 with a pair of sublateral, brown spots near basal margin and a pair of very large, brown spots at apical margin, basal and apical spots sometimes connected and apical spots occasionally fusing on meson; each caudal filament deep yellow at base, alternating pairs of segments dark and light in middle and apical areas.

Known from Alabama, Arkansas, Georgia, Illinois, Indiana, Iowa, Kansas, Michigan, Minnesota, Missouri, New York, Ohio, Oklahoma, Ontario, Pennsylvania, Quebec, Texas, and Wisconsin.

Illinois Records.—Adult specimens, collected April 4 to August 23, are from Alto Pass, Anna, Antioch, Apple River Canyon State Park, Chicago, Cora, Elizabethtown, Evanston, Fox Lake, Golconda, Havana, Herod, Jonesboro, Monticello, Muncie, Oakwood, Parker, Spring Grove, St. Joseph, Wilmington, and Wolf Lake.

11. *Stenonema femoratum* (Say)

Baetis femorata Say (1823:162).
Stenonema femoratum femoratum (Say).
    Spieth (1947:98).

MALE.—Length of body 8–11 mm., of fore wing 10–13 mm. Differs from *tripunctatum* only in the following particulars: compound eyes separated on meson by a space only two-thirds as wide as lateral ocellus; outer margin of hind wing shaded with light brown; in dark specimens, area between inner and outer parapsides of mesonotum lighter brown than rest of mesonotum; genitalia, fig. 342, with penis lobes relatively more truncate laterally; in most specimens, alternating articulations of caudal filaments darker brown than others.

NYMPH.—Length of body 9–12 mm. Very similar to nymph of *tripunctatum,* but differing in having anterior margin of head slightly flattened on meson, rather than evenly rounded from side to side, median, white spot on anterior margin absent; abdominal sternum with sublateral, brown spots usually present on sternites 5– or 6– 8 only, occasionally spots entirely wanting from these sternites; sternite 9 always with sublateral, longitudinal, brown stripe and pair of large, brown spots at posterior margin.

Known from Georgia, Illinois, Indiana, New York, Ohio, Ontario, Quebec, and South Carolina.

Illinois Records. — ASHLEY: Little Muddy River, April 29, 1946, Mohr & Burks, 1 ♂. CARBONDALE: June 13, 1944, Frison & Sanderson, 1 ♂. DIXON SPRINGS: at light, April 3, 1946, Burks and Sanderson, 5 ♂. EDDYVILLE: Lusk Creek, April 4, 1946, Burks & Sanderson, 1 ♂ ; Lusk Creek, May 24, 1940, Ross & Riegel, 1 ♂ ; Lusk Creek, June 1, 1940, B. D. Burks, 2 ♂ ; Belle Smith Springs, June 7, 1946, Mohr & Burks, 9 ♂ ; Belle Smith Springs, July 16, 1946, Mills & Ross, 1 ♂. FOX LAKE: July 1, 1931, Frison, Betten, & Ross, 1 ♂. GIANT CITY STATE PARK: Aug. 22, 1944, Sanderson & Leighton, 12 ♂. GOLCONDA: May 30, 1928, T. H. Frison, 1 ♂. HEROD: May 2, 1946, Burks & Sanderson, 1 ♂ ; May 23, 1946, Ross & Mohr, 1 ♂ ; May 29, 1939, Burks & Riegel, 1 ♂ ; July 8, 1935, DeLong & Ross, 1 ♂. MOMENCE: June 22, 1938, Ross & Burks, 1 ♂. OAKWOOD: May 7, 1936, Ross & Mohr, 1 ♂. PITTSFIELD: at light, Aug. 11, 1948, Sanderson & Stannard, 1 ♂. QUINCY: June 24, 1948, L. J. Stannard, 2 ♂. RUDEMENT: Blackman Creek, May 14, 1946, Mohr & Burks, 2 ♂. WAUKEGAN: Aug. 15, 1938, Ross & Burks, 1 ♂.

BIPUNCTATUM Group

12. *Stenonema bipunctatum* (McDunnough)

Ecdyonurus bipunctatus McDunnough (1926:191).

MALE.—Length of body 6–8 mm., of fore wing 8–10 mm. Face below antennal sockets white, vertex red- or orange-brown; eyes in life pearl-gray; each antenna light yellow, flagellum usually slightly darkened in basal half. Thoracic notum dull gray-tan or gray-brown, with apex of mesoscutellum white; semimembranous area of pleuron anterior to base of fore wing, and dorsal to spiracle, pink, balance of pleuron white except for brown-shaded spot dorsal to each middle coxa; thoracic sternum light yellow; fore leg light tan, with red-brown shading in middle and at apex of fore femur, and with black shading at apex of fore tibia and at apexes of tarsal segments; first tarsal segment two-thirds as long as second segment; middle and hind legs light yellow, each femur usually with a median and an apical red crossband, apical segment of each tarsus shaded with brown; wings hyaline, brown shading always present in humeral cell of fore wing, this brown often extending almost across base of wings; veins and crossveins of fore wing very dark brown, those of hind wing usually hyaline, sometimes light brown; stigmatic area of fore wing not colored, crossveins not crowded in region of bulla. Abdomen mostly white, apical three tergites with pinkish-tan shading, each tergite 3–7 with a pair of black, transverse, submedian dashes, these marks occasionally present also on tergite 2; genitalia, fig. 341, white or light yellow; caudal filaments white, basal articulations usually shaded with orange-tan or brown.

NYMPH.—Length of body 7–8 mm. Pale median spot present on anterior margin of head. Each tarsal claw with two minute ventral denticles near apex. Abdominal gills borne by segments 1–6 truncate at apexes, seventh pair of gills without tracheae; abdominal dorsum solid brown, interrupted by light-colored areas on meson of segments 5, 8, and 9, and with a pair of lateral, light-colored spots near anterolateral angles of tergites 2–7, these lateral spots normally covered by gills; venter white, with a pair of brown, sublateral spots on sternites 5–8; posterolateral angles of segments 3–9 spinelike; sternite 9 with a pair of lateral, longitudinal, brown stripes and with two large, brown spots at posterior margin; caudal filaments white, the middle area of each filament with alternating pairs of segments brown and white.

Known from Illinois and Ontario.

Illinois Records.—AROMA PARK: June 4, 1947, B. D. Burks, 2 ♂. AURORA: July 9, 1925, T. H. Frison, 1 ♂ ; at light, July 17, 1927, Frison & Glasgow, 2 ♂. CASEY: Catfish Creek, April 29, 1942, H. H. Ross, 1 ♂. DIXON: June 27, 1935, DeLong & Ross, 3 ♂. EXLINE: Kankakee River, June 4, 1947, B. D. Burks, 1 ♂. GRAND TOWER: May 30, 1935, Ross & Mohr, 1 ♂. HARRISBURG: at light, Aug. 16, 1937, Ross & Ritcher, 1 ♂. HAVANA: June 1, 1938, C. O. Mohr, 1 ♂. KEITHSBURG: July 4, 1946, Burks & Sanderson, 1 ♂. MOMENCE: Aug. 16, 1938, Ross & Burks, 1 ♂. MONTICELLO: May 7, 1936, Ross & Burks, 1 ♂. OAKWOOD: June 5, 1948, Burks & Sanderson, 2 ♂ ; June 23, 1948, B. D. Burks, 5 ♂ ; Aug. 4, 1939, Burks

& Riegel, 1 ♂. OREGON: at light, July 2, 1946, Burks & Sanderson, 5 ♂ ; July 9, 1925, T. H. Frison, 1 ♂ ; Rock River, Aug. 5, 1948, Burks & Stannard, 1 ♂. PITTSFIELD: at light, Aug. 11, 1948, Sanderson & Stannard, 1 ♂. PONTIAC: Aug. 22, 1938, H. H. Ross, 4 ♂. PROPHETSTOWN: Rock River, July 24–25, 1947, Burks & Sanderson, 1 ♂. ROCKFORD: May 13, 1942, Ross & Burks, 2 ♂ ; May 15, 1946, Ross & Burks, 1 ♂ ; at light, June 29, 1938, B. D. Burks, 1 ♂ ; July 12, 1938, Burks & Boesel, 2 ♂. ROCK ISLAND: June 7, 1938, Burks & Riegel, 1 ♂. ROCKTON: Rock River, June 25, 1947, B. D. Burks, 3 ♂. SERENA: Indian Creek, May 12–16, 1938, Ross & Burks, 2 ♂. SOUTH BELOIT: July 2, 1931, Frison, Betten, & Ross, 1 ♂. URBANA: June 22, 1947, H. H. Ross, 1 ♂. WILMINGTON: at light, Aug. 6, 1947, Burks & Sanderson, 1 ♂.

13. *Stenonema ares* new species

This species is most closely related to *Stenonema bipunctatum* (McDunnough) in having a medially interrupted, black crossline at the posterior margin of each abdominal tergite 2–7, but differs in having the mesonotum bright Mars orange rather than gray-brown or gray-tan; each of abdominal tergites 1–7 in *ares* has a relatively broad, Mars orange crossband at the posterior margin, and the first fore tarsal segment is one half as long as the second, rather than two-thirds as long, as in *bipunctatum*. Specimens of *ares* and *bipunctatum* which have remained very long in alcohol are quite difficult to separate to species.

The nymph of *ares* differs from that of *bipunctatum* in abdominal color pattern and in having no ventral denticles on the claws.

MALE.—Length of body 6–9 mm., of fore wing 8–11 mm. Head below level of antennal sockets white, vertex chrome orange, shaded on meson and laterally with Mars orange; each antenna with scape Mars orange, pedicel yellow, flagellum white, somewhat grayed near base; eyes in life pearl-gray. Thoracic notum Mars orange, apex of meso- and metascutellum white, area on meson of mesonotum extending anteriorly from apex of scutellum to outer parapsides usually chrome orange, occasionally becoming grenadine pink; pleuron mostly pinkish-tan or testaceous, semimembranous area anterior to fore wing base grenadine pink or orange-brown, area of pleuron ventral to fore wing base light yellow; thoracic sternum pale yellow, anterior end of meso-basisternum shaded with testaceous; coxae of all legs shaded with light Mars orange, apical segment of tarsus of each leg gray; fore leg light tan, middle and apex of femur shaded with red-brown, apex of tibia and apexes of tarsal segments dark brown, first tarsal segment one-half as long as second; middle and hind legs light yellow, apex of hind femur and middle and apex of middle femur shaded with red-brown; wings hyaline, humeral cell of fore wing shaded with light brown, proximal part of stigmatic area faintly stained with brown, veins of fore wing yellow-brown, crossveins dark brown, anterior veins and crossveins of hind wing pale yellow, more posterior ones hyaline. Abdomen pale yellow or white, each tergite 2–7 with a medially interrupted, narrow, black crossline at posterior margin and, in addition, a fairly broad, Mars orange crossband at posterior margin of each tergite 1–7; spiracular markings absent; apical three tergites Mars orange, shading to chrome orange at apex of tergite 10, these tergites sometimes also with a pink suffusion; genitalia, fig. 343, white; caudal filaments white, basal articulations Mars orange, more distal articulations white.

NYMPH.—Length of body 10 mm. Head light brown in area anterior to compound eyes and on vertex, this dark area freckled with numerous, relatively large, white dots; three large, white spots on lateral margin near each compound eye; median, white spot anterior to median ocellus and on vertex near posterior margin; base of each antennal flagellum dark; rest of antenna white. Pronotum light brown, with many irregular, white spots; each tibia with a basal and a median, brown band; each tarsus brown except at base and apex; tarsal claws without ventral denticles. Abdominal tergites 1 and 2 white, 3–5 mostly white, each with a longitudinal, median, brown mark on basal half, a brown spot at posterolateral angle, and a pair of large, submedian, vague, light brown spots at posterior margin; tergites 6–8 mostly brown, a pair of small, anterolateral and submedian, white marks at anterior margin of each; tergite 9 brown on meson and near lateral margins; and 10 brown except for two submedian, basal spots; sternites 1–8 white, 9 with vague, lateral and basal, brown

markings; posterolateral angles of segments 7–9 spinelike; gills borne by segments 1–6 truncate at apexes, seventh pair without tracheae; caudal filaments light yellow, apical articulations slightly darkened with tan.

Holotype, male.—Rockford, Illinois, at light, June 11, 1948, Burks, Stannard, & Smith. Specimen dry, on pin.

Paratypes.—ILLINOIS: Same data as for holotype, 18 ♂. DIXON: at light, June 25, 1947, B. D. Burks, 5 ♂. ELIZABETHTOWN: at light, July 14, 1948, Mills & Ross, 1 ♂. FREEPORT: at light, June 10–11, 1948, Burks, Stannard, & Smith, 11 ♂; Aug. 4, 1948, Burks & Stannard, 2 ♂. GREENVILLE: Shoal Creek, April 12, 1946, Mohr & Burks, 1 ♂. OREGON: July 4, 1946, Burks & Sanderson, 8 ♂; July 9, 1925, T. H. Frison, 1 ♂; Aug. 5, 1948, Burks & Stannard, 4 ♂. PITTSFIELD: at light, Aug. 11, 1948, Sanderson & Stannard, 3 ♂. PROPHETSTOWN: at light, June 25, 1947, B. D. Burks, 1 ♂; July 24–25, 1947, Burks & Sanderson, 10 ♂. ROCKTON: June 25, 1947, B. D. Burks, 17 ♂; Aug. 4, 1948, Burks & Stannard, 6 ♂. ROSCOE: June 25, 1947, B. D. Burks, 4 ♂. SHAWNEETOWN: July 14, 1948, Mills & Ross, 2 ♂. STERLING: at light, June 26, 1947, B. D. Burks, 5 ♂. URBANA: at light, May 29, 1947, H. H. Ross, 1 ♂. All specimens dry, on pins; genitalia on microscope slides.

Additional Illinois specimens preserved in alcohol, and not included in the type series, are from the following: ALTON: May 18, 1932, Ross & Mohr, 1 ♂. BILLETT: Wabash River, May 15, 1942, Mohr & Burks, 4 ♂. ROCKFORD: May 22, 1941, Ross & Burks, 31 ♂. STERLING: May 22, 1941, Ross & Burks, 23 ♂.

VICARIUM Group

14. *Stenonema pudicum* (Hagen)

Ephemera pudica Hagen (1861:39).

Eaton (1885:280) placed this species as a synonym of *vicarium*, but McDunnough (1925b:191) studied the type and was able to show that *pudicum* was not synonymous with *vicarium*. The type specimen is a female subimago in poor condition, now in the collection of the Museum of Comparative Zoology. I have studied it and agree with McDunnough.

MALE.—Length of body 10–12 mm., of fore wing 12–14 mm. Head brown, usually a narrow, black line extending across face, from eye to eye, below antennal bases; eyes in life brown, each antenna brown, becoming hyaline toward tip of flagellum. Thoracic notum dark olive-brown, often with a reddish cast toward anterior margin of mesonotum, pleuron anterior to wing bases light yellow or clay color, elsewhere light clay-brown; thoracic sternum chestnut brown; wings hyaline, entire stigmatic area of fore wing shaded with dark red-brown, outer margin of hind wing brown; all veins and crossveins brown, fore wing with 3 or 4 crossveins in each interspace crowded together in region of bulla; usually all coxae brown, the fore leg dark yellow-brown to olive-brown, middle and hind legs light yellow-brown, each femur with a middle and an apical, dark-brown color band, apex of fore tibia black, first fore tarsal segment from one-half to two-thirds as long as second segment. Abdomen dark yellow-brown to medium brown, with a broad, somewhat diffuse, dark brown, transverse band at posterior margin of each tergite 2–8 and, usually, double, longitudinal, dark brown line on meson extending the length of these tergites; apical tergites uniformly very dark brown; sternum gray-brown, lateral and posterior margins of middle sternites usually dark brown; genitalia, fig. 345, yellow-brown; caudal filaments light gray-tan, articulations brown.

NYMPH.—Length of body 12–14 mm. Entire dorsum of head anterior to ocelli and vertex between eyes dark brown, freckled with pale dots, areas lateral to eyes almost completely light. Pronotum variegated with fairly large, light spots near lateral margins, balance of thoracic notum dark brown; pronotum as wide as head; tarsal claws without denticles. Abdominal tergites 6 and 8–10 usually uniformly dark brown, others variegated with light spots; gills borne by segments 1–6 truncate at apexes, gills of seventh pair without tracheae; slender, spinelike projections borne by posterolateral angles of abdominal segments 3– or 4–9; sternum pale yellow, sternites 3– or 4–8 each with a broad, dark brown crossband on median two-thirds of anterior margin; sternite 9 with a broad, longitudinal, dark brown band near each lateral margin, these two bands sometimes almost or quite joined at anterior margin of sternite; caudal filaments usually uniformly yellow or tan.

Known from District of Columbia, Illinois, New York, North Carolina, Tennessee, and Virginia.

Illinois Record. — EDDYVILLE: Lusk Creek, May 16, 1947, B. D. Burks, 1 ♂.

15. *Stenonema vicarium* (Walker)

Baetis vicaria Walker (1853:565).
Baetis tesselata Walker (1853:566).
Ecdyonurus rivulicolus McDunnough (1933a:40). New synonymy.

Spieth (1940:336) gives notes made from a study of the type of this species in the British Museum. The identity of the species now is firmly established. McDunnough's *rivulicolus* differs from *vicarium* only in being slightly smaller and in having the tibiae more tan than red-brown; I have found those characters to intergrade.

MALE.—Length of body 10–14 mm., of fore wing 12–16 mm. Head brown, vertex often tinged with red; eyes in life gray; each antenna tan, shaded with brown at apex of pedicel and on basal half of flagellum, dorsum of thorax dark brown, apex of scutellum red-brown; pleuron yellow-brown, dark red-brown to almost black shading present at wing bases and dorsal to mid-coxa; thoracic sternum brown, anterior and posterior margins of mesobasisternum yellow to light brown, wings hyaline, proximal part of stigmatic area of fore wing shaded with dark red; all veins and crossveins dark brown, crossveins in region of bulla in the fore wing usually not greatly crowded, usually only two or three in each interspace; all coxae brown, with black shading on outer side of each, fore femur light brown, tibia gray-yellow, with apex black, fore tarsus brown, with apexes of segments black, first segment from one-fourth to two-fifths as long as second segment; middle and hind legs yellow-brown, with tarsi darkened; each femur with a median and an apical, broad, dark red-brown band; some or all of the tibiae may be faintly stained with red, especially near bases. Abdomen dark yellow-brown, heavily shaded with blackish brown at posterior margins of tergites 1–7 and on median longitudinal line; apical three tergites lighter, red-brown, with lateral margins often salmon-pink; sternites 1–7 yellow-brown or red-brown, in lighter specimens dark brown shading usually visible near posterolateral angles and on meson; geni-

talia, fig. 344, smoky yellow; caudal filaments gray-brown or tan, articulations
brown.

NYMPH.—Length of body 14–18 mm.
Head anterior to ocelli brown, freckled with
pale dots, area lateral to eyes and at posterior margin of head mostly pale. Pronotum
with large pale spots at lateral and anterior
margins, disc with pale dots, rest of thoracic
notum usually uniform brown; tarsal claws
without ventral denticles. Gills borne by
abdominal segments 1–6 truncate at apexes,
gills of seventh pair without tracheae; abdominal tergites 1–10 with broad, dark
brown crossband at posterior margin of
each, these bands sometimes obsolescent or
wanting entirely on meson of basal three or
four tergites; brown, mid-dorsal band extending from base to apex of abdomen, tergites 5 and 6 often almost completely shaded
with brown, occasional specimens with almost entire abdominal dorsum brown; venter white, with a broad, dark brown crossband at posterior margin of each sternite,
entire apical half to two-thirds of terminal
segment dark brown; posterolateral angles
of segments 3– or 4–9 produced, spinelike;
caudal filaments uniformly tan or yellow-
brown in basal and middle areas, alternating pairs of segments usually dark and
light in apical areas of filaments.

Known from Illinois, Indiana, Kentucky,
Michigan, New Hampshire, New York,
Ontario, Pennsylvania, Quebec, and Wis-
consin.

Illinois Records. — ALTO PASS: Union
Springs, May 15, 1946, Mohr & Burks,
1 ♂. EDDYVILLE: Belle Smith Springs, April
29, 1949, Sanderson & Stannard, 1 ♂.

16. *Stenonema fuscum* (Clemens)

Heptagenia fusca Clemens (1913:254).

MALE.—Length of body 9–11 mm., of
fore wing 12–14 mm. Head yellow-brown,
face with dark red-brown, transverse stripe
below antennal sockets; eyes gray in living
insect; each antenna tan, scape and base of
flagellum shaded with red-brown. Thoracic
notum dark chestnut brown, apex of scutellum white or yellow; area of each pleuron
anterior to base of fore wing and dorsal to
middle and hind coxae red-brown, pleuron
otherwise tan or yellow; thoracic sternum
yellow-brown, with mesobasisternum usually
entirely yellow; all coxae red-brown, fore

leg dark yellow-brown, apex of fore tibia
and apexes of all tarsal segments dark red-
brown, first tarsal segment one-third to one-
half as long as second segment; middle and
hind legs light yellow-brown; each femur
with a median and an apical, dark red-brown
crossband; wings hyaline, entire stigmatic
area of fore wing washed with yellow-
brown, crossveins in region of bulla often
not at all crowded, occasionally two or three
crossveins in each interspace at this point;
all veins and crossveins of both wings very
dark yellow-brown. Abdominal segments
with ground color yellow, tergites usually
almost entirely shaded with brown, occasionally this darkening confined to posterior
margins of tergites and to median dorsal
line; apical three tergites always tinged
with bright Mars orange, with white or pale
yellow on lateral margins; abdominal
sternum dull yellow, rarely with vague,
brown shading either side of median line;
genitalia, fig. 346, grayish yellow; caudal
filaments pale gray-yellow, articulations
brown.

NYMPH.—Length of body 10–12 mm.
Head and thorax mostly brown, pale spots
on lateral margins of head lateral to compound eyes and near margins of pronotum;
tarsal claws without ventral denticles. Gills
borne by abdominal segments 1–6 truncate
at apexes, gills of seventh pair without
tracheae; abdominal tergites with rather
vague, transverse, darkened area at each
posterior margin, tergites 6 and 7 often almost completely brown; posterolateral
angles of segments 3– or 4–9 produced,
spinelike; sternites 1–8 each with a broad,
transverse, brown crossband at posterior
margin, sternite 9 with a large, brown spot
near each posterolateral angle, these spots
usually extending almost to anterior margin
of sternite; each caudal filament uniformly
tan in basal area, alternating pairs of segments dark and light in more distal area.

Known from Michigan, New Brunswick,
New York, Ohio, Ontario, Pennsylvania,
Quebec, and Tennessee.

17. *Stenonema ithaca* (Clemens & Leonard)

Heptagenia ithaca Clemens & Leonard
(1924:17).

MALE.—Length of body 9–10 mm., of
fore wing 11–12 mm. Head and thoracic

notum dark red-brown, thoracic pleuron mostly yellow-brown, with darker shading at wing bases and dorsal to coxae; thoracic venter red-brown; stigmatic area of fore wing washed with yellow-brown, crossveins usually not crowded in region of bulla; first fore tarsal segment about one-half as long as second segment. Abdominal tergites 1–7 almost entirely dark brown, each tergite slightly lighter only near anterior margin; apical tergites uniformly dark yellow-brown, often with an added reddish suffusion; genitalia, fig. 347, yellow; the three well-developed caudal filaments gray-yellow, with articulations brown.

NYMPH. — Length of body 10–11 mm. Tarsal claws without ventral denticles. Gills borne by abdominal segments 1–6 truncate at apexes, posterolateral angles spinelike on segments 6– or 7–9; tergites 6 and 8–10 mostly brown, the others mostly light; middle and apical sternites each with a medianly sinuate, transverse, brown band crossing middle of sternite; the three caudal filaments usually uniformly yellow, with the apical segments sometimes alternately dark and light.

Known from Georgia, Michigan, New York, North Carolina, Ohio, Quebec, South Carolina, Tennessee, and West Virginia.

MEDIOPUNCTATUM Group

18. *Stenonema mediopunctatum* (McDunnough)

Ecdyonurus mediopunctatus McDunnough (1926:191).

MALE.—Length of body 9 mm., of fore wing 10 mm. Face below antennal bases white, vertex sepia brown. Dorsum of thorax entirely blackish brown, except that apex of mesoscutellum is light brown; thoracic pleura very dark brown, each with small, white areas at wing bases; thoracic sternum dark brown; first fore tarsal segment two-thirds as long as second segment, each femur with a median and an apical, red crossband. Abdominal segments white, with a small, black spot on meson of posterior margin of each of tergites 2–8, black spiracular dots usually present on segments 4–7, apical three tergites shaded with brown; caudal filaments entirely white.

Known from Ontario.

19. *Stenonema metriotes* new species

This species resembles *mediopunctatum* in having the markings at the posterior margins of abdominal tergites 2–7 reduced to very short, median dashes; the two differ in that the mesonotum is dark brown in *mediopunctatum*, while it is white in *metriotes*; in *mediopunctatum*, usually the spiracular dots on the abdomen and the articulations of the caudal filaments are not darkened, but, in *metriotes*, the spiracular dots always are wanting and the articulations of the caudal filaments are red-brown. *S. metriotes* may eventually prove to be only a variant of *integrum*, although the abdominal markings characteristic of *integrum* are wanting in this species.

MALE.—Length of body 5–6 mm., of fore wing 7–8 mm. Face below antennal sockets white; vertex flesh colored, with a pair of short, oblique, submedian, tan streaks between eyes; eyes in life white; each antenna white, flagellum slightly grayed near base. Mesonotum chalky white, a pair of faint, tan marks near posterior ends of outer parapsides; pleuron white, a very faint, tan, oblique streak ventral to each wing base; venter of thorax white; fore coxa shaded with tan, fore femur tan, with middle and apex red-brown, tibia white, with brown or tan shading at base and apex; apexes of tarsal segments tan, first tarsal segment three-fifths to two-thirds as long as second segment; middle and hind legs white, apexes of femora shaded with red-brown; wings hyaline, stigmatic area of fore wing stained with tan, anterior veins yellow-brown, crossveins dark brown, posterior veins of fore wing and all veins and crossveins of hind wing hyaline, outer margin of hind wing shaded with brown. Abdomen white, tergites 2–6 each with a minute, black, median dash at posterior margin, tergites 7 and 8 each with a medianly interrupted, black crossline at posterior margin, tergite 9 with a continuous, black crossline at posterior margin, tergites 8 and 9 shaded with tan on meson; spiracular markings absent on abdomen; genitalia, fig. 348, white; caudal filaments white, articulations red-brown.

Holotype, male. — East Dubuque, Illinois, at light, July 3, 1946, Burks & Sanderson. Specimen dry, on pin.

Paratypes. — ILLINOIS: Data same as for holotype, 3 ♂. OREGON: July 4, 1946, Burks

& Sanderson, 1 ♂. PROPHETSTOWN: July 24–25, 1947, Burks & Sanderson, 3 ♂. QUINCY: June 24, 1948, L. J. Stannard, 2 ♂. SHAWNEETOWN: July 14, 1948, Mills & Ross, 3 ♂. All specimens dry, on pins.

TERMINATUM Group

20. Stenonema luteum (Clemens)

Heptagenia lutea Clemens (1913:252).

MALE.—Length of body 9–11 mm., of fore wing 10–12 mm. Face below level of antennae light yellow, vertex yellow, with orange-red shading; eyes in life light green; a red-brown ring surrounding each antennal socket, antenna pale yellow, flagellum darkened in basal half. Thoracic notum light yellow, usually gray-brown shading present on meson of pronotum, mesoscutellum white; pleuron pale yellow, dark shading present around all coxae, area anterior to base of fore wing usually faintly tinged with orange-brown; sternum pale yellow, anterior part of mesobasisternum usually shaded with gray-tan; fore leg dark yellow, middle and hind legs light yellow, each femur with a median and an apical, dark red-brown crossband, apex of each tibia and apexes of all tarsal segments dark brown or black, first fore tarsal segment slightly less than one-half as long as second segment; wings hyaline, stigmatic area of fore wing stained with red-tan; veins and crossveins of fore wing dark brown or black, those of hind wing hyaline. Abdomen light yellow, posterior margin of each tergite 1–7 with a narrow, black cross-stripe; spiracular markings absent; apical three tergites pinkish tan; genitalia, fig. 349, pale yellow; caudal filaments pale yellow, articulations faintly darkened with gray, sometimes three or four basal articulations of each filament light brown.

NYMPH.—Length of body 10 mm. Head with area anterior to eyes brown, freckled with pale dots, large, pale spots lateral to each compound eye and at posterior margin of vertex. Each lateral margin of pronotum with a large, light spot, disc of pronotum brown, with numerous minute, pale dots; each tarsal claw with two ventral denticles near apex. Abdominal tergites 3, 6, and 7 almost entirely dark brown, others with a contrasting pattern of dark and light markings, tergite 9 almost entirely light; gills borne by segments 1–6 truncate at apexes, gills of segment 7 without tracheae; abdominal venter white, sternites 2–8 each with brown shading near lateral margins and at posterior margin, sternite 9 with two large, dark brown, sublateral spots; each caudal filament uniformly tan in basal half, apically with alternating pairs of segments dark and light.

Known from Illinois, Ontario, and Quebec.

Illinois Records.—KANKAKEE: June 17, 1939, B. D. Burks, 1 ♂. MOMENCE: June 15, 1938, Ross & Burks, 3 ♂; Aug. 5, 1938, Burks & Boesel, 1 ♂; Aug. 22, 1939, B. D. Burks, 1 ♂. MOUNT CARMEL: June 25, 1936, DeLong & Ross, 1 ♂. OAKWOOD: June 5, 1948, Burks & Sanderson, 1 ♂; June 9, 1926, Frison & Auden, 1 ♂. ROCKFORD: May 22, 1941, Ross & Burks, 1 ♂; June 12, 1938, Ross & Burks, 1 ♂. SAVANNA: June 29, 1935, DeLong & Ross, 1 ♂. WHITE HEATH: Sangamon River, Aug. 5, 1939, Ross & Riegel, 2 ♂.

21. Stenonema terminatum (Walsh)

Palingenia terminata Walsh (1862:376).
Heptagenia placita Banks (1910:199).
    New synonymy.

The lectotype of Walsh's species and the type of placita Banks are in the Museum of Comparative Zoology.

MALE.—Length of body 6–8 mm., of fore wing 8–11 mm. Head below antennal sockets light yellow, vertex orange-rufous or clay color, with lateral, orange shading; eyes in life light yellow-green; each antenna white or faint yellow, flagellum darkened in basal half. Thoracic notum light clay color to ochraceous-tawny, apex of mesoscutellum white; pleuron light yellow, area anterior to fore wing base often tinged with light buff-brown, a pale orange-rufous spot dorsal to each coxa; fore leg deep yellow, middle and hind legs light yellow, fore femur with a median and an apical red-brown spot, apex of fore tibia and apexes of all tarsal segments dark brown or black; first fore tarsal segment two-fifths as long as second segment; wings hyaline, stigmatic area faintly stained with yellow-brown, veins of fore wing light yellow-brown, crossveins brown, veins and crossveins of hind wing usually entirely hyaline. Abdomen pale yellowish, tergites 1–7 each with a fairly broad, but rather vague, dark brown or black crossband at posterior margin; spiracular markings absent; apical three tergites ochraceous-tawny; genitalia, fig. 350, light yellow; caudal filaments white or pale yellowish, basal two or three articulations of each filament sometimes darkened with brown.

Known from Illinois, Manitoba, New York, Ontario, and Quebec.

Illinois Records. — MONTICELLO: May 24, 1947, B. D. Burks, 1 ♂. OREGON: July 9, 1925, T. H. Frison, 2 ♂. ROCKFORD: May 22, 1941, Ross & Burks, 5 ♂ ; June 29, 1938, B. D. Burks, 2 ♂ ; Sept. 4, 1940, Frison & Ross, 2 ♂. ROCK ISLAND: 14 ♂, 7 ♀ (Walsh 1862:376). ROCKTON: Aug. 4, 1948, Burks & Stannard, 3 ♂. SHAWNEETOWN: July 14, 1948, Mills & Ross, 2 ♂.

22. *Stenonema lepton* Burks

Stenonema lepton Burks (1946:614).

In the male of this species, the first fore tarsal segment often is as much as five-sixths as long as the second segment. In previously used keys to the genera of the Heptageniidae, this character would refer the species to the genera *Cinygmula, Cinygma, Epeorus,* or *Iron.* The male genitalia of *lepton* are, however, typical for the genus *Stenonema.* I have, accordingly, placed it here.

Several recent collections of this species have been made, and the specimens have been preserved dry, on pins. Study of these specimens necessitates some change in the color description of the species, which was drawn from alcoholic material.

MALE.—Length of body 7–9 mm., of fore wing 8–10 mm. Face ventral to antennae white; four minute, gray spots on posterior margin of vertex between eyes: a pair at margins of compound eyes and a submedian pair; vertex near ocelli yellow, shaded with orange-brown posteriorly; each antenna white, flagellum slightly darkened in basal half; eyes in life chalky white. Pronotum very pale yellow, mesonotum chalky white, scutellum white, pleuron white, a faintly darkened streak ventral to base of fore wing between middle and hind coxae, and dorsal to hind coxa; thoracic sternum white; fore leg very pale yellow, red-brown shading present in middle and at apex of fore femur, apex of fore tibia and apexes of all fore tarsal segments darkened with brown, first

fore tarsal segment from two-thirds to five-sixths as long as second segment; middle and hind legs white, femora not (or very faintly) shaded with red-brown in the middle; wings hyaline, stigmatic area washed with very light red stain, anterior veins and all crossveins of fore wing yellow-brown, those of hind wing hyaline; posterior margin of hind wing slightly darkened with brown. Abdomen white, each of tergites 1–9 with a narrow, black crossline at posterior margin, spiracular markings absent, apical three tergites shaded with faint tan on meson; genitalia, fig. 351, white; caudal filaments entirely white.

Known only from Illinois.

Illinois Records. — AROMA PARK: July 8, 1948, Ross & Burks, 1 ♂ ; Aug. 6, 1947, Burks & Sanderson, 1 ♂. KANKAKEE: at light, July 9, 1948, Ross & Burks, 3 ♂. MOMENCE: at light, June 22, 1938, Ross & Burks, 6 ♂ ; June 24, 1939, Burks & Ayars, 23 ♂. WILMINGTON: at light, Aug. 6, 1947, Burks & Sanderson, 10 ♂.

PULCHELLUM Group

23. *Stenonema integrum* (McDunnough)

Heptagenia (*Ecdyonurus?*) *integer* McDunnough (1924a:9).
Stenonema bellum Traver (1933a:202).
    New synonymy.
Stenonema wabasha Daggy (1945:378).
    New synonymy.

The concept of *integrum* followed here is derived entirely from a study of the holotype of the species.

MALE.—Length of body 5–7 mm., of fore wing 6–8 mm. Head chalky white, vertex stained with pale yellow, a pair of submedian, light brown dots between compound eyes; each antenna white, flagellum slightly darkened with gray. Pronotum pale yellow, sometimes with a lunate, black streak on either side; meso- and metanotum chalky white, with clay-colored shading along outer parapsides of mesonotum and just dorsal to fore wing bases; pleura and sternum chalky white, occasionally with pale yellow-brown shading on each pleuron near mesocoxa. Fore leg pale yellow, dark brown at apex of tibia and at apexes of tarsal segments, first tarsal segment three-fifths as long as second; middle and hind legs white; each femur of all legs with a prominent, red-brown crossband in middle and at apex, each middle and hind tibia with a subbasal, red-brown spot; wings hyaline, stigmatic area stained with brown, outer margin of hind wing shaded with brown, veins of fore wing a faint yellow-brown, crossveins dark brown, veins and crossveins of hind wing hyaline. Abdomen chalky white, each tergite 1–9 with a narrow, black crossline at posterior margin, those on tergites 8 and 9 often interrupted on the meson; a longitudinal, dark gray line on meson of tergites 3 and 6, sometimes also on tergites 2 and 7; in each spiracular area of segments 3–8 an oblique, dark brown streak present, occasionally these markings becoming obsolete on anterior and posterior segments but always persisting on at least segments 5 and 6; genitalia, fig. 352, white; caudal filaments white, articulations dark red-brown.

Known from Georgia, Illinois, Indiana, Kansas, Kentucky, Michigan, Minnesota, Missouri, and North Carolina.

Illinois Records.—Adult specimens, collected June 7 to August 27, are from Alton, Aroma Park, Dixon, East Dubuque, Elizabethtown, Foster (Mississippi River), Freeport, Fort Kaskaskia State Park, Kankakee, Keithsburg, Momence, Monmouth, Monticello, Oregon, Poplar Bluff, Prophetstown, Quincy, Rockton, Shawneetown, Urbana, and Wilmington.

24. *Stenonema nepotellum*
(McDunnough)

Ecdyonurus nepotellus McDunnough (1933a:20).

Although the nymphs of this species and those of *rubromaculatum* are, as pointed out by McDunnough (1933a:20), quite different, the adults of the two species are very similar. Freshly collected, dry specimens of the adult males can, however, be separated by color characters.

Male.—Length of body 8–9 mm., of fore wing 8–11 mm. Face below antennal sockets light yellow, vertex yellow, shaded with orange-brown; eyes in life pearl-gray; each antennal scape orange-brown, pedicel tan, flagellum tan at base, hyaline toward apex. Thoracic notum yellow-brown, mesoscutellum and lateral margins of posterior half of mesonotum white; pleuron pale yellow, with brown shading dorsal to coxae, semimembranous area anterior to fore wing base red-brown; sternum pale yellow, mesofurcisternum shaded with tan. Fore leg pale yellow-brown, apex of tibia and apexes of tarsal segments very dark brown or black, first fore tarsal segment one-half as long as second segment; middle and hind legs pale yellow, each femur of all legs with a median and an apical, red-brown crossband; wings hyaline, basal part of stigmatic area red, entire stigmatic area also stained with light brown; all veins and crossveins of fore wing brown, anterior ones of hind wing brown, posterior ones hyaline; crossveins in bullar area of fore wing not crowded. Abdomen yellow, a black crossline at posterior margins of tergites 1– or 2–7; large spiracular dots present; apical tergites shaded with pinkish brown, almost all of tergite 8 suffused with this color, tergite 10 with narrow, white area at posterior margin; abdominal venter yellow; genitalia, fig. 353, white; caudal filaments white, articulations red-brown.

Nymph.—Length of body 7–9 mm. Anterior border of head rather truncate, lacking a median, pale spot; large, pale spot on lateral margin of head lateral to each compound eye, this pale spot sometimes divided by a brown crossbar. Pronotum with a broad, pale area at either lateral margin; tarsal claws without ventral denticles. Abdominal dorsum with rather vague color pattern of light and dark spots, tergites 5 and 7 predominantly pale, others mostly dark; gills borne by segments 1–6 truncate at apexes, gills of seventh pair without tracheae; abdominal venter white, each of sternites 2– or 3–8 with a curved, brown crossbar borne near anterior margin, these bars wider and more intensely colored on posterior segments; sternite 9 with a U–shaped, brown mark, the open end directed posteriorly, occasional specimens with basal crossbar of this U–shaped mark faint or obsolete; posterolateral angles of abdominal segments 3–9 spinelike, those borne by segment 9 long and slender; caudal filaments light brown near bases, apically alternating pairs of segments dark and light.

Known from Illinois, Indiana, Ohio, Ontario, Quebec, and Wisconsin.

Illinois Records. — Oakwood: July 14, 1939, Burks & Riegel, 1 ♂; July 30, 1939, Burks & Riegel, 1 ♂; Aug. 4, 1939, Burks & Riegel, 1 ♂; Aug. 10–14, 1939, B. D. Burks, 3 ♂, 3 ♀, 5 N. Spring Grove: June

15, 1938, B. D. Burks, 1 ♂ ; Nippersink Creek, June 12–29, 1938, B. D. Burks, 3 ♂, 11 ♀, 30 N. ST. CHARLES: at light, June 9, 1948, Burks & Stannard, 1 ♂. STERLING: June 7, 1939, Burks & Riegel, 1 ♂.

25. *Stenonema rubromaculatum*
(Clemens)

Heptagenia rubromaculata Clemens (1913:256).

MALE.—Length of body 8–9 mm., of fore wing 9–10 mm. Face below antennal sockets white, vertex pinkish yellow except at posterior margin, where it is white; eyes in life pearl-gray; each antennal scape and pedicel light brown, flagellum light yellow. Thoracic notum light clay colored, occasionally tinged with olive-gray; each pleuron white or pale yellow, with a brown spot at base of each coxa, semimembranous area of mesopleuron anterior to base of fore wing and dorsal to spiracle pale flesh color; sternum light yellow; all legs yellow, fore coxa shaded with brown, femur with a median and an apical red-brown band, apex of fore tibia and apical segment of each tarsus shaded with brown; wings hyaline, stigmatic area of fore wing shaded with yellow-brown, basal area suffused with red; all veins and crossveins of fore wing brown, anterior ones of hind wing tan, others hyaline; crossveins near bulla, in fore wing, usually not crowded. Abdomen light yellow or white; fine, black, transverse line at posterior margin of each tergite 1–7, black or dark brown spiracular dots present; meson of tergite 8, all of tergite 9, and all but white, posterior margin of tergite 10 pinkish brown; abdominal venter light yellow to white; genitalia, fig. 354, light yellow to white; caudal filaments white, articulations dark brown.

NYMPH.—Length of body 8–10 mm. Head dark brown, freckled with pale dots, a pair of large, pale spots lateral to each compound eye. Pronotum with a broad, pale area at each lateral margin and, usually, a sublateral, pale spot near either margin; each tarsal claw with two ventral denticles near tip. Abdominal dorsum without conspicuous color pattern, nearly uniform dark brown, sometimes with minute, light-colored spots on meson and near lateral margins of middle and apical tergites; venter white, with variable, dark brown color pattern: usually two pairs of submedian dots on each sternite, as

well as a third pair near anterolateral angles, the areas between these dots filled in on some darker specimens to produce a mushroom-like figure on each of sternites 3– or 4–8; sternite 9 usually with lateral and basal margins brown, so as to make a somewhat rectangular pattern; posterolateral angles of segments 3–9 spinelike, those on segment 9 small; caudal filaments brown.

Known from Illinois, Massachusetts, New Brunswick, Nova Scotia, Ontario, Quebec.

Illinois Records. — OAKWOOD: June 5, 1948, Burks & Sanderson, 1 ♂ ; June 6, 1925, T. H. Frison, 10 ♂ ; June 9, 1926, Frison & Auden, 4 ♂ ; July 8, 1946, B. D. Burks, 1 ♂.

26. *Stenonema rubrum* (McDunnough)

Ecdyonurus ruber McDunnough (1926:192).

MALE.—Length of body 7–8 mm., of fore wing 8–9 mm. Face below level of ocelli white, vertex deep orange-brown; antennae pale yellow; eyes in life pearl-gray. Thoracic pro- and mesonotum red-brown, apex of mesoscutellum white, with a red area just anterior to this; metanotum yellow, with a red stain on meson; semimembranous area of each mesopleuron anterior to base of fore wing and dorsal to spiracle orange-brown, sometimes with a pinkish cast; pleuron pale yellow, except for a pinkish brown stain dorsal to each fore and middle coxa; sternum pale yellow; fore leg yellow-tan, with brown shading in middle and at apex of femur, at apex of tibia, and at apexes of tarsal segments; first tarsal segment three-fifths as long as second segment; middle and hind legs yellow, with red-brown shading in middle and at apex of each femur, and dark brown shading at apexes of tibiae and tarsal segments; wings hyaline, faint pink shading in stigmatic area of each fore wing, humeral cell shaded with dark brown, veins and crossveins orange-brown, no crowding of crossveins at bulla. Abdomen light yellow, a rather broad, dark, orange-brown crossband at posterior margin of each tergite 1–7, large spiracular dots present; apical tergites bright orange-brown with, usually, red shading overlying ground color; abdominal venter light yellow; genitalia, fig. 355, pale yellow; caudal filaments pale yellow or white, articulations dark brown.

NYMPH.—Length of body 8–9 mm. Each tarsal claw with two minute denticles on ventral side near apex. Abdominal gills borne by segments 1–6 truncate at apexes, gills of seventh pair without tracheae; abdominal tergites usually uniformly brown, sometimes tergites with median, lighter spots faintly indicated; abdominal sternum white, sternite 8 with a median, dark brown spot at anterior margin and sternite 9 with a median, U–shaped, dark brown mark; caudal filaments uniformly tan in basal and middle areas, but usually with alternating pairs of light and dark segments toward apexes.

Known from Connecticut, Georgia, Illinois, New York, Ontario, and Quebec.

Illinois Records.—MOMENCE: May 26, 1936, H. H. Ross, 1 ♂ ; June 15, 1938, Ross & Burks, 1 ♂ . WILMINGTON: at light, Aug. 6, 1947, Burks & Sanderson, 5 ♂ .

27. *Stenonema pulchellum* · (Walsh)

Palingenia pulchella Walsh (1862:375).

The male lectotype of this species, now in the Museum of Comparative Zoology, is considerably bleached, due probably to exposure to sunlight sometime in the past, and the genitalia are missing. The remaining parts of this specimen, however, agree well with recently collected material from near the type locality. Also in the M.C.Z. is another male specimen, taken by Walsh at Rock Island in 1863 and identified by him as of this species. This specimen is in good condition and unquestionably agrees with specimens at present being identified as of this species. Specimens collected very early in the season are very large and deep yellow in color; specimens taken in midsummer are smaller and predominately white.

MALE.—Length of body 6–9 mm., of fore wing 8–11 mm. Face below antennal sockets white, vertex yellow, shaded with tan; each antennal scape tan, pedicel white, flagellum faintly gray in basal part; eyes in life pearl-gray. Dorsum of thorax blackish brown, meson of mesoscutellum with a white area, this area usually extending anteriorly to ends of outer parapsides and along lateral scutellar ridge toward wing bases; metanotum with broad, white area on meson; pleuron mostly tan, white, or faint pink on semimembranous area anterior to fore wing base; mesosternum tan, metasternum white; front leg light tan, apexes of tibia and tarsal segments dark brown, first tarsal segment three-fifths as long as second; middle and

hind legs white; each femur of all legs with a median and an apical red-brown band; wings hyaline, pale brown stain in stigmatic area; fore wing with dark brown shading

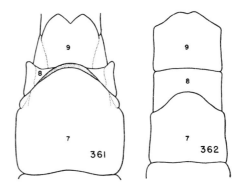

Fig. 361. — *Epeorus namatus*, terminal abdominal sternites of female.
Fig. 362.—*Heptagenia diabasia*, terminal abdominal sternites of female.

at base; longitudinal veins of fore wing yellow-brown, crossveins dark brown; veins and crossveins in costal area of hind wing light yellow, others hyaline. Abdominal segments white, a narrow, black or dark red-brown crossline at posterior margin of each tergite 1–7; dark spiracular dots present; terminal three abdominal tergites bright orange-brown, sometimes with pink suffusion also; genitalia, fig. 356, white; caudal filaments white, articulations dark red-brown.

NYMPH.—Length of body 7–9 mm. Head anterior to eyes, and on vertex between eyes, dark brown, with numerous pale, freckle-like dots, large, pale spot lateral to each compound eye and, on posterior margin of head, at inner eye margins. Pronotum with two or three large, pale spots near either lateral margin; each tarsal claw with two minute denticles; gills borne by abdominal segments 1–6 truncate at apexes, gills of seventh pair without tracheae; abdominal tergites 6, 8, and 10 almost entirely dark brown, tergites 1–5 white, with brown markings, tergite 7 brown near lateral margins and on meson, white elsewhere, and tergite 9 brown with large, submedian white areas; abdominal sternites 1–8 entirely white, sternite 9 white, with longitudinal, brown mark near either lateral margin, also sometimes with median, brown spot at anterior margin; postero-lateral angles of segments 7 and 8 produced as spines, these angles of segment 9 slightly produced, spines obsolescent; caudal filaments brown in basal half, apically alternating pairs of segments dark and light.

Known from Illinois, Indiana, Iowa, New York, Ontario, and Wisconsin.

Illinois Records.—Adult specimens, collected April 24 to August 13, are from Aroma Park, Aurora, Dixon, Elizabethtown, Erie, Havana, Kankakee, Keithsburg, Milan, Momence, Mount Carmel, New Boston, Oakwood, Oregon, Prophetstown, Rockford, Rock Island, St. Charles, Savanna, Shawneetown, Sterling, White Heath, Wichert, and Wilmington.

42. *HEPTAGENIA* Walsh

Heptagenia Walsh (1863:197).

In the males of this important and widely distributed genus, the compound eyes are large, but are not contiguous on the meson, except in the *lucidipennis* group of species; each fore leg is slightly longer than the body, the fore tarsus is from one and one-quarter to one and one-half times as long as the fore tibia, and the first tarsal segment varies from one-fifth to nearly one-half as long as the second tarsal segment. In both sexes, the wing venation is typical for the family, fig. 318, with the costal crossveins in the basal area usually well developed and the stigmatic crossveins not, or sometimes very slightly, anastomosed; in the hind wing, vein M_2 diverges from M_1 slightly basad of the center of the wing. In the male genitalia, the forceps are four-segmented, the second segment being as long as or longer than the two apical segments combined; the penis lobes, figs. 363–380, are fused on the meson two-thirds the distance from the base to the apex, each lobe typically bears spines or teeth, and a single posterolateral spine often is present on each penis lobe. The posterior margin of the terminal abdominal sternite in the females, fig. 362, is either evenly rounded from side to side or has a small median indentation.

In the nymphs, figs. 383–385, the frontal margin of the head is entire; the apical segment of the maxillary palp is relatively slender, with the apex acute, bears a dense row of hair along the outer margin, and lacks pectinate spines; the crown of each galealacinia of the maxilla bears a row of small,

hooklike teeth; the apical segment of each labial palp is extremely broad, with the apex truncate; the outer margin bears a dense row of hair, below which is a bank of pectinate spines. In the legs, the femora are only moderately flattened, and the posterior margin of each bears a dense row of hair and a sparse row of short, stout spines; the tarsal claws are long, slightly enlarged at the bases, and are either edentate or have a short row of ventral denticles; there may be a prominent ventral tooth in the basal area of each claw. Gills are borne by abdominal segments 1–7, with all gills of the same form but not same size. Each gill, fig. 325, is composed of a dorsal, platelike element and a ventral, filamentous tuft; in some species this tuft of filaments is greatly reduced or wanting on the gills of segment 7. None of the gills is extended beneath the abdominal venter. The three caudal filaments are equal in length or the median one is slightly the longer. Each of the cerci bears fairly prominent setae on the mesal side in the basal area; otherwise the cerci are virtually bare.

The nymphs of this genus occur under stones and among debris in shallow water near the banks of brooks, creeks, and rivers. They cannot be reared through to maturity in stagnant water.

The species *Heptagenia quebecensis* (Provancher) (1876:267; 1878:127) is shown by an examination of the lectotype, now in the Provincial Museum in Quebec, to have been based on a female specimen of the genus *Ephemerella*. This female specimen cannot be identified specifically.

The species *Heptagenia manifesta* (Eaton) remains unknown. It was originally the species identified as *Baetis debilis* Walker by Walsh (1862:371), using specimens collected at Rock Island, Illinois. Eaton (1871:130) transferred this species to the genus *Siphlurus* without seeing specimens of Walsh's material. Later, Eaton (1885:253) transferred it to *Rhithrogena* and renamed it *manifesta* (as he considered it to have been originally misidentified as Walker's species). There is no evidence that Eaton ever saw Walsh's material. I have been unable to locate specimens determined as *debilis* by Walsh; there is none in the Hagen collection at the Museum of Comparative Zoology.

Reliable characteristics for the specific separation of the females of this genus have not yet been found.

KEY TO SPECIES

Adult Males

1. Genitalia of the *persimplex* type, fig. 363: penis lobes short, somewhat rounded at apexes; discal, posterolateral, and subapical spines absent....**1. persimplex**
   Genitalia not of the *persimplex* type, but of the types shown in figs. 364, 365, 368, 372, or 374...............2
2. Genitalia of the *elegantula* type, fig. 364: each penis lobe semiquadrate, and with a prominent, medioapical, hooked spine ....**2. diabasia**
   Genitalia not of the *elegantula* type, but of the types shown in figs. 365, 368, 372, or 374....................3
3. Genitalia of the *pulla* type, fig. 365: penis lobes stout, divergent; each lobe bearing two stout, median spines and a minute, posterolateral spine..**3. pulla**
   Genitalia not of the *pulla* type, but of the types shown in figs. 368, 372, or 374..4
4. Genitalia of the *flavescens* type, figs. 366–369: penis lobes relatively long and slender, with apexes strongly diverged 5
   Genitalia not of the *flavescens* type, but of the types shown in figs. 372 or 374...................................8
5. Abdominal segments white or faintly smoky, each tergite with a narrow, black, transverse line at posterior margin and a pair of oblique, sublateral, black lines...........**4. marginalis**
   Abdominal segments light yellow or golden tan, tergites shaded on broad, median area with red-brown or orange-brown.........................6
6. Median spines of penis lobes slender throughout, fig. 367.......**5. patoka**
   Median spines of penis lobes broadened at apexes, figs. 368, 369.............7
7. Basal costal crossveins extremely weak or entirely wanting......**6. flavescens**
   Basal costal crossveins well developed, black................**7. cruentata**
8. Genitalia of the *lucidipennis* type, figs. 370–373: each penis lobe short and broad, with a heavy, usually recurved lobe at both apical and lateral margins and bearing a large, spinulose, mesal spine; discal spines sometimes present 9
   Genitalia of the *maculipennis* type, figs. 374–380: each penis lobe relatively elongate and narrow, and bearing one subapical, one posterolateral, one mesal, and always at least one discal spine..12
9. Penis lobes without discal spines, fig. 370 ....**8. rusticalis**
   Each penis lobe with at least one discal spine, figs. 371–373..............10
10. Abdominal tergites uniformly shaded with

light orange-brown; fore wing not over 6 mm. long.........**9. inconspicua**
Abdominal tergites shaded with tan, red-brown, or dark orange, but with prominent, light yellow spots or stripes on meson of basal and middle tergites; fore wing at least 7 mm. long......11
11. Apex of each penis lobe with a large, rounded, projecting lobule, and apical margin of forceps base with a pair of obscure, sublateral projections, fig. 372 ....**10. lucidipennis**
Apex of each penis lobe with a broad, recurved lobule, and apical margin of forceps base with a pair of conspicuous, sublateral projections, fig. 373 ....**11. perfida**
12. Costal and subcostal crossveins stained with brown, but this dark staining not extending to membrane surrounding these crossveins......**12. umbratica**
Costal and subcostal crossveins dark, and with this shading extending to membrane surrounding these crossveins, fig. 318........................13
13. Discal spines of penis lobes extremely long, fig. 375; abdominal tergites usually entirely dark brown..........**13. hebe**
Discal spines of penis lobes shorter, figs. 376–380; basal and middle abdominal tergites partly or almost entirely white 14
14. Basal and middle abdominal tergites entirely white, without darker markings ....**14. maculipennis**
Basal and middle abdominal tergites white, with posterior margins marked with black, or these tergites white or yellow, with dark red-brown spots near lateral margins..................15
15. Basal and middle abdominal tergites white, with a narrow, black, transverse line at posterior margin of each tergite 16
Basal and middle abdominal tergites white or yellow, with a pair of large, dark brown, triangular marks on each tergite at posterolateral angles......17
16. Thorax dark brown dorsally and on each pleuron..................**15. walshi**
Thorax entirely light yellow or white.... ....**16. juno**
17 Fore femur with a median, reddish streak and a black mark at apex; mesonotum shaded with dark brown on meson only near anterior margin....**17. minerva**
Fore femur with a black mark at apex only; mesonotum with a broad, dark brown-shaded area on meson which extends from anterior to posterior margin.... ....**18. aphrodite**

Mature Nymphs

1. Each gill borne by seventh abdominal segment with a well-developed ventral, filamentous tuft; each tarsal claw with a large basal tooth, but with ventral denticles absent, fig. 382...........2
Each gill borne by seventh abdominal

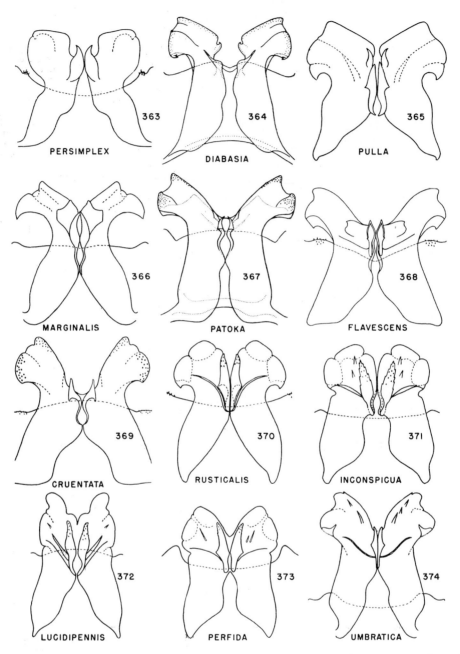

Fig. 363.—*Heptagenia persimplex*, male genitalia.
Fig. 364.—*Heptagenia diabasia*, male genitalia.
Fig. 365.—*Heptagenia pulla*, male genitalia.
Fig. 366.—*Heptagenia marginalis*, male genitalia.
Fig. 367.—*Heptagenia patoka*, male genitalia.
Fig. 368.—*Heptagenia flavescens*, male genitalia.

Fig. 369.—*Heptagenia cruentata*, male genitalia.
Fig. 370.—*Heptagenia rusticalis*, male genitalia
Fig. 371.—*Heptagenia inconspicua*, male genitalia.
Fig. 372.—*Heptagenia lucidipennis*, male genitalia.
Fig. 373.—*Heptagenia perfida*, male genitalia.
Fig. 374.—*Heptagenia umbratica*, male genitalia.

segment lacking ventral, filamentous tuft; each tarsal claw with a prominent basal tooth and a row of minute ventral denticles, fig. 381..............5
2. Venter of abdomen light yellow to white, and with a longitudinal, dark brown bar extending the length of the abdomen near either lateral margin.......3
Venter of abdomen light yellow to white, and entirely without darker markings, or with dark shading present only at posterior margin of ninth sternite ..4
3. Abdominal tergites almost entirely dark brown, with only a pair of submedian and a pair of anterolateral, small, white marks on each tergite, fig. 383........
.......................**2. diabasia**
Abdominal tergites with dark brown shading more restricted: submedian and lateral marks relatively large and

round, first tergite almost entirely white, fourth and eighth each with a large, median, white mark...**3. pulla**
4. Venter of abdomen entirely white or light yellow, without darker markings......
......................**4. marginalis**
Posterior margin of ninth sternite with a broad, dark brown border...........
......................**6. flavescens**
5. Caudal filaments entirely white, articulations not darkened...**10. lucidipennis**
Caudal filaments white, with articulations brown, or these filaments tan, gray, or brown.............................6
6. Caudal filaments white, with articulations brown......................7
Caudal filaments tan, brown, or gray...8
7. Abdominal sternites 8 and 9 each with a short, transverse, brown mark on meson of anterior margin..........**16. juno**

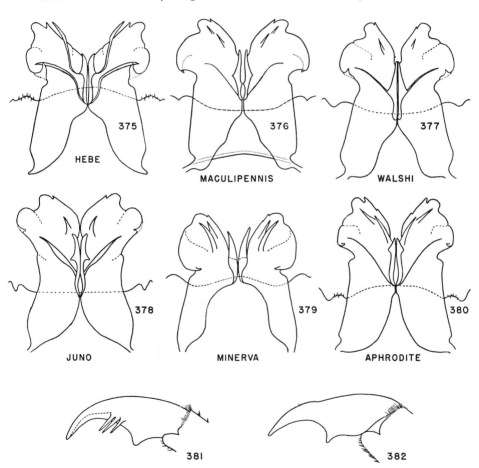

Fig. 375.—*Heptagenia hebe*, male genitalia.
Fig. 376. — *Heptagenia maculipennis*, male genitalia.
Fig. 377.—*Heptagenia walshi*, male genitalia.
Fig. 378.—*Heptagenia juno*, male genitalia.
Fig. 379.—*Heptagenia minerva*, male genitalia.

Fig. 380.—*Heptagenia aphrodite*, male genitalia.
Fig. 381.—*Heptagenia maculipennis*, middle tarsal claw of mature nymph.
Fig. 382.—*Heptagenia diabasia*, middle tarsal claw of mature nymph.

Abdominal sternites 8 and 9 usually entirely without brown, median marks at anterior margins, occasionally a small, median, brown dot at anterior margin of sternite 9.......**14. maculipennis**
8. Dorsum of abdomen uniformly dark brown, entirely or virtually without light markings; caudal filaments gray.
.....................**12. umbratica**
Dorsum of abdomen brown, with well-marked pattern of fairly large, white spots; caudal filaments tan or brown..9
9. Abdominal sternites 1–8 uniformly light tan, sternite 9 with lateral and posterior margins shaded with dark brown.....
.....................**18. aphrodite**
Abdominal sternum white, with brown markings on sternites 2–9.........10
10. Abdominal sternites 2–8 each with an irregular, brown crossband at anterior margin...................**11. perfida**
Abdominal sternites 2–8 each with a pair of fairly large, sublateral, brown spots
.........................**13. hebe**

PERSIMPLEX Group

1. *Heptagenia persimplex* McDunnough

Heptagenia persimplex McDunnough (1929:179).

This species often has been confused with *Anepeorus simplex*. As McDunnough has pointed out (1929:179), even Walsh's type material of *simplex* included specimens of *persimplex*. I have seen specimens of *persimplex* in several collections identified as *simplex*.

MALE.—Length of body 6–7 mm., of fore wing 7–8 mm. Compound eyes separated on meson by a space as wide as one compound eye; head very light yellow-brown; eyes in life light gray. Thorax very light cream color, almost white; legs light yellow, with apexes of tibiae and tarsi darkened with dirty tan; fore tibia as long as fore femur, fore tarsus one and one-third times as long as tibia, first tarsal segment one-third to almost one-half as long as second segment; wings hyaline, with veins and crossveins in costal half of fore wing light brown. Abdomen light cream colored, without darker markings; genitalia, fig. 363, light cream colored, forceps segments 3 and 4 of the same length, and their combined lengths slightly less than one-half as great as length of second segment; caudal filaments almost white.

NYMPH.—Unknown.

This species is known from Illinois, Iowa, Missouri, Nebraska, and Ohio.

Illinois Records. — HAVANA: Matanzas Beach, June 20, 1947, Ross & Stannard, 1 ♂; at light, June 25, 1898, C. A. Hart, 1 ♂. MOUNT CARMEL: June 18, 1947, B. D. Burks, 1 ♂. QUINCY: Mississippi River, June 7, 1939, Burks & Riegel, 1 ♂; July 6, 1939, Mohr & Riegel, 2 ♂. SHAWNEETOWN: at light, June 21, 1927, Frison & Glasgow, 3 ♂.

ELEGANTULA Group

2. *Heptagenia diabasia* Burks

Heptagenia diabasia Burks (1946:610).

MALE.—Length of body 9–13 mm., of fore wing 8–12 mm. Compound eyes pearl-gray, with a faint, yellow-green tint, the two eyes separated on meson by a space almost as wide as median ocellus; head yellow, shaded with light red on vertex and with a minute, black dot at base of frontal shelf at either eye margin. Thorax yellow, with tan shading on dorsal meson; legs yellow, with red-brown shading at apexes of middle and hind femora and at middle and apex of fore femur; all tibiae shaded at bases and apexes with brown; fore tibia slightly longer than fore femur, fore tarsus one and one-fifth times as long as tibia, first tarsal segment one-fourth as long as second; wings hyaline, with veins and crossveins brown, and crossveins in costal and subcostal interspaces slightly thickened; stigmatic crossveins of fore wing occasionally partly anastomosed. Abdomen yellow, with a longitudinal, black spiracular line on either side, a narrow, black, transverse line at posterior margin of each tergite, and apical three tergites shaded with golden- or carmine-brown; genital forceps light yellow, penis lobes, fig. 364, tan; third forceps segment one-third longer than fourth, second segment four times as long as third; caudal filaments white, with brown articulations.

FEMALE.—Length of body 9–14 mm., of fore wing 10–15 mm. Color lighter than in male, almost white; black markings of head and abdomen as in male; apical abdominal tergites without brown shading; apical abdominal sternites shaped as shown in fig. 362; caudal filaments white, articulations brown.

NYMPH.—Fig. 383. Length of body 8–15 mm. Each tarsal claw with large basal tooth, apical denticles wanting. Dorsum of

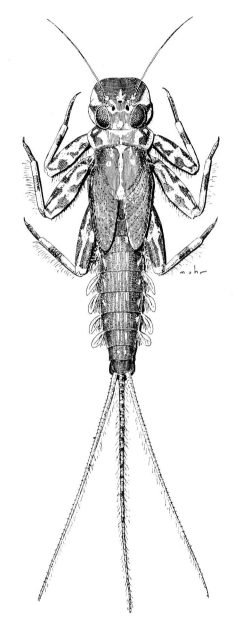

Fig. 383. — *Heptagenia diabasia*, mature nymph, dorsal aspect.

abdomen uniformly dark brown, with minute, submedian and anterolateral, light spots on tergites; abdominal sternites white, with a wide, brown, sublateral, longitudinal band near either lateral margin; gills, fig. 325, relatively slender, with filamentous tuft prominent and bushy, gills borne by seventh segment with well-developed, filamentous

tuft; caudal filaments rather vaguely marked with alternating, narrow, dark and light color bands.

Known from Illinois, Iowa, and Minnesota.

Illinois Records.—CAIRO: July 12, 1948, L. J. Stannard, 1 ♂. DIXON: June 25, 1947, B. D. Burks, 1 ♂. EAST DUBUQUE: July 3, 1946, Burks & Sanderson, 1 ♀. FREEPORT: June 10–11, 1948, Burks, Stannard, & Smith, 1 ♂. HAVANA: June 17, 1909, 1 ♀; June 1, 1933, at light, C. O. Mohr, 1 ♂; White Oak Creek, near Matanzas Lake, June 2, 1940, B. D. Burks, 4 ♂, 1 ♀; Matanzas Beach, June 25, 1947, Ross & Stannard, 5 ♂, 2 ♀. HOMER: July 4, 1943, H. H. Ross, 1 ♂. HOMER PARK: June 30, 1925, T. H. Frison, 1 ♀. JERSEYVILLE: June 2, 1938, T. H. Frison, 1 ♂. KANKAKEE: Aug. 3, 1938, Burks & Boesel, 1 ♂. KAPPA: June 22, 1943, H. H. Ross, 1 ♂. MAHOMET: June 8, 1940, H. H. Ross, 1 ♂. MAZON: Mazon Creek, June 25–27, 1938, B. D. Burks, 3 ♂, 2 ♀. MILAN: Rock River, June 4, 1940, Mohr & Burks, 5 ♂, 2 ♀. MOMENCE: June 22, 1938, Ross & Burks, 1 ♂. MOUNT CARMEL: June 18, 1947, B. D. Burks, 1 ♂, 1 ♀. MUNCIE: May 22, 1942, Ross & Burks, 1 ♂. PEORIA: June 23, 1938, F. F. Hasbrouck, 1 ♂. ROCKFORD: June 12, 1938, at light, Ross & Burks, 1 ♂; June 29, 1938, at light, B. D. Burks, 1 ♂; June 11, 1948, Burks, Stannard, & Smith, 13 ♂, 1 ♀. ST. JOSEPH: July 29, 1922, T. H. Frison, 1 ♀.

PULLA Group

3. *Heptagenia pulla* (Clemens)

Ecdyurus grandis Clemens (1913:147).
    Nomen nudum.
Ecdyurus pullus Clemens (1913:330).

MALE.—Length of body 10–11 mm., of fore wing 11–12 mm. Head light yellow, with a transverse, brown mark crossing face below antennal bases, and vertex shaded with brown. Pronotum dark brown; mesonotum and thoracic sternum light red-brown, pleura yellow. Legs yellow, with apexes of femora and bases of tibiae red-brown, tarsi with articulations stained with red-brown; wings hyaline, veins and crossveins of each fore wing brown. Abdomen yellow, tergites shaded with red-brown on broad, longitudinal, median area, this dark area interrupted by a narrow, yellow, median line and a pair of lunate, submesal marks on each tergite; posterior margins of tergites shaded with dark brown; sternites yellow, with a broad, light red, shaded area at posterior margin of each; genitalia, fig. 365, yellow, shaded with red at apexes; caudal filaments almost white, articulations brown.

NYMPH.—Length of body 11–13 mm. Head dark brown, with a white area on either side between compound eye and lateral margin of head. Each tarsal claw with a large basal tooth, ventral denticles wanting. Dorsum of abdomen mostly dark brown, with prominent, white markings: tergites 1 and 2 mostly white, tergites 4 and 8 each with a large, quadrate, white spot on meson at posterior margin, other tergites each with a pair of submesal and a pair of posterolateral, round, white spots; gills relatively small and oval, with filamentous, ventral tufts well developed and present in all gills; caudal filaments alternately brown and white throughout.

Known from Manitoba, New York, Ohio, and Ontario.

FLAVESCENS Group

4. Heptagenia marginalis Banks

Heptagenia marginalis Banks (1910:198).

MALE.—Length of body and of fore wing 9–10 mm. Head tan, with black markings along margin of frontal shelf, just below each antennal base, at either side of median ocellus, on vertex just posterior to each lateral ocellus, and along posterior margin of head; compound eyes separated on meson by a space as wide as one lateral ocellus. Thorax cream colored, with narrow, black lines on pleura; legs light yellow, almost white, each femur with a prominent, dark brown ring at apex, fore tibia and fore tarsal segments shaded with brown at apex; first fore tarsal segment one-sixth as long as second; wings hyaline, with veins and crossveins brown. Abdomen white, each tergite with a narrow, black band across posterior margin and a pair of oblique, sublateral, black marks; genitalia, fig. 366, white; caudal filaments very light gray, with articulations slightly darker.

FEMALE.—Length of body 9–10 mm., of fore wing 10–11 mm. Color pattern very similar to that of male, as all black marks of male are apparent in female, ground color

of body entirely white; caudal filaments white.

NYMPH.—Length of body 10–11 mm. Head light brown, with small, white spots. Each tarsal claw with basal tooth, but ventral denticles wanting. Dorsum of abdomen light brown, with a narrow, black crossband at posterior margin and a pair of sublateral, black streaks on each tergite; tergite 1 mostly white, following tergites each with a pair of sublateral, round, white spots; tergite 8 with a large, median, white blotch; abdominal sternum white to cream colored; gills semiovate, with fibrillar ventral tuft well developed on all segments;

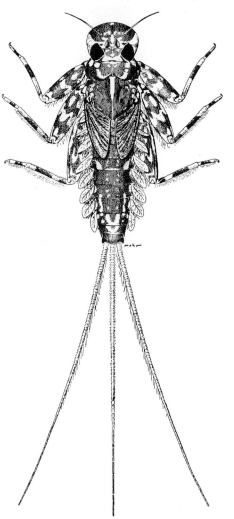

Fig. 384. — *Heptagenia flavescens*, mature nymph, dorsal aspect.

caudal filaments with alternating brown and white banding.

Known from New York, North Carolina, Ohio, Pennsylvania, Virginia, and West Virginia.

5. *Heptagenia patoka* Burks

Heptagenia patoka Burks (1946:612).

MALE.—Length of body and of fore wing 9 mm. Head tan; compound eyes separated on meson by a space as wide as lateral ocellus. Thorax brown, pleura yellow; each fore leg light tan, fore femur with red-brown shading in middle and at apex, brown shading at apex of fore tibia and at articulations of tarsus, first fore tarsal segment one-third as long as second; wings hyaline, veins and crossveins brown, with crossveins in costal and subcostal interspaces of fore wing broader and darker than others. Abdomen red-brown on dorsum, light yellowish tan on venter; tergites 1–3 each with a pair of yellow, submesal, round spots near anterior margin, following tergites with these spots more elongate and progressively darker and more vague, becoming scarcely distinguishable from dark ground color; a faint, median, yellow line present on tergites 1–5; sternites unmarked; apical three segments shaded with tan; genitalia, fig. 367, tan, with brown shading at edges; caudal filaments tan, articulations brown.

NYMPH.—Unknown.

The species is known only from Illinois.

Illinois Record.—PATOKA: July 19, 1945, Ross & Sanderson, 1 ♂.

6. *Heptagenia flavescens* (Walsh)

Palingenia flavescens Walsh (1862:373).

MALE.—Length of body 8–11 mm., of fore wing 10–13 mm. Head yellow, vertex shaded with tan; compound eyes grayish yellow-green, the two eyes separated on meson by a space slightly wider than a lateral ocellus. Thorax mostly red-tan, with pleura mostly yellow. Legs yellow, each fore femur shaded with red-brown in middle and at apex, fore tibia and tarsal segments shaded with gray-brown at apexes, first tarsal segment one-fifth as long as second segment; middle and hind femora shaded with brown at apexes; wings hyaline, veins and crossveins brown, those of hind wing slightly lighter brown, basal costal cross-

veins of fore wing weak or virtually wanting. Abdomen yellow, with a broad, median, gray-brown band on tergites 1–7, a narrow, dark gray, transverse band at posterior margin of tergites 1–7 or –8; apical three tergites orange-tan, apical two sternites shaded with orange-tan; genitalia, fig. 368, with forceps tan or yellow and penis lobes rose-pink; caudal filaments white or pale yellow, articulations brown or tan.

FEMALE.—Length of body 9–12 mm., of fore wing 11–15 mm. Head and thorax entirely yellow, or mesonotum faintly stained with tan; legs as in male; abdomen yellow, without dark, dorsal shading or with median, dorsal area faintly tan stained, a narrow, transverse, gray line present at posterior margins of tergites 1–8 or –9; terminal abdominal sternite more deeply cleft on meson of posterior margin than in *diabasia*; caudal filaments white.

NYMPH.—Fig. 384. Length of body 12–16 mm. Head mostly dark brown, with a prominent, triangular, white mark extending from anterolateral angle of each compound eye to lateral margin of head. Thorax brown, with small, irregular, white spots, each tarsal claw with a prominent basal tooth, but no ventral denticles. Abdomen dorsally dark brown with, typically, a pair of submedian, a pair of sublateral, and a pair of anterolateral, white spots on each of tergites 2–7; tergite 1 mostly white; submedian spots of tergite 4 and of 5 enlarged so as to coalesce at posterior margin; tergites 8 and 9 each white in median area, brown laterally; tergite 10 usually entirely brown; gills borne by seventh segment with well-developed ventral tuft of filaments; entire venter white, except that posterior margin of sternite 9 has a brown border; caudal filaments alternately banded brown and white.

Known from Georgia, Iowa, Illinois, Kansas, Manitoba, Minnesota, and Texas.

Illinois Records.—BILLETT: May 6, 1942, Burks & Mohr, 2 ♀. CALVIN: May 26, 1942, Mohr & Burks, 1 ♂. DIXON: June 27, 1935, DeLong & Ross, 1 ♂. FREEPORT: June 10, 1948, Burks & Stannard, 1 ♂. HAVANA: May 18, 1894, Hart, 1 ♂. MOUNT CARMEL: April 22, 1946, Mohr & Burks, 2 ♂, 1 ♀. OREGON: July 9, 1925, T. H. Frison, 14 ♂, 17 ♀; July 13, 1926, Frison & Hayes, 2 ♂; July 19, 1927, Frison & Glasgow, 3 ♂. PROPHETSTOWN: July 19,

1927, Frison & Glasgow, 1 ♂ ; Rock River, June 26, 1947, B. D. Burks, 1 ♂ ; July 24–25, 1947, Burks & Sanderson, 1 ♂. QUINCY: June 7, 1939, Burks & Riegel, 1 ♂. ROCKFORD: May 22, 1941, Ross & Burks, 1 ♂, 1 ♀ ; May 15, 1942, Ross & Burks, 1 ♂. ROCK ISLAND: 12 ♂, 4 ♀ (Walsh 1862:374). ROCKTON: Rock River, June 25, 1947, B. D. Burks, 1 ♂. SHAWNEETOWN: July 14, 1948, Mills & Ross, 1 ♂. STERLING: at light, May 21, 1925, D. H. Thompson, 1 ♂ ; May 22, 1941, Ross & Burks, 1 ♂.

7. Heptagenia cruentata Walsh

Heptagenia cruentata Walsh (1863:205).

The types of this species are lost, but there are a male and female in the collection at the Museum of Comparative Zoology which were determined as of this species by Walsh. They were collected a year after the description was published.

MALE.—Length of body 7–8 mm., of fore wing 9–11 mm. Head usually entirely shaded with red; compound eyes pearl-gray, the two eyes separated on meson by a space slightly narrower than a lateral ocellus. Thorax with notum light red-brown, pleura and sternum deep yellow. Legs yellow, femora and tibiae extensively stained with red, this red shading more intense in middle and at apex of each femur and on basal half of each tibia; tarsi grayed toward apexes; first fore tarsal segment one-fourth as long as second; wings hyaline, with membrane in costal and subcostal interspaces washed with yellow; all longitudinal veins and crossveins in hind wing golden, crossveins dark brown in fore wing, those in costal and subcostal interspaces broadened and darker. Abdomen deep yellow, with a broad, longitudinal, median, red-brown stripe on dorsum, this interrupted on each of tergites 2–8 by a longitudinal, median, yellow line and a pair of lunate, submedian, light streaks; genitalia, fig. 369, with forceps yellow and penis lobes red-tan; caudal filaments light yellow, with articulations light red-brown.

FEMALE.—Length of body 8 mm., of fore wing 10 mm. General color as in male, but yellow of body lighter, and red- and brown-shaded areas less conspicuous; legs colored as in male except that red shading is obscure on tibiae; dorsum of abdomen only lightly shaded with red-tan; posterior margin of terminal abdominal sternite only slightly incised on meson; caudal filaments light yellow, articulations sometimes faintly darkened.

NYMPH.—Unknown.

The species is known from Illinois, Manitoba, and Nebraska.

Illinois Records.—ANNA: at light, July 22, 1938, Burks & Boesel, 1 ♂. DIXON: June 27, 1935, DeLong & Ross, 1 ♂. PROPHETSTOWN: Rock River, July 24–25, 1947, Burks & Sanderson, 5 ♂, 1 ♀ ; June 26, 1947, B. D. Burks, 1 ♂. QUINCY: June 8, 1939, Burks & Riegel, 11 ♂, 13 ♀ ; July 13, 1937, Mohr & Burks, 1 ♂, 3 ♀. ROCK ISLAND: 4 ♂, 3 ♀ (Walsh 1863:205) ; June 7, 1937, Burks & Riegel, 1 ♂. ROCKTON: Rock River, June 25, 1947, B. D. Burks, 4 ♂.

LUCIDIPENNIS Group

8. Heptagenia rusticalis McDunnough

Heptagenia rusticalis McDunnough (1931b:92).

MALE.—Length of body and of fore wing 5–6 mm. Head yellow on face, shading to brown on vertex; compound eyes contiguous on meson. Dorsum of thorax brown, blending into deep yellow on pleura and sternum; legs dull, deep yellow, femora suffused with brown; wings hyaline, veins and crossveins colorless. Abdominal tergites dull brown, venter deep yellow, anterior tergites each with a faint, longitudinal, median, yellow line and a pair of submesal, longitudinal, yellow streaks; genitalia, fig. 370, smoky yellow; caudal filaments gray.

FEMALE.—Length of body 5–6 mm., of fore wing 6–7 mm. In appearance, similar to male, but head entirely light red-brown; entire thorax dull yellow; abdominal tergites lightly washed with red-brown, sternites yellow; posterior margin of terminal abdominal sternite evenly rounded from side to side; caudal filaments pale yellow.

NYMPH.—Unknown.

The species is known from New York, Ohio, and Quebec.

9. Heptagenia inconspicua McDunnough

Heptagenia inconspicua McDunnough (1924b:118).

MALE.—Length of body 4–5 mm., of fore wing 5–6 mm. Head tan; compound eyes light gray, contiguous on meson. Thoracic notum dull, light brown, pleura and sternum yellow. Legs yellow, apex of each fore femur and fore tibia, and fore tarsus shaded with dull brown, first fore tarsal segment one-third as long as second; wings hyaline, veins and crossveins colorless. Abdominal tergites shaded with orange- or red-brown, this shading more intense at posterior margins of tergites; genitalia, fig. 371, yellow, apical two forceps segments slightly smoky; caudal filaments white or faint yellow, with basal articulations red-brown.

FEMALE.—Length of body 4–5 mm., of fore wing 6–7 mm. Head tan, stained with orange; legs, thorax, and abdomen colored as in male; posterior margin of terminal abdominal sternite produced on meson, truncate; caudal filaments white.

NYMPH.—Unknown.

This species is known from Illinois, Indiana, Manitoba, Missouri, Ohio, and Wisconsin.

Illinois Records.—DIXON: June 27, 1935, 2 ♂, 2 ♀ ; at light, June 25, 1947, B. D. Burks, 1 ♂. LAKE GLENDALE: May 16, 1947, B. D. Burks, 1 ♂. OAKWOOD: June 6, 1925, T. H. Frison, 2 ♂ ; June 9, 1926, T. H. Frison, 6 ♂ ; June 23, 1948, B. D. Burks, 4 ♂ ; June 24, 1948, Mills & Ross, 1 ♂. OREGON: July 18, 1927, Frison & Glasgow, 1 ♂. RICHMOND: June 20, 1938, B. D. Burks, 1 ♂ ; Aug. 15, 1938, Ross & Burks, 3 ♂, 5 ♀. ROCKFORD: June 29, 1938, B. D. Burks, 1 ♂. SOUTH BELOIT: July 2, 1931, Frison, Betten, & Ross, 1 ♂. ST. CHARLES: at light, July 8, 1948, Ross & Burks, 1 ♂, 1 ♀. URBANA: at light, July 5, 1907, 1 ♂ ; July 8, 1931, H. H. Ross, 1 ♂. WAUKEGAN: Aug. 15, 1938, Ross & Burks, 5 ♂.

10. *Heptagenia lucidipennis* (Clemens)

Ecdyurus lucidipennis Clemens (1913:329).

MALE.—Length of body 6 mm., of fore wing 7 mm. Head with face yellow, shaded with brown, vertex dark brown or red-brown; compound eyes contiguous on meson. Thoracic notum brown, pleura and sternum yellow. Legs yellow, fore femur shaded with brown, first fore tarsal segment one-fourth as long as second; wings hyaline, costa and subcosta of fore wing slightly grayed, other longitudinal veins and all crossveins colorless. Dorsum of abdomen dark brown, venter yellow; each basal tergite with dark-shaded area vaguely interrupted by a

lighter, longitudinal, median line and a pair of obscure, submedian streaks; genitalia, fig. 372, yellow; caudal filaments light yellow, almost white, with faint, gray shading in basal halves.

FEMALE.—Length of body 6 mm., of fore wing 8 mm. Color pattern very similar to that of male, but slightly lighter, the yellow areas almost white and the brown areas tan or reddish; caudal filaments white.

NYMPH.—Length of body 7–8 mm. Head as wide as pronotum, dark brown, with numerous small, white spots. Thorax predominantly white, with numerous dark brown spots on dorsum; each tarsal claw with large basal tooth and several ventral denticles. Abdominal dorsum predominantly brown, each tergite typically white along posterior margin, two submedian and two anterolateral, white spots in brown area near anterior margin; gills of anterior segments each with ventral, filamentous tuft well developed; pair of gills on segment 6, each with filamentous tuft small and reduced; pair on segment 7 without ventral, filamentous tufts; caudal filaments white, unmarked.

Known from New York, Ohio, and Ontario.

11. *Heptagenia perfida* McDunnough

Heptagenia perfida McDunnough (1927b:301).

MALE.—Length of body 5.5–6.5 mm., of fore wing 6.5–7.5 mm. Head yellow, with red-orange shading; compound eyes bluish gray, contiguous on the meson. Thoracic notum chestnut brown; pleura tan, paling to yellow ventrally, sternum yellow. Legs yellow, apical half of fore femur red-brown, fore tibia and fore tarsus shaded with faint gray; wings hyaline, veins and crossveins colorless except that costa and subcosta of fore wing sometimes show faint, yellow staining. Abdominal tergites a deep, rich brown, with anteromesal area of tergites 2–8 yellow-brown, a longitudinal, median, yellow line and a pair of submedian, yellow streaks often visible within this yellow-brown area; venter yellow; genitalia, fig. 373, yellow, apical margin of forceps base with a pair of sublateral, prominent, setose projections; caudal filaments light yellow, basal articulations brown.

FEMALE.—Length of body 5.5–6.5 mm., of fore wing 7.5–8.5 mm. Head yellow, with

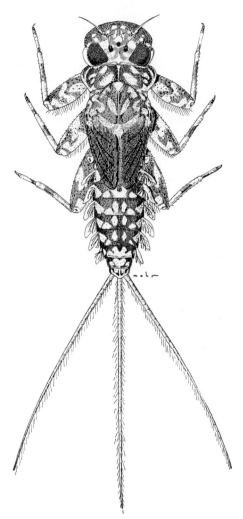

gin of head. Thorax brown, marked with obscure, paler spots; each tarsal claw with basal tooth virtually wanting and only three to five ventral denticles. Dorsum of abdomen brown, tergites 1–7 each with a pair of submedian, white streaks, usually a pair of small, white dots at posterior margin lateral to submedian streaks, and a pair of posterolateral, white, triangular marks; median area of tergites 8 and 9 each with a large, white blotch; filamentous tuft absent in gills borne by seventh segment, tuft borne by sixth pair of gills reduced in size; abdominal sternites each with a slightly irregular, brown crossband at anterior margin; caudal filaments tan, unmarked.

Known from Illinois and Ontario.

Illinois Records.—ALTO PASS: May 31, 1940, B. D. Burks, 2 ♂ ; Union Springs Church, May 12, 1939, Burks & Riegel, 2 ♂ ; June 1, 1940, B. D. Burks, 1 ♂, 1 N ; May 3, 1946, Burks & Sanderson, 2 ♂, 1 ♀ ; May 15, 1946, Mohr & Burks, 2 ♂, 2 ♀ ; branch of Clear Creek, May 19, 1946, Mohr & Burks, 2 ♂, 1 ♀. CORA: May 3, 1946, Burks & Sanderson, 1 ♂, 1 ♀. EDDYVILLE: Lusk Creek, June 6, 1946, Mohr & Burks, 1 ♂. GOLCONDA: May 13, 1939, Burks & Riegel, 2 ♂, 2 N. JONESBORO: branch of Clear Creek, May 15, 1946, Mohr & Burks, 7 ♂. OAKWOOD: Salt Fork River, May 22, 1942, Ross & Burks, 1 ♂. WOLF LAKE: Hutchins Creek, May 15–June 1, 1940, B. D. Burks, 4 ♂, 2 N.

MACULIPENNIS Group

12. Heptagenia umbratica McDunnough

Heptagenia umbratica McDunnough (1931b:92).

MALE.—Length of body 7–8 mm., of fore wing 8–9 mm. Frontal shelf of head stained with brown, and with a transverse, black streak across base, face above frontal shelf yellow, vertex brown; compound eyes separated on meson by a space as wide as a lateral ocellus. Thoracic notum and pleura dull brown, sternum yellow. Legs light yellow, fore femur, tibia, and tarsus shaded with brown; first fore tarsal segment one-fourth to one-fifth as long as second segment; wings hyaline, veins and crossveins in costal half of fore wing brown, others colorless. Abdomen yellow, tergites with posterior margin black and dorsal area

Fig. 385. — *Heptagenia perfida,* mature nymph, dorsal aspect.

orange shading. Thorax dull brown dorsally, light yellow ventrally. Abdomen dull brown dorsally, darker at posterior margins of tergites, sternites light yellow; posterior margin of terminal sternite produced and truncate on meson; caudal filaments light yellow, basal articulations tan.

NYMPH.—Fig. 385. Length of body 6–8 mm. Head slightly wider than thorax, dorsal side of head brown, with a pair of submesal, white spots on anterior margin, a second small pair of white spots just posterior to these, three fairly large spots between eyes and posterior to antennal bases, and a white streak extending from anterolateral angle of each compound eye to mar-

shaded with brown; tergites 2–6 each with a transverse, black streak at posterior margin and a pair of large, posterolateral, brown triangles; apical three segments brown; genitalia, fig. 374, with forceps pale yellow, shaded with brown toward apexes, penis lobes tan; caudal filaments white.

FEMALE.—Size as in male; color generally identical with that of male except that brown shading is less intense; posterior margin of terminal abdominal sternite produced, and evenly rounded from side to side.

NYMPH.—Length of body 9 mm. Head and dorsum of body extremely dark brown, with lighter markings obscure or virtually wanting; each tarsal claw with a small basal tooth and a few ventral denticles; gills uniformly dark purplish gray, those of seventh pair lacking ventral, filamentous tufts; caudal filaments dark gray.

Known from Quebec.

13. Heptagenia hebe McDunnough

Heptagenia hebe McDunnough (1924*b*:122).

MALE.—Length of body 6–7 mm., of fore wing 7–8 mm. Frontal shelf of head stained with brown, a black, transverse streak across base of shelf; face yellow, vertex brown; the compound eyes pearl-gray, separated on meson by a space slightly wider than a lateral ocellus. Thorax yellow-brown dorsally, pleura yellow, with a longitudinal, brown streak just dorsad of each coxa; sternum yellow. Legs yellow, apexes of femora shaded with tan; an obscure, dark brown, ventral streak usually present near apex of each femur; fore tibia shaded with tan in apical half, fore tarsus tan; first fore tarsal segment one-fifth as long as second segment; wings hyaline, all veins of fore wing tan, crossveins dark brown, those in costal and subcostal interspaces surrounded by a dark brown cloud, another cloud at base of outer fork; often these clouds surround virtually all crossveins in fore wing; veins and crossveins of hind wing hyaline. Abdomen yellow, tergum heavily shaded with dark brown, this shading sometimes uniformly and completely covering tergites, but usually tergites 2, 3, and 8 each with a pair of submedian, yellow streaks, tergites 4–7 each with large, round, median, yellow spot; genitalia, fig. 375, yellow, with median, discal spines extremely long; caudal filaments light yellow.

FEMALE.—Length of body 5–6 mm., of fore wing 7–8 mm. Head almost entirely yellow. Thorax yellow, with dark brown, longitudinal stripe on each pleuron above coxae; legs light yellow, fore femur lightly shaded with tan; wings with crossveins in costal and subcostal interspaces surrounded by dark brown clouds, other crossveins in fore wing dark brown, veins white or a faint yellow. Abdomen light yellow, with a pair of broad, submarginal, lunate, dark brown streaks on each of tergites 1–7; posterior margin of terminal abdominal sternite produced, evenly rounded from side to side; caudal filaments white.

NYMPH.—Length of body 5–7 mm. Head brown, with dark brown freckles and three white spots on margin just anterior to either eye, and a pair of submedian, white spots on posterior margin of vertex. Thorax brown, with numerous small, white spots; each tarsal claw with a small, acute basal tooth and 3–5 ventral denticles. Dorsum of abdomen brown, with prominent, white markings; tergite 1 almost entirely white, tergite 2 white on median area and on posterolateral triangles, tergites 3 and 6 each with four white dots, tergites 4 and 5 white on broad, median area and on lateral triangles at posterior margins, tergite 7 white across posterior half, tergite 8 almost entirely white, tergite 9 with a white stripe across anterior margin and with a pair of circular, posterolateral, white spots, last tergite with a median, white spot at posterior margin; venter entirely white except for a brown mark at each posterolateral angle of terminal abdominal sternite; gills borne by seventh segment without ventral, filamentous tufts; caudal filaments tan, alternating articulations brown.

Known from Connecticut, Illinois, Indiana, Iowa, Maryland, Michigan, Minnesota, New York, Ohio, Ontario, Pennsylvania, Quebec, and Tennessee.

Illinois Records. — DOWNS: Kickapoo Creek, June 22, 1943, H. H. Ross, 1 ♂. EICHORN: June 6, 1946, Mohr & Burks, 1 ♂. HOMER: Salt Fork River, June 30, 1925, T. H. Frison, 4 ♀ ; July 19, 1924, T. H. Frison, 1 ♀. KANKAKEE: Kankakee River, June 17, 1939, B. D. Burks, 1 ♂ ; July 10, 1925, 1 ♂ ; July 10, 1925, T. H. Frison, 3 ♂, 2 ♀. LA GRANGE: June 17, 1938, J. S. Ayars, 2 ♂. MAZON: Mazon Creek, June 23–28, 1938, Ross & Burks,

1 ♂, 9 ♀, 5 N. OAKWOOD: Salt Fork River, June 6, 1925, T. H. Frison, 6 ♂, 1 ♀ ; June 14, 1930, T. H. Frison, 2 ♂ ; Camp Drake, June 24, 1948, Mills & Ross, 1 ♂. QUINCY: Mississippi River, June 7, 1939, Burks & Riegel, 1 ♂ ; June 8, 1939, T. E. Musselman, 1 ♂. URBANA: at light, July 29, 1947, L. J. Stannard, 1 ♀ ; July 5, 1907, 1 ♀. WILMINGTON: at light, Aug. 6, 1947, Burks & Sanderson, 5 ♂.

14. *Heptagenia maculipennis* Walsh

Heptagenia maculipennis Walsh (1863:206).

The male lectotype of this species is in the Museum of Comparative Zoology. It is badly broken, but the remaining fragments indicate that the present concept of the species is correct.

MALE.—Length of body 4.5–6.0 mm., of fore wing 6.5–8.0 mm. Head yellow; vertex stained with red-brown and having a brown stripe at posterior margin; compound eyes greenish gray, with a brown stripe across middle of each, eyes separated on meson by a space almost twice as wide as a lateral ocellus. Thorax cream colored, almost white, with a longitudinal, mesal, brown stripe on mesonotum; a dark brown stripe on each pleuron extending from mesocoxa to pronotum and a lighter brown stripe extending ventrally from fore wing base to sternum. Legs pale yellow to white, fore femur yellow, ventral side with a black mark at apex, apexes of fore tibia and fore tarsal segments shaded with gray; first fore tarsal segment one-sixth to one-fifth as long as second segment; wings hyaline, anterior veins of fore wing tan, all crossveins brown, those in costal and subcostal interspaces surrounded by brown clouds, fig. 318; posterior veins of fore wing and all veins and crossveins of hind wing colorless. Abdomen white, with tergites 7–9 and basal half of 10 shaded with bright orange- or red-brown; genitalia, fig. 376, white; caudal filaments white.

FEMALE.—Length of body 5.0–6.5 mm., of fore wing 7.0–8.5 mm. Head as in male. Thorax much as in male except that brown shading is reduced in area, the longitudinal, mesal, brown shading of mesonotum is usually confined to anterior half of sclerite; legs as in male, but fore tibia and tarsus with no dark shading; wings hyaline, all veins colorless, anterior crossveins of fore wing brown. Entire abdomen light yellow to white; posterior margin of terminal abdominal sternite produced on meson, evenly rounded; caudal filaments white.

NYMPH.—Length of body 5.0–6.5 mm. Head wider than pronotum, anterior, dorsal portion gray-brown, freckled with dark spots, three white spots on either margin just anterior to compound eye, three large, round, white spots on vertex between eyes. Thoracic notum gray-brown, with numerous white spots; each tarsal claw with a small, acute basal tooth and four or five ventral denticles. Abdomen dorsally gray-brown, with large, white markings; tergite 1 almost entirely white; tergites 2, 3, and 6 each with a pair of submedian, white streaks, a pair of posterior, white triangles, and a pair of lateral triangles; tergites 4 and 5 each with a large, median, white area, a pair of posterior triangles, and a pair of lateral triangles; tergites 7 and 8 with a confluent, median, white blotch, and each with a pair of posterior and a pair of lateral triangles; tergite 9 brown on posterior two-thirds, white anteriorly, three marginal, white dots in brown portion; tergite 10 brown, with a pair of large, submedian, white marks; sternites entirely white; gills borne by seventh segment lacking ventral, filamentous tufts; caudal filaments white, articulations brown, these latter alternately of lighter and darker tones.

Known from Illinois, Manitoba, Missouri, Ohio, Ontario, and Tennessee.

Illinois Records.—Specimens, collected June 3 to September 8, are from Antioch, Aurora, Beardstown, Dixon, Effingham, Elizabethtown, Hardin, Harrisburg, Havana, Homer, Hoopeston, Illini State Park (La Salle County), Kankakee, Lewistown, Mahomet, Momence, Monmouth, Monticello, Mount Carmel, Muncie, Oakwood, Oregon, Ottawa, Pontiac, Prophetstown, Quincy, Rockford, Rock Island, Rockton, Rossville, St. Charles, Shawneetown, Shelbyville, South Beloit, Sterling, Urbana, Waukegan, and Wilmington.

15. *Heptagenia walshi* McDunnough

Heptagenia walshi McDunnough (1926:193).

MALE.—Length of body and of fore wing 6 mm. Head brown, a light tan or yellow crossband just below antennae; compound eyes separated by a space almost as wide as one compound eye. Thorax brown on dorsum and pleura, sternum tan. Legs yellow, each femur with a black streak at apex, fore tibia and fore tarsus shaded with brown; first fore tarsal segment one-fourth as long as second segment; wings hyaline, veins and all crossveins but those posterior to Cu_1 in fore wing brown, with brown clouds surrounding crossveins in costal and subcostal interspaces as well as those posterior to bulla; stigmatic area stained with brown; veins and crossveins of hind wing almost or quite colorless. Abdomen white, each tergite tinged with tan in posterior half; a black crossline at posterior margin of each tergite; genitalia, fig. 377, light tan to white; caudal filaments white.

NYMPH.—Unknown.

This species is known from Ohio and Ontario.

16. Heptagenia juno McDunnough

Heptagenia juno McDunnough (1924b:121).

MALE.—Length of body 5.5–6.5 mm., of fore wing 6.5–7.5 mm. Head white, with a narrow, brown crossline at base of frontal shelf and at posterior margin of vertex; compound eyes separated on meson by a space as wide as one eye. Thorax cream colored, a longitudinal, black line on each pleuron above coxae, and a vague, dark streak at bases of each pair of wings; legs light yellow, each femur with a dark streak at apex, first fore tarsal segment one-fourth as long as second; wings hyaline, all veins except anterior ones of fore wing hyaline, crossveins of fore wing anterior to Cu_1 brown, others hyaline, crossveins in costal and subcostal interspaces of fore wing surrounded by brown clouds. Abdomen white, a black line at posterior margin of each tergite 1–7; apical tergites each vaguely darkened in mesal area with brown; venter white; genitalia, fig. 378, white; caudal filaments white.

FEMALE.—Length of body 5 mm., of fore wing 6.5 mm. Coloration similar to that of male, but light yellow of thorax becoming white or almost white; transverse, black marks at posterior margins of abdominal tergites faint or wanting, terminal three tergites not darkened; posterior margin of terminal abdominal sternite produced on meson as an obscure, blunt point; caudal filaments white.

NYMPH.—Length of body 6 mm. Head dark gray-brown, slightly wider than pronotum, three white spots on margin just anterior to either compound eye. Thorax dark grayish-brown, with numerous white spots, each tarsal claw with a small, basal tooth and four or five ventral denticles. Abdomen dark gray-brown, with white spots, tergite 1 mostly white, tergites 7 and 8 with a coalescing, median, white blotch, tergite 9 white at anterior margin, dark posteriorly; sternites 8 and 9 each with a transverse, dark brown mark on meson of anterior margin; caudal filaments white, articulations brown.

This species is known from Kentucky, New York, Pennsylvania, Quebec, and Tennessee.

17. Heptagenia minerva McDunnough

Heptagenia minerva McDunnough (1924:121).

MALE.—Length of body 6 mm., of fore wing 7 mm. Head yellow, with a black, transverse line at base of frontal shelf and at posterior margin of vertex; compound eyes separated on meson by a space as wide as one eye. Thorax cream colored, with a median, brown streak on anterior third of mesonotum and a very dark brown spot on either side of base of mesoscutellum; a longitudinal, black line on each pleuron above coxae. Legs yellow, fore femur with a median, reddish spot and an apical, black streak; wings hyaline, veins of fore wing brown, crossveins darker brown, crossveins of costal and subcostal interspaces surrounded by brown clouds, as are those posterior to bulla; outer fork with brown cloud at base. Abdomen faint yellow or white, each of tergites 1–7 with a transverse, black line at posterior margin and a pair of sublateral, elongate, brown triangles; apical three tergites shaded with red-brown; venter, genitalia, fig. 379, and caudal filaments white.

FEMALE.—Length of body 7 mm., of fore wing 8 mm. Coloration similar to that of male, but generally lighter, with light yellow of thorax becoming white; dark shading of abdominal tergites much reduced, terminal tergites white.

NYMPH.—Unknown.

The species is known from Maryland and Ontario.

18. *Heptagenia aphrodite* McDunnough

Heptagenia aphrodite McDunnough
(1926a:194).

MALE.—Length of body 6 mm., of fore
wing 7 mm. Head with face yellow, vertex
red-brown, with black shading around ocelli
and at posterior margin; compound eyes
separated on meson by a space as wide as
one eye. Thorax with pronotum yellow,
mesonotum chestnut brown, with yellow
shading laterally; pleuron yellow, with a
broad, red-brown, longitudinal stripe above
coxae; sternum yellow; legs yellow, each
femur shaded with tan at apex, fore femur
also with a black, apical streak; wings hyaline, all longitudinal veins but C, Sc, and
R_1 of fore wing colorless, crossveins of fore
wing anterior to Cu_1 brown, those in costal
and subcostal interspaces surrounded by
brown clouds. Abdomen yellow, tergites
1–7 each with a black, transverse line at
posterior margin and a pair of large, brown,
elongate, sublateral triangles; terminal three
tergites very dark brown; genitalia, fig. 380,
light yellow; caudal filaments light yellow
in color.

FEMALE.—Length of body 6–7 mm., of
fore wing 7–8 mm. Coloration similar to
that of male, but generally lighter. Thoracic
notum yellow, with vague, brown darkening
on meson at anterior margin of mesonotum.
Abdomen yellow, tergites 1–9 each with a
minute, transverse, black line at posterior
margin, tergites 2–7 each with a pair of
sublateral, oblique, dark brown streaks; terminal abdominal sternite with posterior
margin produced on meson in form of a
Tudor arch; caudal filaments white.

NYMPH.—Length of body 7 mm. Head
dark brown, slightly wider than pronotum,
three white spots at margin on either side
just anterior to compound eye. Thorax
dark brown, with numerous white spots;
each tarsal claw with small, acute basal
tooth and 4–6 ventral denticles. Abdomen
dorsally mostly brown, with white markings;
tergite 1 mostly white, tergites 7 and 8 with
a median, coalesced, white blotch, tergite 9
with a white, median triangle based on anterior margin; abdominal sternum uniformly tan, entire margin of sternite 9
brown; caudal filaments tan, articulations
brown.

This species is known from Georgia, Illi-
nois, Indiana, New York, North Carolina,
Ontario, Tennessee, West Virginia, and
Wisconsin.

Illinois Record.—MOMENCE: at light,
July 17, 1914, 1 ♂.

43. *EPEORUS* Eaton

Epeorus Eaton (1881:26).

The species of the genus *Epeorus* (s. lat.)
can be grouped into four subgenera: *Epeorus*
Eaton (s. s.), *Iron* Eaton (1883, pl. 24, fig.
44), *Ironodes* Traver (1935b:32), and
Ironopsis Traver (1935b:36). The last two
subgenera include only western species in
North America.

In the adult males of this genus, the compound eyes are contiguous on the meson;
the fore leg is approximately as long as the
body, the fore tarsus is one and one-sixth
to one and one-half times as long as the
fore tibia, and the first segment of the fore
tarsus is as long as, or slightly longer than,
second segment; the two fore tarsal claws
are either of equal size and blunt, or unequal, with the larger claw blunt and the
slightly smaller claw hooked at the apex.
In both sexes, the first segment of the hind
tarsus is as long as, or slightly longer than,
the second segment. The wing venation is
typical for the family, fig. 319, with the
crossveins in the basal area of the costal
interspace of the fore wing extremely weak
or absent and the stigmatic crossveins anastomosed or not. In the hind wing, vein M_2
diverges from M_1 slightly basad of the
center of the wing. In the male genitalia,
the forceps are four-segmented, with the
second segment longer than the third and
fourth segments combined; the penis lobes,
fig. 387, are fused on the median line about
one-half of the distance to the apexes. These
penis lobes are simple or are provided with
median hooks or spines; there may also be
lateral prongs on the lobes. In the females,
the posterior margin of the terminal abdominal sternite, fig. 361, is cleft on the
meson.

In the nymphs, fig. 386, the frontal margin
of the head is not incised on the meson; the
apical segment of the maxillary palp is enlarged, but relatively acute at the apex, and
has a dense row of hairs on the outer margin; the apex of the galea-lacinia of the
maxilla bears three stout, curved teeth. The
apical segment of the labial palp has a dense
row of hairs on the outer margin and, be-

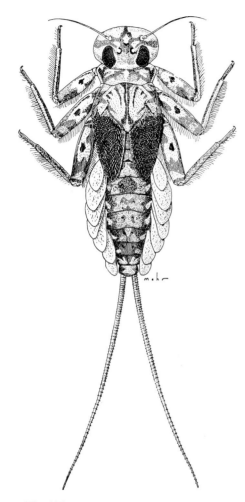

Fig. 386.—*Epeorus namatus*, mature nymph, dorsal aspect.

The nymphs of *Epeorus* inhabit shallow, cool or cold, rapidly flowing water.

The genus *Epeorus* is well represented in the western and northeastern states, but is extremely rare in the Midwest. *Epeorus namatus* (Burks) (1946:607), figs. 386, 387, was described from Indiana and *vitrea* (Walker) (1853:555) is known from Manitoba, Michigan, New York, Ontario, Pennsylvania, and Quebec. Either or both of these might eventually be collected in Illinois.

44. *CINYGMULA* McDunnough

Cinygmula McDunnough (1933b:75).

In the adult males of *Cinygmula,* the compound eyes are large, but are not quite contiguous on the meson; each fore leg is about as long as the body, with the femur three-fourths as long as the tibia and the tarsus one and one-fourth to one and one-half times as long as the tibia; the first tarsal segment is from three-fifths to three-fourths as long as the second segment. The wing venation in both sexes is typical for the family, with the stigmatic crossveins of the fore wing not at all, or only slightly, anastomosed; the wing membrane often is suffused with a gray or yellow tint. Vein M_2 in the hind wing diverges from M_1 in the center of the wing. In the male genitalia, the forceps have four segments, the second of which is the longest and is about as long as the third and fourth segments combined; the penis lobes, fig. 388, are rather long and slender, have conspicuous lateral and mesal spines, and are either entirely separate or fused on the meson at the base only. The terminal abdominal sternite of the female has a V-shaped, median indentation on the posterior margin.

In the nymphs, the frontal margin of the head is emarginate on the meson, exposing a small portion of the labrum when viewed in dorsal aspect. The femora are relatively long, narrow, and flattened, and the tarsal claws are short, stout, and bear only three to five minute, ventral denticles. A pair of gills is present on each of abdominal segments 1–7; these gills are similar on all segments, each being platelike, with the fibrillar portion either entirely wanting or reduced to two or three filaments, fig. 328; the gills normally are held against the sides of the abdomen and project over the tergites but do not

low this, a bank of pectinate spines. In the legs, the femora are relatively long and slender, with a closely set row of bristles along the posterior margin of each; each tarsal claw is short, rather slender at the base, and has two to six extremely minute, ventral denticles near the tip. Gills are borne by abdominal segments 1–7; each gill, fig. 327, is composed of a ventral, platelike element and a much-reduced, dorsal tuft of filaments; the anterior and posterior pairs of gills may or may not project beneath the abdominal venter to form, with the intermediate gills, a partial or complete, ventral, adhesive disc. There are only two long caudal filaments; the median one is atrophied.

extend beneath the venter at all. The median caudal filament is slightly longer than the lateral ones.

with the frontal margin of the head entire. The tarsal claws are relatively short and stout at the bases; each claw has only two

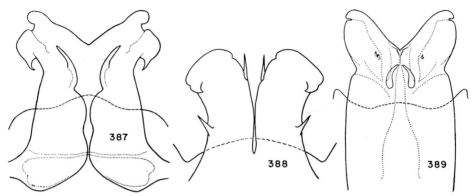

Fig. 387.—*Epeorus namatus*, male genitalia.   Fig. 388.—*Cinygmula atlantica*, male genitalia.
Fig. 389.—*Cinygma integrum*, male genitalia.

The species of *Cinygmula* are principally western and northern in distribution. *C. altantica* (McDunnough) (1924b:131) occurs in New York and Nova Scotia.

45. CINYGMA Eaton

Cinygma Eaton (1885:247).

In the adult males of this genus, the compound eyes are large, but do not quite meet on the meson; each fore leg is as long as the body, the femur and tibia almost or quite equal in length, the tarsus twice as long as the tibia, and the first tarsal segment from three-fifths to three-fourths as long as the second tarsal segment. In both sexes, the wing venation is typical for the family, with the stigmatic crossveins, fig. 322, numerous and anastomosed in such a way that a fine, slightly irregular line, parallel with vein C, divides the costal interspace into two parts; the costal crossveins of the fore wing in the basal area are quite weak. Vein M_2 in the hind wing diverges from M_1 in the center of the wing. The male genital forceps have four segments, the second of which is longer than the other three combined; the penis lobes, fig. 389, are broad and fused on the meson almost to the tips, and there is a small, median spine on each penis lobe at a point midway from the base to the apex. In the females, the posterior margin of the apical abdominal sternite is broadly rounded, and has a median notch.

The nymphs are typical for the family,

or three ventral denticles near the tip. A pair of gills is borne by each of abdominal segments 1–7; each gill, fig. 329, is composed of a broad and platelike, dorsal element and a small ventral tuft of filaments; the gills do not extend over the abdominal sternites; the three caudal filaments are all of practically the same length.

The species of *Cinygma* known at present occur only in the western states. *C. integrum* Eaton (1885:248) is known from British Columbia, Oregon, and Washington.

45. RHITHROGENA Eaton

Rhithrogena Eaton (1881:23).

In the adult males of this genus, the compound eyes are contiguous on the meson; each fore leg is slightly longer than the body, with the tibia one and one-third times as long as the femur, and the tarsus one and one-fourth to one and one-half times as long as the tibia; the first segment of the fore tarsus is only one-fifth to one-third as long as the second segment. In both sexes, the wing venation is typical for the family, fig. 320, but the costal crossveins in the basal area of the fore wing are very weak, and, in Nearctic species, the costal, stigmatic crossveins are anastomosed, fig. 323. In the hind wing, vein M_2 diverges from vein M_1 slightly basad of the middle of the wing. In the male genitalia, the four-segmented forceps arise from a base which has a shallow, median indentation with a broadly

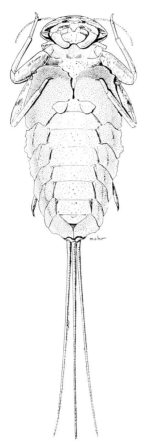

Fig. 390.—*Rhithrogena* sp., mature nymph, ventral aspect.

rounded projection on either side of this indentation, fig. 391. The penis lobes vary considerably in shape, but are always fused on the meson at the base; apical teeth and lateral spines are often present on the penis lobes. The posterior margin of the apical abdominal sternite of the female is broadly rounded, without a median emargination.

In the nymphs, fig. 390, the frontal margin of the head is very slightly emarginate on the meson, exposing a small part of the labrum when viewed in dorsal aspect; each maxillary palp has the distal segment broad and clavate, with the apex obliquely truncate and bearing multiple rows of pectinate spines and teeth on the inner surface and a dense row of hairs on the outer margin; the crown of each maxillary galea-lacinia bears a row of minute, stout teeth or spines; the apical segment of each labial palp is greatly enlarged and has a dense row of fine hairs

along the outer, apical margin, below which is a bank of pectinate spines. In the legs, the femora are only moderately flattened, and have a row of closely set bristles along the posterior margin of each; each tarsal claw is short, broad at the base, and bears one or two ventral spines near the base and two or three ventral denticles near the tip. The gills are borne by segments 1–7, each gill being composed of a ventral, platelike element and a dorsal tuft of filaments, fig. 324. The platelike element of each gill has the outer margin irregularly fissured. The gills are held spread out beneath the venter of the abdomen in such a way as to form an adhesive disc, the first and last pairs of gills meeting on the meson to complete this disc at the anterior and posterior ends of the abdomen. The median caudal filament is somewhat longer than the cerci; all filaments are bare except for minute setae on the mesal margin of each cercus.

The species of *Rhithrogena* are relatively numerous and common in the western states and in the Northeast. Only one species has been collected in Illinois.

Rhithrogena pellucida Daggy

Rhithrogena pellucida Daggy (1945:383).

MALE.—Length of body and of fore wing 6–7 mm. Compound eyes in life dark gray-green; scape and pedicel of each antenna light yellow, flagellum shaded with gray; vertex and frontal shelf of head dark brown, except for a light yellow, transverse band extending from eye to eye at level of bases of antennae. Pronotum dark red-brown; mesonotum dark brown, almost black, with faint, greenish tinge, apex of scutellum red-brown; pleura brown, with yellow markings. Legs yellow or tan-yellow, each femur with a vague, brown, median band, and each tarsal segment shaded with gray-brown at apex; first fore tarsal segment one-third as long as second; wings hyaline, veins C, Sc, and R of fore wing shaded with gray-brown at bases, all other veins and all crossveins hyaline; stigmatic crossveins only slightly anastomosed. Abdominal tergites 2–6 each with a pair of large, brown, lateral spots; tergites 7–10 brown, with faint orange cast; sternites dirty white or yellow; genitalia, fig. 391, greatly reduced, the penis lobes lacking all spines, teeth, or tubercles; caudal filaments white, articulations not darkened.

FEMALE.—Size as in male. Color similar to that of male, but generally lighter, the dark brown of the male being replaced by lighter brown, and light yellow being replaced by white.

NYMPH.—Length of body 6.5–7.5 mm. Head, thoracic dorsum, and abdominal ter-

HILLSDALE: Rock River, July 29, 1925, D. H. Thompson, 1 N. LYNDON: Rock River, July 8, 1925, D. H. Thompson, 28 N; July 15, 1925, 5 N; Aug. 5, 1924, 1 N. NEW MILFORD: Rock River, at mouth Kishwaukee River, July 14, 1927, D. H. Thompson, 5 N. OREGON: Rock River, below bridge,

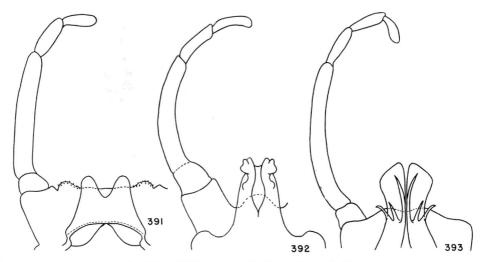

Fig. 391.—*Rhithrogena pellucida,* male genitalia.
Fig. 392.—*Anepeorus simplex,* male genitalia. (After McDunnough.)
Fig. 393.—*Arthroplea bipunctata,* male genitalia.

gites 1–7 and 10 dark chestnut brown, tergites 8 and 9 yellow except at lateral margins. Thoracic sternum white, with narrow, brown lines at edges of sclerites; each femur light brown, variegated with white in middle and at either end, and with scattered, dark brown dots in basal two-thirds; tarsal claws brown at tips. Abdominal venter tan or yellowish, sternites 2–8 each with a brown, median patch and a dark brown, transverse line at the posterior margin, and a brown-shaded area near lateral margins; lamellate portion of each gill white, fibrillar portion faintly stained with tan; caudal filaments tan.

This swift-water species is known from Illinois, Michigan, and Minnesota.

Illinois Records.—COMO: Elkhorn Creek, June 18, 1925, D. H. Thompson, 2 N; Rock River, July 6, 1925, 2 N; Aug. 12, 1924, 1 N. DIXON: Rock River, May 12, 1925, D. H. Thompson, many exuviae; May 22, 1925, 1 N. ERIE: Rock River, July 23, 1925, D. H. Thompson, 1 N. GRAND DETOUR: Rock River, May 27, 1927, 9 N.

May 24, 1927, D. H. Thompson, 1 N; July 11, 1929, T. H. Frison, 1 N. PORTLAND: Rock River, July 21, 1925, D. H. Thompson, 1 N. ROSCOE: Rock River, Aug. 20, 1925, D. H. Thompson, 7 N. STERLING: Rock River, Aug. 7, 1924, D. H. Thompson, 1 N.

47. *ANEPEORUS* McDunnough

Anepeorus McDunnough (1925b:190).

In the males of *Anepeorus,* the compound eyes are only moderately large and are separated on the meson by a space at least as great as the width of one eye; each fore leg is only slightly longer than the middle or hind leg; the fore tibia is one and one-third times as long as the fore femur, and the fore tarsus is only two-thirds as long as the fore tibia; the second tarsal segment is one and one-half times as long as the first, slightly longer than the third, and one and one-half times as long as the fourth, the fifth segment being slightly shorter than the fourth segment. In both sexes, the wing venation is typical for the family, with the stigmatic crossveins in the costal interspace relatively few in number and sometimes partly anastomosed. In the hind wing, vein R_5 diverges from R_4 at a point the same distance proximad of the outer wing margin that R_3 diverges from R_2; vein M_2 diverges from M_1 at a point only one-third the distance from the base to the outer wing margin. The male genitalia consist of a pair of four-segmented forceps arising from a medially

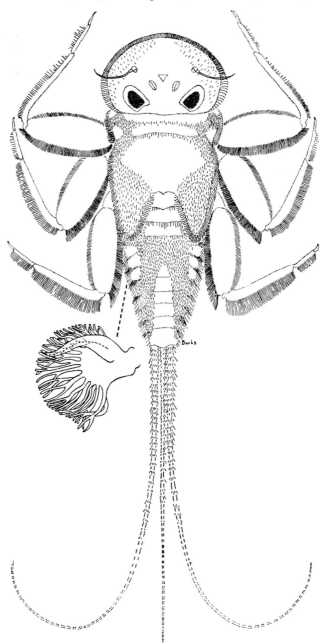

Fig. 394.—Nymph questionably placed as that of *Anepeorus;* nearly mature nymph, dorsal aspect. The small figure at left shows detail of gill borne on venter of fifth abdominal segment.

excavated base, fig. 392; the penis lobes are fused at the meson, on the ventral side, and bear a cluster of thin, somewhat convoluted laminae at the apexes.

The nymphs are not known with certainty, but fig. 394 shows an extremely rare form that may be a nymph of this genus. Only three specimens of this form have been collected in Illinois over a period of more than 20 years. It has not yet been possible to rear it.

This supposed nymph of *Anepeorus* has the head extremely broad and flat, with the compound eyes located near the posterior margin; the anterior and lateral margins bear a dense fringe of hair; the mouth-parts are evidently those of a predator, as the mandibles have long, slender incisors, and the broad, grinding, molar surfaces are absent; the fanglike structure of the maxillae and labium clearly fits them for predacity. The pronotum has lateral, flange-like projections; the legs are flattened and have a dense fringe of long hairs along the posterior margins of all femora and the middle and hind tibiae; the anterior margins of the femora bear dense fringes of shorter hairs; the claws are long, slender, and edentate; the wingpads show the heptageniid pattern of veins. The abdomen is flattened dorsoventrally and is relatively broad; the tergites are clothed with dense, somewhat woolly hairs; each of the first seven abdominal segments has a pair of ventral gills, fig. 394, which are all of the same general shape; the apex of the terminal abdominal segment is produced as a broad lobe; the three caudal filaments are all of approximately equal length, and their length is greater than that of the entire head and body; the caudal filaments are virtually bare, each clothed with short, sparse setae only in the basal area. Both male and female specimens have been collected, the male nymph showing the rudimentary forceps at the posterolateral angles of the projecting lobe on the posterior margin of the terminal abdominal sternite.

The short, thickset abdomen and the shape of the terminal abdominal sternite eliminated this nymph from consideration as the possible naiad of the genus *Pseudiron*. The nymph here considered to be that of *Anepeorus* bears a superficial resemblance to the nymph of the Russian *Behningia ulmeri* Lestage (Behning 1924; Chernova 1938).

Anepeorus simplex (Walsh)

Heptagenia simplex Walsh (1863:204).

Mayflies of this species are almost identical in appearance with those of two other heptageniid species: *Heptagenia persimplex* and *Stenonema integrum*. In both *persimplex* and *integrum*, however, the fore tarsus in the males is much longer than the fore tibia. *A. simplex* is actually an extremely rare species, and some of the published records of it have been based on misidentified specimens.

I agree with McDunnough (1929:179) that the Walsh specimen in the Museum of Comparative Zoology now labeled lectotype of this species is not *simplex* as defined by Walsh himself. There is another male specimen in the type lot at the M.C.Z. which is, in my opinion, the true *simplex*, as it agrees closely with Walsh's original description. The abdomen is, unfortunately, lacking from this specimen; so I was not able to examine the genitalia when I saw the specimen in 1942. I am basing my treatment of the species on McDunnough's redescription (1929:179), based on a single pair, now in the Illinois Natural History Survey collection, which was collected on the Rock River a few miles upstream from Rock Island, and on a single male subimago from Mount Carmel, on the Wabash River.

MALE.—Length of body 6–8 mm., of fore wing 7–9 mm. Head faintly suffused with pink, almost white; antennae white; eyes greenish yellow in life, according to Walsh. Entire thoracic notum a very faint yellowish pink; pleura and sternum white. Legs white, with yellow or brown shading on entire fore femur, at apexes of middle and hind femora, at bases of all tibiae, at apex of fore tibia, and at apexes of all tarsal segments; wings hyaline, with crossveins in costal half of each fore wing stained tan; three to five costal crossveins basad of bulla in fore wing, seven or eight stigmatic crossveins present in each fore wing. Abdomen white; genitalia, fig. 392, faintly yellow, caudal filaments white.

FEMALE.—Length of body 7–9 mm., of fore wing 8–10 mm. Coloration identical with that of male, except that yellow-brown shading of each fore femur is confined to a middle stripe and small area at apex; abdomen yellowish (due to color of eggs), posterior margin of terminal abdominal sternite produced as a broadly rounded lobe, without median emargination or excavation; caudal filaments faintly tan stained on basal articulations.

Known from Illinois and Iowa.

Illinois Records. — MOUNT CARMEL: Wabash River, June 10, 1947, Burks & Sanderson, 1 ♂. OREGON: Rock River, July 9, 1925, T. H. Frison, 1 ♂, 1 ♀. ROCK ISLAND: 10 ♂, 9 ♀ (Walsh 1863:204). Records of supposed nymph: DIXON: Rock River, May 22, 1925, D. H. Thompson, 1 N. MOUNT CARMEL: Wabash River, May 25, 1942, Mohr & Burks, 1 N; May 28, 1942, Mohr & Burks, 1 N.

48. *ARTHROPLEA* Bengtsson

Arthroplea Bengtsson (1908:239).
Remipalpus Bengtsson (1908:242).
Haplonia Blair (1929:254).

In the adult males of *Arthroplea*, the compound eyes are almost contiguous on the meson; the first fore tarsal segment is two-thirds as long as the second segment, and the entire tarsus is twice as long as the fore tibia. In both sexes, the venation of the fore wing is typical for the family, with the basal crossveins in the costal interspace rather weak and the stigmatic costal crossveins well developed but relatively few in number and not anastomosed. The hind wing, fig. 321, has the vein R_{4+5} unbranched throughout its length. The male genitalia, fig. 393, consist of a pair of five-segmented forceps and a pair of semirectangular penis lobes; each lobe bears three long, filamentous appendages. In the females, the apical abdominal sternite has the posterior margin evenly rounded from side to side, not indented on the meson.

The nymphs, fig. 395, are unique among American mayflies in that the labial and maxillary palps are considerably lengthened. Each maxillary palp has two segments and is as long as the head and thorax combined; each labial palp has two segments and is one-half as long as the maxillary palp. Normally, the maxillary palps are held extended posteriorly, over the thoracic notum, but the labial palps are concealed beneath the head. Each tarsal claw is short, stout at the base, and bears a row of minute bristles on the

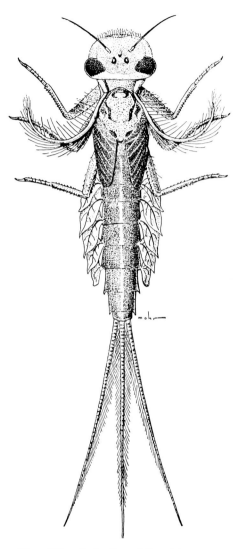

Fig. 395.—*Arthroplea bipunctata*, mature nymph, dorsal aspect.

ventral side in the basal area. A pair of single gills is present on each of abdominal segments 1–7; each gill is platelike, with a point at the apex. There are three long caudal filaments.

Arthroplea is represented by only one North American species, *bipunctata* (McDunnough) (1924c:76), known from Connecticut, Maine, New Hampshire, Ontario, and Quebec.

LITERATURE CITED

Albarda, Herman
    1878.   Descriptions of three new European Ephemeridae. Ent. Monthly Mag. 15:128–30.

Aro, J. E. *in* J.-A. Lestage
    1924.   Les Éphémères finnoises de M. le Docteur J. E. Aro. Bulletin de la Société Ento-mologique de Belgique 6:33–6.

Banks, Nathan
    1900.   New genera and species of Nearctic Neuropteroid insects. Am. Ent. Soc. Trans. 26:239–59.
    1903.   A new species of *Habrophlebia*. Ent. News 14:235.
    1908.   Neuropteroid insects—notes and descriptions. Am. Ent. Soc. Trans. 34:255–67. Pls. 17–9.
    1910.   Notes on our eastern species of the May-fly genus *Heptagenia*. Can. Ent. 42:197–202. Figs. 13–6.
    1914.   New Neuropteroid insects, native and exotic. Acad. Nat. Sci. Phila. Proc. 66:608–32. Pl. 28. (Ephemeridae:612–6.)
    1924.   Descriptions of new Neuropteroid insects. Harvard Univ. Mus. Compar. Zool. Bul. 65:421–55. Pls. 1–4. (Ephemeridae:423–6. Pls. 2, 4.)

Barnard, K. H.
    1932.   South African May-flies (Ephemeroptera). Roy. Soc. So. Africa, Cape Town, Trans. 20:201–59. 48 figs.
    1940.   Additional records and descriptions of new species of South African Alder-flies (Megaloptera), May-flies (Ephemeroptera), Caddis-flies (Trichoptera), Stone-flies (Perlaria), and Dragonflies (Odonata). S. African Mus. Ann. 32:609–61. 19 figs.

Behning, A. L.
    1924.   Zur Erforschung der am Flussboden der Wolga lebenden Organismen. (German translation of Russian title.) Monographien, der Biologischen Wolga-Station der Naturforscher-Gesellschaft zu Saratow 1(4):121-289. (Ephemeroptera, *in* chapter 4:246–53.)

Bengtsson, Simon
    1908.   Berättelse öfver en resa i entomologiskt syfte till mellersta Sverige sommaren 1907. Kungliga Svenska Vetenskapsakademien Årsbok 6:237–46. Stockholm.
    1909.   Beiträge zur Kenntnis der paläarktischen Ephemeriden. Lunds Universitets Årsskrift Ny Följd Afdelningen 2, 5(4):1–19. Lund.
    1912.   Neue Ephemeriden aus Schweden. Entomologisk Tidskrift 33:107–17.
    1915.   Eine Namensänderung. Entomologisk Tidskrift 36:34.
    1917.   Weitere Beiträge zur Kenntnis der nordischen Eintagsfliegen. Entomologisk Tidskrift 38:174–94.

Berner, Lewis
    1940.   Baetine mayflies from Florida (Ephemeroptera). Fla. Ent. 23:33–45, 49–62. 2 pls.
    1941.   Ovoviviparous mayflies in Forida. Fla. Ent. 24:32–4.
    1946.   New species of Florida mayflies (Ephemeroptera). Fla. Ent. 28:60–82. 1 pl.
    1950.   The mayflies of Florida. Fla. Univ. Studies, Biol. Sci. Ser. 4(4):xii + 1–267. 24 pls., 88 figs., 19 maps.

Billberg, Gustav J.
    1820.   Enumeratio insectorum in Museo Billberg. 138 pp. Stockholm.

Blair, K. G.
    1929.   Two new British mayflies (Ephemeroptera). Ent. Monthly Mag. 65:253–5. 4 figs.

Burks, B. D.
    1946.   New heptagenine mayflies. Ent. Soc. Am. Ann. 39:607–15. 10 figs.
    1949.   New species of *Ephemerella* from Illinois (Ephemeroptera). Can. Ent. 79:232–6. 7 figs. (Publication dated 1947.)

Burmeister, H. C. C.
    1839.   Handbuch der Entomologie. Neuroptera 2:757–1050. Berlin.

Campion, Herbert
    1923.   On the use of the generic name *Brachycercus* in Plectoptera and Orthoptera. Ann. and Mag. Nat. Hist. (ser. 9) 11:515–8.

Carpenter, F. M.
1933.   The Lower Permian insects of Kansas. Part 6. Delopteridae, Protelytroptera, Plectoptera and a new collection of Protodonata, Odonata, Megasecoptera, Homoptera, and Psocoptera. Am. Acad. Arts and Sci. Proc. **68**:411–504. 1 pl., 29 text figs. (Plectoptera:487–503. Figs. 26–9.)

Chernova, O. A.
1938.   Sur une nouvelle famille Ephemeroptera. (French translation of Russian title.) Akademiia Nauk S.S.S.R., Izvestiia – Otdelenie Matematicheskikh i Estestvennykh Nauk, Seriia Biologicheskaia 1938(1):129–37. Leningrad. (Bulletin de l'Académie des Sciences de l'URSS – Classes des Sciences Mathématiques et Naturelles, Série Biologique.)
1940.   Keys to the nymphs of Ephemeroptera of USSR with descriptive notes. *In* Freshwater Life in USSR, by V. I. Zhadin. Pt. 1:127–57. Moscow Acad. Sci. U.S.S.R.

Chopra, B.
1927.   The Indian Ephemeroptera (Mayflies). Part I.—The sub-order Ephemeroidea: Families Palingeniidae and Polymitarcidae. Indian Mus. Rec. **29**:91–138. Pls. 8–10., 18 figs.

Clemens, W. A.
1913.   New species and new life histories of Ephemeridae or mayflies. Can. Ent. **45**:246–62, 329–41. Pls. 5–7.
1915*a*.   Mayflies of the *Siphlonurus* group. Can. Ent. **47**:245–60. Pls. 9–11.
1915*b*.   Rearing experiments and ecology of Georgian Bay Ephemeridae. Contr. Can. Biol., Sessional Paper No. 39*b*:113–28. Pls. 13–4 (misnumbered 15–6).
1915*c*.   Life-histories of Georgian Bay Ephemeridae of the genus *Heptagenia*. Contr. Can. Biol., Sessional Paper No. 39*b*:131–43. 1 text fig., pls. 17–8 (pl. 17 misnumbered 15).
1917.   An ecological study of the mayfly *Chirotenetes*. Toronto Univ. Studies, Biol. Ser. **17**:1–43. 5 pls.
1922.   A parthenogenetic mayfly (*Ametelus ludens* Needham). Can. Ent. **54**:77–8.

Clemens, W. A., and A. K. Leonard
1924.   On two species of mayflies of the genus *Heptagenia*. Can. Ent. **56**:17–8.

Curtis, John
1834.   Descriptions of some nondescript British species of May-flies of anglers. London and Edinburgh Phil. Mag. and Jour. Sci. (ser. 3) **4**:120–5; 212–8.

Daggy, Richard H.
1945.   New species and previously undescribed naiads of some Minnesota mayflies (Ephemeroptera). Ent. Soc. Am. Ann. **38**:373–96. 4 figs., 2 pls.

Eaton, A. E.
1866.   Notes on some species of the orthopterous genus *Cloëon*, Leach (as limited by M. Pictet). Ann. and Mag. Nat. Hist. (ser. 3) **18**:145–8. 5 figs.
1868.   An outline of a re-arrangement of the genera of Ephemeridae. Ent. Monthly Mag. **5**:82–91.
1869.   On *Centroptilum*, a new genus of the Ephemeridae. Ent. Monthly Mag. **6**:131–2.
1871.   A monograph on the Ephemeridae. Ent. Soc. London Trans. **19**:1–164. 6 pls.
1881.   An announcement of new genera of Ephemeridae. Ent. Monthly Mag. **17**:191–7, **18**:21–7.
1882.   An announcement of new genera of the Ephemeridae. Ent. Monthly Mag. **18**:207–8.
1883–   A revisional monograph of recent Ephemeridae or mayflies. Linn. Soc. London
1888.   Trans., Zool. Ser. **3**:1–352. 65 pls.
1901.   Ephemeridae collected by Herr E. Strand in South and Arctic Norway. Ent. Monthly Mag. **37**:252–5. 2 figs.
1907–   [Description of a new species of Ephemeridae]. *In* Bidrag til en Fortegnelse over
1908.   arktisk norges *Neuropter*fauna, by Esben Petersen. Tromsø Museums Aarshefter **25**(1902):119–53. (*Ephemerella aronii* sp. n. :149–51.)

Edmunds, George F., Jr.
1945.   Ovoviviparous mayflies of the genus *Callibaetis* (Ephemeroptera:Baetidae). Ent. News **56**:169–71.
1948*a*.   The nymph of *Ephoron album* (Ephemeroptera). Ent. News **59**:12–4. 2 figs.
1948*b*.   The mayfly genus *Lachlania* in Utah. Ent. News **59**:43.
1948*c*.   A new genus of mayflies from western North America (Leptophlebiinae). Biol. Soc. Wash. Proc. **61**:141–8. Pls. 5–6.

Forbes, Stephen A.
1878.   The food of Illinois fishes. Ill. State Lab. Nat. Hist. Bul. **1**(2):71–86.
1880,   Studies of the food of birds, insects and fishes, made at the Illinois State Laboratory
1883.   of Natural History, at Normal, Illinois. Ill. State Lab. Nat. Hist. Bul. **1** (No. 3, 1880): 1–160; (No. 6, 1883):1–109.

1888*a*.   Studies of the food of fresh-water fishes. Ill. State Lab. Nat. Hist. Bul. 2:433–73.
1888*b*.   On the food relations of fresh-water fishes: a summary and discussion. Ill. State Lab.
           Nat. Hist. Bul. 2:475–538.

Gordon, Eva
1933.   Notes on the ephemerid genus *Leptophlebia*. Brooklyn Ent. Soc. Bul. 28:116–34.
        Pls. 12–4.

Grandi, M.
1941–   Contributi allo studio degli Efemerotti italiani, I–XVI. Bologna Università Istituto
1951.   di entomologia Bollettino 12:1–62, 50 figs.; 179–205, 20 figs.; 13:29–71, 24 figs.;
        137–71, 20 figs.; 14:114–30, 11 figs.; 15:103–28, 22 figs.; 229–32; 16:85–114, 20 figs.;
        176–218, 23 figs.; 17:62–82, 12 figs.; 275–300, 17 figs.; 18:58–92, 21 figs.; 117–27,
        7 figs., 181.

Hagen, Herman
1861.   Synopsis of the Neuroptera of North America, with a list of the South American
        species. Smithsn. Inst. Misc. Collect. xx + 347 pp. (Ephemerina:33–55.)
1863.   *In* Observations on certain N. A. Neuroptera, by H. Hagen, M. D. of Koenigsberg,
        Prussia; translated from the original French MS., and published by permission of
        the author, with notes and descriptions of about twenty new N. A. species of
        Pseudoneuroptera. By Benj. D. Walsh. Ent. Soc. Phila. Proc. 2:182–272.
1868.   On *Lachlania abnormis* a new genus and species from Cuba belonging to the
        Ephemerina. Boston Soc. Nat. Hist. Proc. 11:372–4.

Handlirsch, Anton
1906.   Die fossilen Insekten und die Phylogenie der rezenten Formen. 10 + 1430 pp.
        51 pls., 10 charts, 6 figs. Leipzig. (Plectoptera:600–4, 1228–9.)
1918.   Fossile Ephemeridenlarven aus dem Buntsandstein der Vogesen. Verhandlungen der
        Kaiserlich-Königlichen Zoologisch-Botanischen Gesellschaft in Wien 68:112–4. 1 fig.
1925.   Systematische Übersicht *in* Handbuch der Entomologie, by Christoph Schröder
        3:414–24. 17 figs. Jena.

Ide, F. P.
1930*a*.   The nymph of the mayfly genus *Cinygma* Eaton. Can. Ent. 62:42–5. Pl. 6.
1930*b*.   Contributions to the biology of Ontario mayflies with descriptions of new species.
           Can. Ent. 62:204–13, pl. 17; 218–31, pls. 18–9.
1935*a*.   Life history notes on *Ephoron, Potamanthus, Leptophlebia,* and *Blasturus* with de-
           scriptions (Ephemeroptera). Can. Ent. 67:113–25. Pls. 4–5.
1935*b*.   Post embryological development of Ephemeroptera (mayflies), external characters
           only. Can. Jour. Res. 12:433–78. 13 figs.
1936.   The significance of the outgrowths on the prothorax of *Ecdyonurus venosus* Fabr.
        (Ephemeroptera). Can. Ent. 68:234–8. Pl. 13.
1937*a*.   The subimago of *Ephoron leukon* Will., and a discussion of the imago instar
           (Ephem.). Can. Ent. 69:25–9. Pl. 2.
1937*b*.   Descriptions of eastern North American species of baetine mayflies with particular
           reference to the nymphal stages. Can. Ent. 69:219–31, pls. 8–10; 235–43, pls. 11–12.
1940.   Quantitative determination of the insect fauna of rapid water. Toronto Univ. Studies,
        Biol. Ser. 47, Pub. Ont. Fish. Res. Lab. 59:1–20. 4 pls.
1941.   Mayflies of two tropical genera, *Lachlania* and *Campsurus,* from Canada with
        descriptions. Can. Ent. 73:153–6. 3 figs.
1942.   Availability of aquatic insects as food of the speckled trout, *Salvelinus fontinalis.*
        N. Am. Wildlife Conf. Trans. 7:442–50. 1 fig.

Joly, Émile
1870.   Contributions pour servir a l'histoire naturelle des Éphémérines. Société d'Histoire
        Naturelle de Toulouse Bulletin 4:142–50. Pl. 3.

Kimmins, D. E.
1942.   Keys to the British species of Ephemeroptera with keys to the genera of the nymphs.
        Freshwater Biol. Assn. Brit. Empire, Sci. Pub. No. 7:1–64. 36 figs.

Klapálek, Franz
1905.   Plecopteren und Ephemeriden aus Java. Naturhistorischen Museum Hamburg
        Mitteilungen 22:101–7.
1909.   Ephemerida, Eintagsfliegen. Die Süsswasserfauna Deutschlands (Herausgegeben
        von A. Brauer) Heft 8:1–32. 53 figs.

Knox, Velma
1935.   The body-wall of the thorax; the musculature of the thorax. *In* The biology of mayflies,
        by Needham, Traver, Hsu, *et al.* Pp. 135–78, pls. 20-30.

Leach, William E.
1815.   Entomology *in* Brewster's Edinburgh Encyclopaedia 9:57–172.

Leonard, Justin W.
1949.   The nymph of *Ephemerella excrucians* Walsh. Can. Ent. 81:158–60. 3 figs.

Lestage, J.-A.
1917.   Contribution a l'étude des larves des Éphémères paléarctiques. Annales de Biologie Lacustre 8(fasc. 3–4):213–457. 54 figs. Brussels.
1919.   Contribution a l'étude des larves des Éphémères paléarctiques (sér. 2). Annales de Biologie Lacustre [for 1918] 9:79–182. 13 figs.
1921.   Les Éphémères indo-chinoises. Société Entomologique de Belgique Annales 61:211–22.
1924a. Note sur les Éphémères de la *Monographical Revision* de Eaton. Société Entomologique de Belgique Annales 64:33–59.
1924b. Les Éphémères de l'Indochine française. Faune entomologique de l'Indochine française, fasc. 8:79–93. Säigon.
1925a. Contribution à l'étude des Éphémères. Série 3. Le groupe Éphémérellidien. Annales de Biologie Lacustre [for 1924] 13(fasc. 3–4):227–302. 14 figs.
1925b. Éphéméroptères, Plécoptères et Trichoptères recueillis en Algérie par M. H. Gauthier et liste des espèces connues actuellement de l'Afrique du Nord. Société d'Histoire Naturelle de l'Afrique du Nord Bul. 16:8–18. Algiers.
1928–   Les Éphéméroptères de la Belgique. Société Entomologique de Belgique Bul. et Ann.
1929.   68:251–64; 69:126–30, 217–21.
1930a. La dispersion holarctique de quelques Éphéméroptères. Société Entomologique de Belgique Bul. et Ann. 70:201–7.
1930b. Contribution à l'étude des larves des Éphéméroptères, V–VI. Société Entomologique de Belgique Bul. et Ann. 69:433–40; 70:79–89.
1930c. Contribution à l'étude des larves des Éphéméroptères, VII. Société Entomologique de Belgique Mém. 23:73–146.
1931a. Note à propos l'homonymie de deux Éphéméroptères. Société Entomologique de Belgique Bul. et Ann. 71:119.
1931b. Contribution à l'étude des Éphéméroptères, VIII. Société Entomologique de Belgique Bul. et Ann. 71:41–60. 5 figs.
1935.   Contribution à l'étude des Éphéméroptères, IX–XII. Société Entomologique de Belgique Bul. et Ann. 75:76–139, 11 figs.; 173–83, 2 figs.; 312–4; 346–58.
1938.   Contribution à l'étude des Éphéméroptères, XVI–XXI. Société Entomologique de Belgique Bul. et Ann. 78:155–82, 2 pls.; 246–9; 273–4; 315–9; 320; 381–94, 1 pl.
1939.   Contribution à l'étude des Éphéméroptères, XXII–XXIII. Société Entomologique de Belgique Bul. et Ann. 79:77–85, 135–8.
1940.   Contribution à l'étude des Éphéméroptères, XXIV. Société Entomologique de Belgique Bul. et Ann. 80:118–24.
1945.   Contribution à l'étude des Éphéméroptères, XXVI. Société Entomologique de Belgique Bul. et Ann. 81:81–9.

Linnaeus, Carolus
1758.   Systema Naturae. Tenth edition. 824 + 2 pp. Holmiae.
1761.   Fauna Suecica. Second edition. 48 + 578 pp., 2 pls. Stockholm.

Lubbock, John W.
1863.   On the development of *Chloëon* (*Ephemera*) *dimidiatum*.—Part I. Linn. Soc. London Trans. 24:61–78. Pls. 17–8.

Lyman, F. E.
1944.   Taxonomic notes on *Brachycercus lacustris* (Needham) (Ephemeroptera). Ent. News 55:3–4.

McDunnough, J.
1921.   Two new Canadian may-flies (Ephemeridae). Can. Ent. 53:117–20. Pl. 4.
1923.   New Canadian Ephemeridae with notes. Can. Ent. 55:39–50. 3 figs.
1924a. New Ephemeridae from Illinois. Can. Ent. 56:7–9. 1 fig.
1924b. New Canadian Ephemeridae with notes, II. Can Ent. 56:90–8, pl. 1; 113–22, pl. 3; 128–33.
1924c. New Ephemeridae from New England. Boston Soc. Nat. Hist. Occas. Papers 5:73–6. Pl. 6.
1924d. New North American Ephemeridae. Can. Ent. 56:221–6. Pl. 5.
1925a. The Ephemeroptera of Covey Hill, Que. Roy. Soc. Can. Trans., Ser. 3, Sec. 5, 19:207–24. Pl. 1.
1925b. New Canadian Ephemeridae with notes, III. Can. Ent. 57:168–76, pl. 4; 185–92, pl. 5.
1925c. New *Ephemerella* species (Ephemeroptera). Can. Ent. 57:41–3.
1926.   Notes on North American Ephemeroptera with descriptions of a new species. Can. Ent. 58:184–96.

1927a.   A new *Ephemerella* from Illinois (Ephemeroptera). Can. Ent. **59**:10.
1927b.   New Canadian Ephemeridae with notes IV. Can. Ent. **58**:296–303. Pl. 3. (Publication dated Dec., 1926.)
1927c.   Notes on the species of the genus *Hexagenia* with description of a new species (Ephemeroptera). Can. Ent. **59**:116–20. 1 fig.
1928a.   The Ephemeroptera of Jasper Park, Alta. Can. Ent. **60**:8–10.
1928b.   Ephemerid notes with description of a new species. Can. Ent. **60**:238–40.
1929.   Notes on North American Ephemeroptera with descriptions of new species, II. Can. Ent. **61**:169–80. Pls. 3–4. 4 figs.
1930.   The Ephemeroptera of the north shore of the Gulf of St. Lawrence. Can. Ent. **62**:54–62. Pls. 7–9.
1931a.   The *bicolor* group of the genus *Ephemerella* with particular reference to the nymphal stages (Ephemeroptera). Can. Ent. **63**:30–42, 61–8. Pls. 2–5.
1931b.   New species of North American Ephemeroptera. Can. Ent. **63**:82–93.
1931c.   The genus *Isonychia* (Ephemeroptera). Can. Ent. **63**:157–63. Pl. 8.
1931d.   The eastern North American species of the genus *Ephemerella* and their nymphs (Ephemeroptera). Can. Ent. **63**:187–97, pl. 11; 201–16, pls. 12–4.
1931e.   New North American Caeninae with notes (Ephemeroptera). Can. Ent. **63**:254–68. Pls. 17–8.
1932a.   Further notes on the Ephemeroptera of the north shore of the Gulf of St. Lawrence. Can. Ent. **64**:78–81.
1932b.   New species of North American Ephemeroptera II. Can. Ent. **64**:209–15. 3 figs.
1933a.   Notes on the Heptagenine species described by Clemens from the Georgian Bay region, Ont. (Ephemerop.). Can. Ent. **65**:16–24, figs. 1–2; 33–43, figs. 3–4. Pl. 1.
1933b.   The nymph of *Cinygma integrum* and description of a new Heptagenine genus. Can. Ent. **65**:73–7. Pls. 2–3.
1933c.   New species of North American Ephemeroptera III. Can. Ent. **65**:155–8. 4 figs.
1933d.   New Ephemeroptera from the Gaspé Peninsula. Can. Ent. **65**:278–81. 3 figs.
1934.   New species of North American Ephemeroptera IV. Can. Ent. **66**:154–64, pls. 7–8, 1 text fig.; 181–8, pl. 10, 1 text fig.
1935.   Notes on western species of Ephemeroptera. Can. Ent. **67**:95–104. 4 figs.
1936a.   A new Arctic Baetid (Ephemeroptera). Can. Ent. **68**:32–4. Pl. 1.
1936b.   Further notes on the genus *Ameletus* with descriptions of new species (Ephem.). Can. Ent. **68**:207–11. 2 figs.
1938.   New species of North American Ephemeroptera with critical notes. Can. Ent. **70**:23–34. Pl. 1, 2 figs.
1939.   New British Columbian Ephemeroptera. Can. Ent. **71**:49–54. Pl. 15.
1942.   An apparently new *Thraulodes* from Arizona (Ephemerida). Can. Ent. **74**:117.
1943.   A new *Cinygmula* from British Columbia (Ephemeroptera). Can. Ent. **75**:3. 1 fig.

McLachlan, Robert
1873.   *Oniscigaster Wakefieldi*, a new genus and species of Ephemeridae from New Zealand. Ent. Monthly Mag. **10**:108–11. 1 fig.
1874.   On *Oniscigaster Wakefieldi*, the singular insect from New Zealand belonging to the family Ephemeridae; with notes on its aquatic conditions. Linn. Soc. London Jour., Zool. Ser. **12**:139–46. Pl. 5.

Morgan, Anna H.
1911.   May-flies of Fall Creek. Ent. Soc. Am. Ann. **4**:93–126. 2 figs., pls. 6–12.
1913.   A contribution to the biology of May-flies. Ent. Soc. Am. Ann. **6**:371–441. 3 figs., pls. 42–54.
1930.   Field book of ponds and streams. 16 + 448 pp. 23 pls., 314 figs. New York.

Morgan, Anna H., and Margaret C. Grierson
1932.   The functions of the gills in burrowing mayflies (*Hexagenia recurvata*). Physiol. Zool. **5**:230–45. 1 pl.

Murphy, Helen E.
1922.   Notes on the biology of some of our North American species of May-flies. Lloyd Libr. Bot., Pharm. and Materia Med. Bul. No. 22, Ent. Ser. No. **2**:1–46. 7 pls.

Neave, Ferris
1930.   Migratory habits of the mayfly, *Blasturus cupidus* Say. Ecology **11**:568–76. 3 figs.
1932.   A study of the May flies (*Hexagenia*) of Lake Winnipeg. Contr. Can. Biol. and Fish. **7**(15):177–201. 10 figs.
1934.   A contribution to the aquatic insect fauna of Lake Winnipeg. Internationale Revue der gesamten Hydrobiologie und Hydrographie **31**:157–70. 3 figs.

Needham, James G.
1901.   Ephemeridae. *In* Aquatic insects in the Adirondacks, by J. G. Needham and Cornelius Betten. N. Y. State Mus. Bul. No. **47**:418–29. Pls. 11 and 16.

1903. Food of brook trout in Bone pond. *In* Aquatic insects in New York State, by James G. Needham, Alex. D. MacGillivray, O. A. Johannsen, and K. C. Davis. N. Y. State Mus. Bul. 68:204–17. Pl. 7, 2 figs.
1905. Ephemeridae. *In* May flies and midges of New York, by James G. Needham, K. J. Morton, and O. A. Johannsen. N. Y. State Mus. Bul. 86:17–62. Pls. 4–12, 14 figs.
1908. New data concerning May flies and dragon flies of New York. N. Y. State Mus. Bul. 124:188–94. Pl. 10.
1909. Studies of aquatic insects: a peculiar new May fly from Sacandaga Park. N. Y. State Mus. Bul. 134:71–5. Pl. 2, fig. 22.
1918. A new mayfly, *Caenis*, from Oneida Lake, New York. N. Y. State Col. Forestry, Syracuse Univ. Tech. Pub. 9:249–51.
1921. Burrowing mayflies of our larger lakes and streams. U. S. Bur. Fish. Bul. for 1917–18, 36:265–92. Pls. 70–82.
1924. The male of the parthenogenetic mayfly, *Ameletus ludens*. Psyche 31:308–10.
1927. The Rocky Mountain species of the mayfly genus *Ephemerella*. Ent. Soc. Am. Ann. 20:107–17. 1 fig.
1932. Three new American mayflies (Ephemerop.). Can. Ent. 64:273–6. Figs. A–C.

Needham, James G., and Helen E. Murphy
1924. Neotropical mayflies. Lloyd Libr. Bot., Pharm. and Materia Med. Bul. 24, Ent. Ser. 4:3–79. 13 pls.

Needham, James G., Jay R. Traver, Yin-Chi Hsu, *et al.*
1935. The biology of mayflies. 16 + 759 pp., 40 pls., 168 figs. Comstock Publishing Co., Ithaca, N. Y.

Perrier, Rémy
1934. Faune de la France en tableaux synoptiques illustrés 3:41–50. 35 figs. Second edition. Paris.

Petersen, Esben
1909. New Ephemeridae from Denmark, Arctic Norway and the Argentine Republic. Deutsche Entomologische Zeitschrift 1909:551–6. 12 figs.

Phillips, J. S.
1930. A revision of New Zealand Ephemeroptera. New Zeal. Inst. Trans. and Proc. 61:271–390. Pls. 50–67.

Pictet, François J.
1843– Histoire naturelle générale et particulière des insectes Neuroptères. Seconde Mono-
1845. graphie: Famille des Éphémérines. 10 + 300 pp., 47 pls. Geneva, Switzerland.

Provancher, Léon
1876. Petite faune entomologique du Canada, Névroptères, Fam. III, Éphémérides. Nat. Can. 8:264–8.
1878. Additions et corrections aux Névroptères de la Province de Québec. Nat. Can. 10:124–47.

Say, Thomas
1823. Descriptions of insects belonging to the order Neuroptera Linn., Latr. collected by the expedition authorized by J. C. Calhoun, Secretary of War, under the command of Major S. H. Long. Western Quarterly Reporter 2(2):160–5. Cincinnati.
1824. *From* Narrative of the expedition to the source of the St. Peter's river . . . under the command of Stephen H. Long, Major U.S.T.E., 2:268–378. Philadelphia.
1839. Descriptions of new North American neuropterous insects and observations on some already described. Acad. Nat. Sci. Phila. Jour. 8:9–46.

Schoenemund, Eduard
1930. Eintagsfliegen oder Ephemeroptera. *In* Die Tierwelt Deutschlands, by Friedrich Dahl. Teil 19:1–106. 186 figs. Jena.

Serville, Jean G. A.
1829– Description of *Ephemera limbata*. *In* Iconographie du Règne animal de G. Cuvier . . .
1844. Paris, by F. E. Guérin-Méneville. (2:pl. 60, fig. 7; 3:384.)

Smith, Osgood R.
1935. The eggs and egg-laying habits of North American mayflies. *In* The biology of mayflies, by James G. Needham, Jay R. Traver, Yin-Chi Hsu, *et al.* Pp. 67–89, pls. 15–8, figs. 4–5.

Spieth, Herman T.
1933. The phylogeny of some mayfly genera. N. Y. Ent. Soc. Jour. 41:55–86, 327–91. Pls. 16–29.

1937. An oligoneurid from North America. N. Y. Ent. Soc. Jour. 45:139–45. Pl. 2.
1938a. Two interesting mayfly nymphs with a description of a new species. Am. Mus. Nat. Hist. Am. Mus. Novitates No. 970:1–7. 2 pls.
1938b. Taxonomic studies on Ephemerida, I. Description of new North American species. Am. Mus. Nat. Hist. Am. Mus. Novitates No. 1002:1–11. 8 figs.
1940. The North American ephemeropteran species of Francis Walker. Ent. Soc. Am. Ann. 33:324–38. 1 fig.
1941a. The North American ephemeropteran types of the Rev. A. E. Eaton. Ent. Soc. Am. Ann. 34:87–98. 1 pl.
1941b. Taxonomic studies on the Ephemeroptera. II. The genus *Hexagenia*. Am. Midland Nat. 26:233–80. 6 pls.
1943. Taxonomic studies on the Ephemeroptera. III. Some interesting ephemerids from Surinam and other neotropical localities. Am. Mus. Nat. Hist. Am. Mus. Novitates No. 1244:1–13. 2 pls., 8 figs.
1947. Taxonomic studies on the Ephemeroptera. IV. The genus *Stenonema*. Ent. Soc. Am. Ann. 40:87–122. 2 pls., figs. 29–31.

Sprules, William M.
1947. An ecological investigation of stream insects in Algonquin Park, Ontario. Toronto Univ. Studies, Biol. Ser. 56:6 + 1–81. Ephemeroptera:43–5.

Stephens, James F.
1835. Illustrations of British entomology 6. 240 pp., 7 pls.

Tillyard, R. J.
1917. The biology of dragonflies (Odonata or Paraneuroptera). xii + 396 pp., 4 pls., 188 text figs. Cambridge.
1920. Report on the neuropteroid insects of the Hot Springs region, N. Z. in relation to the problem of trout food. Linn. Soc. N. S. Wales Proc. 45:205–13.
1923. The wing-venation of the order Plectoptera or mayflies. Linn. Soc. London Jour., Zool. Sect. 35:143–62. 10 figs.
1925. Kansas Permian insects. Pt. IV. The order Paleodictyoptera. Am. Jour. Sci., Ser. 5, 9:328–35. 3 figs.
1926. The insects of Australia and New Zealand. 17 + 560 pp. 44 pls. Sidney, N. S. W., Angus and Robertson, Ltd. (Plectoptera, Chap. VIII, pp. 57–64.)
1932. Kansas Permian insects. Pt. 15. The order Plectoptera. Am. Jour. Sci., Ser. 5, 23:97–134, 237–72. 22 figs.

Traver, Jay R.
1931a. A new mayfly genus from North Carolina. Can. Ent. 63:103–9. Pl. 7.
1931b. Seven new southern species of the mayfly genus *Hexagenia*, with notes on the genus. Ent. Soc. Am. Ann. 24:591–621. 1 pl., 7 figs.
1931c. The ephemerid genus *Baetisca*. N. Y. Ent. Soc. Jour. 39:45–67. Pls. 5–6.
1932a. Mayflies of North Carolina. Elisha Mitchell Sci. Soc. Jour. 47:85–161, 163–236. Pls. 5–12.
1932b. *Neocloeon*, a new mayfly genus (Ephemerida). N. Y. Ent. Soc. Jour. 40:365–73. Pl. 14.
1933a. Mayflies of North Carolina. Part III. The Heptageninae. Elisha Mitchell Sci. Soc. Jour. 48:141–206. Pl. 15.
1933b. Heptagenine mayflies of North America. N. Y. Ent. Soc. Jour. 41:105–25.
1934. New North American species of mayflies (Ephemerida). Elisha Mitchell Sci. Soc. Jour. 50:189–254. Pl. 16.
1935a. North American mayflies, a systematic account of North American species in both adult and nymphal stages. *In* The biology of mayflies, by Needham, Traver, Hsu, et al. Pp. 237–739, pls. 34–40, figs. 77–168.
1935b. Two new genera of North American Heptageniidae (Ephemerida). Can. Ent. 67:31–8. 1 pl., 6 figs.
1937. Notes on mayflies of the southeastern states (Ephemeroptera). Elisha Mitchell Sci. Soc. Jour. 53:27–86. Pl. 6.
1938. Mayflies of Puerto Rico. Puerto Rico Univ. Jour. Ag. 22:5–42. 3 pls.
1939. Himalayan mayflies (Ephemeroptera). Ann. and Mag. Nat. Hist. (ser. 11) 4:32–56. 22 figs.
1943. New Venezuelan mayflies. Boletin Entomologica Venezolana 2:79–98. 8 figs.
1944. Notes on Brazilian mayflies. Museu Nacional Boletim Zoologia No. 22:2–53. 20 figs. Rio de Janeiro.
1946– Notes on Neotropical mayflies, Pts. I–III. Revista de Entomologia 17:418–36, 3 pls.,
1947. 18:149–60, 1 pl.; 370–95, 4 pls. Rio de Janeiro.

Uéno, Masuzo
1931. Contributions to the knowledge of Japanese Ephemeroptera. Annotationes zoologicae japonenses 13(3):189–231. Pls. 12–3, 34 figs.

Ulmer, Georg
  1914.   Ephemeroptera. *In* Fauna von Deutschland, by Brohmer, pp. 95–9, 11 figs. 587 pp., 212 figs. Leipzig.
  1920*a.*   Neue Ephemeropteren. Archiv für Naturgeschichte (1919)85(Abt. A, Heft 11):1–80. 56 figs.
  1920*b.*   Über die Nymphen einiger exotischer Ephemeropteren. Festschrift für Zschokke No. 25:1–25. 16 figs. Basel.
  1920*c.*   Übersicht über die Gattungen der Ephemeropteren, nebst Bemerkungen über einzelne Arten. Stettiner Entomologische Zeitung 81:97–144.
  1921.   Über einige Ephemeropteren-Typen älterer Autoren. Archiv für Naturgeschichte 87(Abt. A, Heft 6):229–67. 21 figs.
  1924*a.*   Ephemeroptera, Eintagsfliegen. Biologie der Tiere Deutschlands, Teil 34. 40 pp., 28 figs. Berlin.
  1924*b.*   Einige alte und neue Ephemeropteren. Konowia 3:23–37. 4 figs.
  1924*c.*   Ephemeropteren von den Sunda-Inseln und den Philippinen. Treubia 6:28–91. 58 figs. Buitenzorg.
  1926*a.*   Beiträge zur Fauna sinica. III. Trichopteren und Ephemeropteren. Archiv für Naturgeschichte 91 (Abt. A, Heft 5):19–110. 102 figs.
  1926*b.*   *Baëtis luridipennis* Burm. aus Nord-Amerika ist ein *Siphlonurus* (Ephemeropt.). Entomologische Mitteilungen 15:223–5. 1 fig.
  1930.   Entomological expedition to Abyssinia, 1926–27; Trichoptera and Ephemeroptera. Ann. and Mag. Nat. Hist. (ser. 10) 6:479–511. 28 figs.
  1932*a.*   Die Trichopteren, Ephemeropteren und Plecopteren des arktischen gebietes. Fauna Arctica 6:207–26. Jena.
  1932*b.*   Bemerkungen über die seit 1920 neu aufgestellten Gattungen der Ephemeropteren. Stettiner Entomologische Zeitung 93:204–19.
  1933.   Aquatic insects of China. Art. VI. Revised key to the genera of Ephemeroptera. Peking Nat. Hist. Bul. (1932–33) 7:195–218. 2 pls.
  1938.   Chilenische Ephemeropteren, hauptsächlich aus dem deutschen Entomologischen Institut, Berlin-Dahlem. Arbeiten über Morphologische und Taxonomische Entomologie aus Berlin-Dahlem 5:85–108. 16 figs.
  1939–   Eintagsfliegen (Ephemeropteren) von den Sunda-Inseln. Archiv für Hydrobiologie
  1940.   Supplement-Band 16:443–580, 23 pls.; 581–692, 27 pls. Berlin; Stuttgart.
  1942–   Alte und neue Eintagsfliegen (Ephemeropteren) aus Süd- und Mittelamerika. Stet-
  1943.   tiner Entomologische Zeitung 103:98–128, 3 pls.; 104:14–46, 3 pls.

Van Cleave, Harley J., and Jean A. Ross
  1947.   A method for reclaiming dried zoological specimens. Science 105:318.

Walker, Francis
  1853.   Catalogue of the specimens of Neuropterous insects in the collection of the British Museum 3:477–585.

Walsh, Benjamin D.
  1862.   List of the Pseudoneuroptera of Illinois contained in the cabinet of the writer, with descriptions of over forty new species, and notes on their structural affinities. Acad. Nat. Sci. Phila. Proc. (1862) 13:361–402.
  1863.   Observations on certain N. A. Neuroptera, by H. Hagen, M. D. of Koenigsberg, Prussia; translated from the original French MS., and published by permission of the author, with notes and descriptions of about twenty new N. A. species of Pseudoneuroptera. Ent. Soc. Phila. Proc. 2:167–272. 2 figs.
  1864.   On the pupa of the ephemerinous genus *Baetisca* Walsh. Ent. Soc. Phila. Proc. 3:200–6. 1 fig.

Westwood, John O.
  1840.   An introduction to the modern classification of insects, founded on the natural habits and corresponding organization of the different families 2. 11+587 pp. London. (Generic synopsis, p. 47.)

Williamson, Hugh
  1802.   On the *Ephoron leukon,* usually called the White Fly of Passaick River. Am. Phil. Soc. Phila. Trans. 5:71–3.

INDEX

The page entries in **boldface** type refer to the principal treatment of the families, genera, and species in the text; those in *italic* type refer to illustrations. Names that are synonyms, or of changed generic assignment, are indicated by *italic* type.

Randall Library – UNCW
QL505.2.U6 B87 1975   NXWW
Burks / The mayflies, or Ephemeroptera, of Illinoi

304900205459+